I0073231

GRAMMATICAL COMPLEXITY AND ONE-DIMENSIONAL DYNAMICAL SYSTEMS

DIRECTIONS IN CHAOS

Editor-in-Chief: Hao Bai-lin

Published

Directions in Chaos Vol. 1 (published as Vol. 3 of Directions in Condensed Matter Physics series)
edited by Hao Bai-lin

Directions in Chaos Vol. 2 (published as Vol. 4 of Directions in Condensed Matter Physics series)
edited by Hao Bai-lin

Vol. 3: Experimental Study and Characterization of Chaos
edited by Hao Bai-lin

Vol. 4: Quantum Non-Integrability
edited by Da Hsuan Feng and Jian-Min Yuan

Vol. 5: Bibliography on Chaos
compiled by Zhang Shu-yu

DIRECTIONS IN CHAOS VOL. 6

GRAMMATICAL COMPLEXITY AND ONE-DIMENSIONAL DYNAMICAL SYSTEMS

Huimin Xie

Suzhou Univ. China

World Scientific

Singapore • New Jersey • London • Hong Kong

Published by

World Scientific Publishing Co. Pte. Ltd.
5 Toh Tuck Link, Singapore 596224
USA office: 27 Warren Street, Suite 401-402, Hackensack, NJ 07601
UK office: 57 Shelton Street, Covent Garden, London WC2H 9HE

British Library Cataloguing-in-Publication Data
A catalogue record for this book is available from the British Library.

Directions in Chaos —Vol. 6
GRAMMATICAL COMPLEXITY AND ONE-DIMENSIONAL DYNAMICAL SYSTEMS

Copyright © 1996 by World Scientific Publishing Co. Pte. Ltd.

All rights reserved. This book, or parts thereof, may not be reproduced in any form or by any means, electronic or mechanical, including photocopying, recording or any information storage and retrieval system now known or to be invented, without written permission from the publisher.

For photocopying of material in this volume, please pay a copying fee through the Copyright Clearance Center, Inc., 222 Rosewood Drive, Danvers, MA 01923, USA. In this case permission to photocopy is not required from the publisher.

ISBN-13 978-981-02-2398-4
ISBN-10 981-02-2398-6

Preface

"Complexity" has evolved from a common word to a scientific topic that frequently appeared in books and papers[1] There are different definitions of complexity, and many discussions about what complexity is, and how to adapt to or cope with complexity. The purpose of this book, however, is very restrictive. As shown by its title and the table of contents, this book is concerned exclusively to the grammatical complexity and its application to two simplest models of dynamical systems, the unimodal maps and the circle homeomorphisms.

The study of grammatical complexity is simply meant to use the tools of languages and automata to discuss the degree of complexity of symbolic behaviors of dynamical systems. Here the most important elements are nonlinearity and discreteness. The hierarchy proposed by N. Chomsky and others is used as the base to do this study. Thus it can be seen as a development of symbolic dynamics, which was originated from the early study of Hadamard (1898) and Morse (1921), and used successfully in the discussion of the Smale's horseshoe.

The selection of the material in this book reflects the viewpoint of the author and his colleagues that in order to discuss the complicated behaviors of dynamical systems we may begin from the study of symbolic strings and sequences generated by these systems. Since the definition of language is simply a set of strings, it is quite natural and, indeed, inevitable to use formal languages and automata theory to systematize the symbolic study of dynamical systems.

Part I, including Chapters 1 and 2, is introductory. It provides the necessary knowledge for doing the analysis of grammatical complexity for dynamical systems. Here we show that the free monoid, an algebraic structure, is a proper framework in which the manipulation of strings finds its natural place. In Chapter 2 the notion of dynamical language is developed. Some associated notions, such as the distinct excluded block, the connection between dynamical languages and symbolic flows, graphs, and the topological entropy are discussed.

The content of Part II, including Chapters 3–8, is devoted to the grammatical complexity of unimodal maps. There have been published many books about one-dimensional dynamical systems. It is not our purpose to add another book about the same topic. The discussion in this Part is intended to compare the results obtainable

[1]See, e.g., R. Ruthen, *Scientific American* **268** (1993) 130, J. Horgan, *ibid* **272** (1995) 104, and Çambel (1993), Cramer (1993), Grassberger (1986), Hao (1991), Peliti and Vulpiani (1988), Penrose (1989), Roetzheim (1994), Wackerbauer et al (1994), Weisbuch (1991) in References.

by the new approach with the familiar facts in the simplest model of dynamical systems. Of course, the discussion of grammatical complexity is also a combinatorial study about them. Some new insight, however, is indeed obtained by this approach. One of the most important results in this aspect is to give a rigorous proof in the case of unimodal maps that the most complicated behaviors happen between the purely regular motion and the purely random motion as proposed by many scientists working in the field of complexity. The grammatical study of Feigenbaum attractor gives an evidence that the degree of complexity has a jump at phase transition. The connection between the kneading map and the structure of infinite automata is also revealing.

In Part III, including Chapter 9–11, the grammatical complexity of circle homeomorphisms is presented by the discussion of their languages and automata. The universal structure of infinite automata is discovered that here the continued fraction representation of the rotation number is the most important factor.

The book is self-contained. Since some knowledge about languages and automata is indispensable, three appendices are included to summarizes some of the background material about languages and automata for reading the book. Every basic fact about the manipulation of strings is proved in detail. Most proofs in the book are elementary, but some of them are rather lengthy. The reader is encouraged to read the introductions of Chapters and to concentrate on the meaning of ideas and statements.

The author has used most materials of the book for a graduate course in complexity and dynamical systems.

I would like to express my gratitude to Professor Hao Bai-lin who has encouraged me to do research in the field of complexity, and given suggestions in preparing the manuscript of this book. As a matter of fact, his lecture of "A Talk about Complexity" in Suzhou University during November 1991 is the starting point of our work in this direction. My best thanks also go to my colleagues in the group of dynamical systems in the Mathematics Department of Suzhou University, Professors Liu Zeng-rong, Chen Xi, Lu Qing-he, and Cao Yong-luo, for their comments and suggestions.

I am clearly aware that there are difficulties in language for me to write a book in English. I beg the indulgence of the reader for any mistakes remaining in the text.

This book was typeset by the author using the LaTeX document preparation system written by Leslie Lamport and Donald Knuth with much help and suggestion from Dr. Wong Lock-Yee, the Scientific Editor of World Scientific, and my colleague Wang Yi-Xun.

This work has been jointly supported by the National Basic Research Project "Nonlinear Science" (1991–1995) and the Natural Science Foundation of Jiangsu Province (1993–1995).

Suzhou, Jiangsu Huimin Xie
June 1996

Contents

List of Figures

Part I

STRINGS AND LANGUAGES

CHAPTER 1
FREE MONOIDS

In the symbolic study of dynamical systems scientists have to deal with symbolic strings or symbolic sequences in different ways, which on the whole are of an algebraic or combinatorial nature. In fact, the most important operation, among others, on symbolic strings is the concatenation of two strings. For instance, if $x = x_1 \cdots x_m$ and $y = y_1 \cdots y_n$ are two given strings, where x_i and y_j are symbols or letters taken from an alphabet set, then it is simply a matter of juxtaposing them together to obtain their concatenation, which is denoted by

$$xy = x_1 \cdots x_m y_1 \cdots y_n.$$

This is exactly the binary associative operation taken in a free monoid—a mathematical structure in algebra. This is the reason that our presentation of this book begins from a brief introduction about the free monoids. As a matter of fact, nearly all calculations in this book may be seen as algebraic operations on strings in some free monoid.

General references about the free monoids are Lothaire (1983) and Shyr (1991).

In Section 1.1 the concept of free monoids is introduced. Here the concept of free submonoids will play an important role in Chapter 6.

It turns out that the concept of the primitivity of strings is indispensable for symbolic study. It concerns that whether a string can be written as a concatenation of several copies of a shorter string. A well-known fact (Proposition 1.1.8) about the primitivity of strings is proved. At the end of Section 1.1 we list some operators on strings that are used throughout the whole book. They are σ (shift and cyclic shift), π, $\overline{x}$, and $\underline{x}$.

It is evident that if a free monoid has an additional structure then it may have more properties. In Section 1.2 we introduce into the free monoid $\Sigma^* = \{0,1\}^*$ an ordering relation that reflects the order of points in an interval in study of unimodal maps. Some basic results about strings in this ordered free monoid are collected there with proofs. These results will be widely used in Part II (Chapters 3–8), which is devoted to an analysis of the grammatical complexity of unimodal maps.

1.1 Free Monoids and Strings

1.1.1 Definition of Monoids

A monoid is a nonempty set of elements, together with an associative binary operation called multiplication and a unit element.

More precisely, let S be a nonempty set, and a binary associative operation be defined and denoted by juxtaposition. Then for all elements u and v in S, their product is an element in S and denoted by $uv \in S$. A set S is called a *semigroup* if it is defined thus far. Furthermore, if there exists a unit element, denoted by 1, such that $1u = u1 = u$ for each $u \in S$, then S is called a *monoid*.

Here some notations are in need. If $u = v$, then the concatenation of u and v is denoted by u^2. Similarly, for each natural number i the notation u^i is the concatenation of i copies of u. By definition, $u^0 = 1$. The string u^i is referred to as a *power* of u with the exponent i.

The binary operation of S can be extended to subsets of S by defining for $A, B \subseteq S$:

$$AB = \{ uv \mid u \in A, v \in B \}.$$

Denote A^2 for AA and A^n for the product of n copies of A. For a nonempty subset $A \subseteq S$ we define

$$A^* = 1 \cup A \cup A^2 \cup \cdots \cup A^n \cup \cdots$$

Obviously A^* is a submonoid of S and called the *submonoid* of S generated by A. Similarly the notation $A^+ = A^* - \{1\}$ is called the *semigroup* generated by A.

Example 1.1.1 Consider a special kind of monoids used in study of symbolic dynamics. We know that in symbolic study of dynamical systems an alphabet is taken first, which is a nonempty (finite or infinite) set whose elements are called symbols or letters. We denote this alphabet by Σ throughout the book.

A *symbolic string* over Σ consists of zero or more symbols of Σ, whereby the same symbol may occur several times. The string consisting of zero symbols is called the *empty string*. In this book we use ε to denote the empty string.

Now if we denote by Σ^* the set of all finite strings over the alphabet Σ, then it is clear that the set Σ^* is a monoid, in which the multiplication operation is just the concatenation of strings, and the unit element of the monoid is the empty string ε, that is,

$$\varepsilon u = u \text{ and } u = \varepsilon u \text{ for all } u \in \Sigma^*$$

Some notions and terminology for Σ^* are introduced now.

The *length* of a finite string $u = a_1 a_2 \cdots a_n$ is the number n of the symbols which occur in u. It will be denoted by $|u|$:

$$|u| = |a_1 a_2 \cdots a_n| = n.$$

Of course, the length of the empty string ε is 0.

A string $v \in \Sigma^*$ is said to be a *substring* of string $x \in \Sigma^*$, if there exist strings $u, w \in \Sigma^*$ such that

$$x = uvw,$$

and this substring v of x is also referred to as a *prefix* (*suffix*) of x if $u = \varepsilon$ ($w = \varepsilon$).

A substring v of $x \in \Sigma^*$ is said to be *proper* if $v \neq x$. Proper prefix and proper suffix are similarly defined.

Although a string may consist of infinite symbols, in the sequel we will call an infinite string by the name of sequence, and preserve the name of string to denote a finite one.

1.1.2 Free Monoids

A monoid S is said to be *free* if there exists a subset B of S which satisfies the following conditions:

1. $B^* = S$, where $B^* = 1 \cup B \cup B^2 \cup \cdots \cup B^n \cup \cdots$.
2. If $u_1 \cdots u_n = v_1 \cdots v_m$, where $u_1, \ldots, u_n, v_1, \ldots, v_m \in B$, then $m = n$ and $u_i = v_i, i = 1, \ldots, n$.

The set B is referred to as the *base* of S. It may be shown that for a free monoid its base is uniquely determined.

It is easy to see that the monoid Σ^* in Example 1.1.1 is free and Σ is its base. On the other hand, this special kind of monoids is the most important one for us as the following Proposition shows.

Proposition 1.1.2 A monoid S is free if and only if there exists an alphabet Σ and an isomorphism of the free monoid Σ^* onto S.

Its proof is straightforward and omitted.

By this Proposition we always assume that each free monoid we study in the sequel is a Σ^* for some alphabet Σ if not otherwise mentioned. At the same time we also assume the alphabet set Σ is finite in the book.

A subtle problem is that a submonoid of a free monoid need not be free.

Example 1.1.3 Let $\Sigma = \{0, 1\}$ and $A = \{00, 000\}$. The submonoid A^* of Σ^* is $\{0\}^* \setminus \{0\}$ and not free.

Example 1.1.4 Let $\Sigma = \{0, 1\}$ and $A = \{0, 10, 01\}$. Consider the submonoid A^* of Σ^* generated from A. It is easy to see that if A^* is free, then its base must be a subset of A. But at the same time every proper subset of A cannot generate A^*, so this base can only be A itself. But the equality

$$0(10) = (01)0$$

means that A is not the base of A^* and hence A^* is not free.

Example 1.1.5 Let $\Sigma = \{0, 1\}$ and $A = \{00, 10, 100, 11, 110\}$. It can be verified that A^* is a free submonoid of Σ^*.

An important fact used later in Chapter 6 is as follows.

Proposition 1.1.6 Let Σ^* be a free monoid and $u, v \in \Sigma^*$ be given. If $uv \neq vu$, then the submonoid $\{u, v\}^*$ generated by u, v is free, and $\{u, v\}$ is its base.

For its proof see Corollary 1.2.6 in Lothaire (1983).

1.1.3 Primitive Strings

Definition 1.1.7 A nonempty string $x \in \Sigma^*$ is said to be *primitive* if it is not a repetition of a shorter string of Σ^*, that is to say,

$$x = y^n \text{ for some } y \in \Sigma^* \text{ and an integer } n > 0 \Longrightarrow x = y \text{ and } n = 1.$$

A fact used frequently in the sequel is that each nonempty string $x \in \Sigma^*$ can be expressed as $x = y^n$ by some primitive string $y \in \Sigma^*$ and an integer $n > 0$. We may call y the *primitive root* of x (Salomaa 1981).

The following fact, which appears in many references (see, e.g., Lothaire 1983, Shyr 1991, Salomaa 1981), is indispensable for calculation in free monoids, and hence of constant use in this book.

Proposition 1.1.8 Let $u, v \in \Sigma^*$. If $uv = vu$, $u \neq \varepsilon$, $v \neq \varepsilon$, then both u and v are powers of a common primitive string of Σ^*.

Proof. Proceed inductively on $|uv|$, the length of string uv. It is trivially true if $|uv| = 2$. Assume that our claim is true for the length $|uv| \leq k$ and consider the case of $|uv| = k + 1$. Since Σ^* is free, if it happens that $|u| = |v|$, then we have $u = v$. Otherwise, without loss of generality we may assume that $|u| < |v|$. From the condition $uv = vu$ we have $v = uw$ and $uuw = uwu$. This leads to $uw = wu$. Using the inductive hypothesis there exists a primitive string x and integers $m, n > 0$ such that $u = x^m$ and $w = x^n$. Then we obtain $v = x^{m+n}$ and finish our induction. ∎

Corollary 1.1.9 A string $x \in \Sigma^*$ is not primitive if and only if $x = uv = vu$ for some $u, v \in \Sigma^+ = \Sigma \setminus \{\varepsilon\}$.

There are many results about primitive strings in free monoids (see, e.g., Shyr 1991), but in the sequel we only need a special result from Xie (1995b), which play a basic role in the discussion about Fibonacci sequences in Chapter 6.

Proposition 1.1.10 If $x, y \in \Sigma^*$ and $xy \neq yx$, then the string $xyxxy$ is primitive.

Proof. Assume the contrary that

$$xyxxy = u^p$$

holds for a primitive string u and an integer $p > 1$. Observe that by the condition $xy \neq yx$ neither of x and y is empty. Thus it can be seen that the substring xy cannot be a power of u. Otherwise, from the equality $xyxxy = (xy)x(xy) = u^p$ and xy being a power of u, x, and then y, would also be powers of u and this contradicts the condition of $xy \neq yx$.

Now consider the following two cases separately. (Again by $xyxxy = u^p$ and $xy \neq u$ it is clear that $|xy| = |u|$ is impossible.)

(a) $|xy| < |u|$. Here the exponent p must be 2 and we may write

$$u = xyv_1 = v_2xy, \; x = v_1v_2, \text{ and } |v_1| = |v_2|.$$

Because $u = xyv_1 = (v_1v_2)yv_1 = v_2xy$, both v_1 and v_2 are prefixes of u and of the same length, so $v_1 = v_2$ holds. By Corollary 1.1.9, the equality $u = (xy)v_1 = v_1(xy)$ contradicts to the primitivity of u.

(b) $|xy| > |u|$. Then we can find an integer $q > 0$ such that

$$xy = u^q w_1 = w_2 u^q,$$

where w_1 is a nonempty proper prefix of u and $|w_1| = |w_2|$. Since both u and w_2 are prefixes of xy and $|w_2| = |w_1| < |u|$, so w_2 is also a prefix of u. From $|w_1| = |w_2|$ we obtain $w_1 = w_2$. Again by Corollary 1.1.9, the equality $xy = (u^q)w_1 = w_1(u^q)$ leads to a contradiction with the primitivity of u and $|w_1| < |u|$. ∎

1.1.4 *Cyclic Shifts of Strings*

The *cyclic shift* operator on a nonempty string $s = s_1 \cdots s_n$ is defined by

$$\sigma(s) = s_2 \cdots s_n s_1.$$

We call $\sigma^i(s)$, $i = 0, 1, \ldots, |s| - 1$, the *cyclic shifts* (or *cyclic permutation*) of string s.

It is obvious that s itself is also a cyclic shift of each $\sigma^i(s)$ for $i = 0, 1, \ldots, |s| - 1$. (This concept is called *conjugacy* between strings in Lothaire 1983.) A direct consequence of Proposition 1.1.8 is the following fact.

Proposition 1.1.11 If $s \in \Sigma^*$ is primitive, then each $\sigma^i(s)$ is primitive, and any two distinct cyclic shifts of s are different, that is to say,

$$\sigma^i(s) \neq \sigma^j(s) \text{ for all } i \neq j, 0 \leq i, j \leq |s| - 1.$$

Similarly, the *shift* operator on an infinite sequence $s = s_1 \cdots s_n \cdots$ is defined by

$$\sigma(s) = s_2 \cdots s_n \cdots$$

Although we use the same notation σ for both shift and cyclic shift operators, it is easy to recognize its exact meaning from the context.

1.1.5 *Other Operators and Notations*

We need the following operators on strings in the rest of the book.

(1) For a nonempty string x the operator π may be used on x in two ways. The left-π on x is to remove x's first symbol, and the right-π on x is to remove its last symbol. For instance, if $x = 10110$, then

$$\pi(x) = \pi x = 0110, (x)\pi = x\pi = 1011, \text{ and } \pi x \pi = 011.$$

(2) Operator $\bar{x}$ on a nonempty string $x \in \{0, 1\}^*$ is to change its last symbol, that is, from 0 to 1 and vice versa:

$$\text{if } x = 10110, \text{ then } \bar{x} = (x\pi)\bar{0} = 10111.$$

We call $\overline{x}$ the *dual string* of x. Of course, we have $\overline{\overline{x}} = x$ and $x\pi = \overline{x}\pi$. In Part II (Chapters 3–8) we will see that it is often useful for a given string to consider its dual one or both strings simultaneously.

(3) Operator $\underline{x}$ on a nonempty string $x \in \{0,1\}^*$ is to change its first symbol as follows:

$$\text{if } x = 10110, \text{ then } \underline{x} = \underline{1}(\pi x) = 00110.$$

This operator is used in Part III (Chapter 9–11).

1.2 An Ordered Free Monoid

As a matter of fact, in the symbolic study of unimodal maps we treat symbolic strings in a special monoid Σ^* where $\Sigma = \{0,1\}$ and a total order relation is defined on it. (In the symbolic study of circle homeomorphisms (Part III) we meet the same monoid but with a different order relation, the lexicographical order, defined there.)

The results presented in this section are common knowledge in the free monoid defined below. Some of them were scattered in references (see, e.g., Zheng 1989b, Hao 1989, Xie 1993b, Wang and Xie 1994).

1.2.1 An Order Relation on the Monoid $\{0,1\}^*$

As in ordinary number systems we use notations "=", "<", ">", "≤", and "≥" to denote the order relation defined below between strings of monoid $\{0,1\}^*$.

First a nonempty string s over $\Sigma = \{0,1\}$ is called an *odd (even) string* if the number of symbol 1's in s is odd (even).

Let two strings $s, t \in \Sigma^*$ be $s = s_1 \cdots s_m$, $t = t_1 \cdots t_n$, we call $s < t$ $(s > t)$ if there exists an integer $k \geq 0$ such that $s_1 \cdots s_k = t_1 \cdots t_k$, $s_{k+1} \neq t_{k+1}$, and $t_1 \cdots t_{k+1}$ is odd (even), otherwise we have $s \geq t$ $(s \leq t)$. Of course if $m = n$ and $s_i = t_i$ for $i = 1, \ldots, m$ then we have $s = t$. But here some caution is needed. It we find that both $s \leq t$ and $s \nless t$ are true, then we have $s = t$, but, if $|s| \neq |t|$, here its exact meaning is that the shorter string between s and t is a proper prefix of the other string. It is obvious that the order relation defined thus far is total, that is, for any two $s, t \in \Sigma^*$ either $s \leq t$ or $t \leq s$ holds.

We can extend this order relation naturally either between infinite sequences or between an infinite sequence and a finite string.

1.2.2 Maximal Strings

Definition 1.2.1 A finite string x is said to be *maximal* if

$$\sigma^i(x) \leq x \text{ for each } i = 1, \ldots, |x| - 1.$$

Similarly an infinite sequence s is said to be *maximal* or *shift-maximal* if

$$\sigma^i(s) \leq s \text{ for each } i \geq 0.$$

The notation

$$M(x) = \max\{x, \sigma(x), \sigma^2(x), \ldots, \sigma^{|x|-1}(x)\}$$

denote the *maximal cyclic shift* of a string x.

Here we collect some basic results concerned with maximal strings.

Lemma 1.2.2 Let $x = us$ and $y = ut$ be given. If u is odd, and inequalities $y \leq x$ and $t \leq s$ hold, then either x is a prefix of y or vice versa.

Proof. From $ut \leq us$ and u being odd we obtain $t \geq s$. Combining it with $t \leq s$ leads to the conclusion. ∎

The following two facts are often used in symbolic study of unimodal maps.

Proposition 1.2.3 If $z = uuv$ is a maximal finite string, in which u is an odd string, then z is a prefix of a power of u.

Proof. Since $uv \leq uu$ and $v \leq u$, by taking $x = uu$, $y = uv$, $s = u$, $t = v$, and using Lemma 1.2.2, we see that if $|v| \leq |u|$ then our proof finishes. Otherwise, we have $v = uv_1$ and $z = uuuv_1$. From $uv_1 \leq uu$ and $v_1 \leq u$ we can again argue as above. Proceeding inductively we obtain the result required. ∎

For infinite sequences we have the following strong result by the same argument. A direct proof, however, is given below.

Proposition 1.2.4 If $x = uuy$ is a maximal infinite sequence, in which u is an odd string, then $x = u^\infty$.

Proof. Assume the contrary that the conclusion is not true, then the sequence x must be of the form $x = u^n u' \cdots$, where $n \geq 2$, $u' \neq u$, and $|u'| = |u|$. Since the sequence x is maximal, we have

$$u' \leq u, \text{ and } uu' \leq uu.$$

Since u is odd, the latter inequality equals to

$$u' \geq u,$$

hence $u' = u$, a contradiction with $u' \neq u$. ∎

1.2.3 Maximal and Primitive Strings

Here we consider those strings which are simultaneously maximal and primitive. They have special properties that are indispensable for the symbolic study of unimodal maps of Part II.

The first result is a direct consequence of Propositions 1.1.11 and 1.1.8.

Lemma 1.2.5 A nonempty string x is maximal and primitive if and only if

$$\sigma^i(x) < x \text{ for all } 1 \leq i < |x|.$$

First we introduce a notion of prefix-suffix of a string (to itself). Its abbreviation PS will be used in the rest of the book.

Definition 1.2.6 Let a string x be given. If a string y is x's prefix and suffix at the same time, then y is called a *prefix-suffix* of x to itself, or simply a PS of x.

Definition 1.2.7 Three basic properties that a nonempty string x may have are:

(1) each suffix of x, say y, satisfies the condition $y \leq x$, (P1)

(2) there exists no (nonempty) even proper PS of x to itself, (P2)

(3) the string x is either maximal and primitive or $x = yy$ where the string y is odd, maximal and primitive. (P3)

We will study relations between (P1), (P2), and (P3), and then obtain some basic facts about maximal and primitive strings.

Lemma 1.2.8 A string x has the property (P1) $\Longrightarrow$ its dual string $\bar{x}$ has the property (P2).

Proof. Assume the contrary that $\bar{x}$ has y as an even proper PS. Then x will have y as its even prefix, and $\bar{y}$ its odd suffix. From $\bar{y} > y \Rightarrow \bar{y} > x$ we find a contradiction with the property (P1) which x satisfies. ∎

Remark. The converse need not be true. For instance, if $x = 101001$, then $\bar{x}$ has (P2), while x do not satisfy (P1).

Lemma 1.2.9 x satisfies (P1), (P2) $\Longleftrightarrow$ $\bar{x}$ satisfies (P1), (P2).

Proof. Since $\bar{\bar{x}} = x$, it suffices to prove that x has (P1), (P2) $\Rightarrow$ $\bar{x}$ has (P1), (P2).

By Lemma 1.2.8 it suffices to prove $\bar{x}$ has (P1). Assume the contrary that $\bar{x}$ has a (proper) suffix y such that $y > \bar{x}$. Since $|y| < |x|$ this leads to $y > x$ too. On the other hand, the string x will have $\bar{y}$ as its proper suffix. By (P1) for x we obtain

$$\bar{y} \leq x < y.$$

These inequalities mean two things: y is odd, and x has $\bar{y}$ as its prefix. Therefore, x has an even proper PS $\bar{y}$. This contradicts the property (P2) which x satisfies. ∎

The main result in this subsection is the following Theorem.

Theorem 1.2.10 A string x has the properties (P1), (P2) if and only if x has the property (P3).

Proof. (P3) $\Longrightarrow$ (P1) and (P2).

The condition of x being maximal already implies (P1). If x do not have (P2), then there are decompositions

$$x = \alpha\beta = \beta'\alpha, \text{ where } \alpha \text{ is even, and } 0 < |\alpha| < |x|.$$

Since x is maximal, we have $\alpha\beta' \leq x = \alpha\beta$, and, since α is even, $\beta' \leq \beta$. On the other hand, by the property (P1) we have $\beta \leq x$, and thus $\beta \leq \beta'$. Combining these inequalities we obtain $\beta' = \beta$ and $x = \alpha\beta = \beta\alpha$. By applying Proposition 1.1.8, there are two natural integer a, b and a primitive string u such that $\alpha = u^a$, $\beta = u^b$, and hence $x = u^{a+b}$. By (P3) we can only have $a = b = 1$ and u being odd, maximal and primitive. But then $\alpha = u$, here α is even, while u is odd, a contradiction.

(P1) and (P2) $\Longrightarrow$ (P3). Assume the contrary that x is not a maximal and primitive string, then by Lemma 1.2.5, there exists a decomposition of x such that $x = \alpha\beta$ where $0 < |\alpha| < |x|$ and

$$\sigma^{|\alpha|}(x) = \beta\alpha \geq x = \alpha\beta. \tag{1.1}$$

If ">" holds it means that x is not maximal, and if "=" holds then x is not primitive.

By the property (P1) we have $\beta \leq x = \alpha\beta$. Hence from (P2) β is an odd PS of x. Writing $x = \beta\alpha'$ and using (1.1) we obtain $\alpha \leq \alpha'$. But by (P1) we also have $\alpha' \leq \alpha$, and hence $\alpha = \alpha'$. Thus $x = \alpha\beta = \beta\alpha$. Here we have proved that x is maximal.

Furthermore, by applying Proposition 1.1.8 to x there exist two natural integers a, b and a primitive string y such that

$$\alpha = y^a, \beta = y^b, \text{ and hence } x = y^{a+b}$$

If $a + b > 2$ then x would have y^2 as its even proper PS which contradicts (P2). Thus we have but $a = b = 1$ and $x = y^2$ with $y = \beta$ being odd. Since x is maximal, the string y in $x = yy$ is maximal. ∎

The following Corollaries are often used later, and each of them is a direct consequence of Theorem 1.2.10.

Corollary 1.2.11 A maximal and primitive string cannot have even proper prefix-suffix to itself.

Corollary 1.2.12 An odd string x is maximal and primitive if and only if x has the properties (P1) and (P2).

Corollary 1.2.13 An odd string x is maximal and primitive if and only if its dual string $\bar{x}$ has property (P3), that is, $\bar{x}$ is either maximal and primitive or $\bar{x} = yy$, where y is an odd maximal and primitive string.

The following result is also a consequence of Theorem 1.2.10 and will be used later.

Proposition 1.2.14 If a string $x = ab$ is maximal and primitive, where the prefix a is odd and the suffix b is even, then $b < a$.

Proof. Since x is maximal we have $b \leq ab$. Since b is even, by applying Corollary 1.2.11 we see that b cannot be a prefix of x, and hence $b < ab$ holds. Thus $b > a$ is impossible. Now assume the contrary that $b < a$ is false, then we have $a = b$, which means that one of a, b is a prefix of the other one. Since their parity is different, $a = b$ is impossible. If b is a proper prefix of a, then b becomes a proper even prefix-suffix of x. This contradicts to Corollary 1.2.11 again. Therefore, a must be a proper prefix of b. Writing $b = ab_1$, $x = a^2b_1$, and using Proposition 1.2.3 to x, we have $x = a^n b'$ for some $n > 1$ and b' being a prefix of a. Since x is primitive, we have $0 < |b'| < |a|$. But now both ab' and b' are proper prefixes of x. Since a is odd, one substring between b' and ab' must be even, hence we obtain a contradiction with Corollary 1.2.11 again. ∎

CHAPTER 2
DYNAMICAL LANGUAGES

In order to understand symbolic behaviors of dynamical systems it is usually not enough to analyze finite strings or infinite sequences one by one. On the contrary, people have often to consider some set of strings or sequences collectively. Recalling from the theory of formal languages that a language is simply a set of symbolic strings over an alphabet set, it is thus natural to use the theory of formal languages to analyze problems in symbolic dynamics (Hopcroft and Ullman 1979, Günther, Schapiro and Wagner 1994).

For reader's convenience a brief introduction about regular languages, non-regular languages, and languages generated by Lindenmayer's L systems is given in Appendices A, B, and C.

In this chapter a special class of languages generated from general dynamical systems is discussed and the name of dynamical languages is proposed for them. Some basic notions, including distinct excluded blocks, symbolic flows, graphs, and *topological entropy, associated with dynamical languages are presented in Sections.* These notions are main tools used throughout in Parts II and III.

Section 2.1 gives the definition of dynamical languages. Other names used in references are briefly reviewed.

In Section 2.2 a theory about distinct excluded blocks (or forbidden words) is developed. This is a notion used widely but discussed rarely in references.

Section 2.3 is used to present the connection between symbolic flows and dynamical languages, including a brief discussion about subshifts of finite type and sofic systems in this aspect.

In Section 2.4 graphic representations of dynamical languages, including transition diagrams of automata, are discussed. For the transitive case a theory of minimal graphs due to Fischer is presented.

In Section 2.5 the notion of topological entropy is presented from the viewpoint of languages.

Remark. In this book we focus our attention on the complexity of sets of strings, namely, of languages. There exist, however, many works devoted to the complexity of individual string or sequence. Here the notion of complexity proposed by Kolmogorov, Chaitin and Solomonoff occupies the central position. Readers who are interested in this field can see, e.g., Kolmogorov (1965), Chaitin (1966), Solomonoff (1965), Lempel and Ziv (1976), Brudno (1983), Kaspar and Schuster (1987), Herzel (1988), Chapter 10 of Rasband (1990), Keller (1991) Li and Vitányi (1993), Chapter 7 of Xie (1994).

2.1 Definition of Dynamical Languages

Let Σ be an alphabet, that is, a finite set of symbols, and Σ^* be the set of all finite strings over Σ. Every subset of Σ^* is called a *formal language* (or simply *language*) over Σ.

It is natural that languages generated from different fields will have different features. An early example is as follows. We know that the use and study of symbolic dynamics have a long history. It began with the study of geodesic flows by Hadamard (1898) and Morse (1921a, 1921b). (See references in Morse and Hedlund 1938, 1940, Gottschalk and Hedlund 1955.) In Morse and Hedlund (1938 pp. 822–823) it was explicitly said that in the applications to dynamical theory the sequences admitted are not formed freely from the generating symbols, but are subject to certain limitations. Four conditions of admissibility were listed for the problem presented there.

It is evident that the languages generated from the symbolic study of dynamical systems have some common features. Different names have appeared in references to reflect these features. Here we prefer using the name of dynamical languages to denote them. It seems that this name appeared first in Coven and Paul (1977 p.265), where a language L is called dynamical if it is the set of words of some symbolic flow. (The symbolic flows and their associate languages will be discussed later in Section 2.3.)

Definition 2.1.1 A nonempty language $L \subset \Sigma^*$ is called *dynamical* if it satisfies the following conditions:

D1. a string $z \in L$ if and only if every substring of z belongs to L.
D2. if $z \in L$, then there exists symbols $a, b \in \Sigma$ such that $azb \in L$.

It seems that the conditions D1 and D2 are ubiquitous in most studies of dynamical systems. In Nasu (1985) and Hurd (1988) the names of *admissible languages* and *subshift language* are used for the same object. In de Luca and Varricchio (1990), the condition D1 is called *factorial* and the condition D2 *prolongable*. In Hanson and Crutchfield (1992), the subclass of regular languages that satisfy the condition D1 is called the class of *process languages*. In Troll (1993) similar ideas were proposed for the study of truncated horseshoes.

Remark. For some applications the condition D2 could be replaced by the right-prolongable (left-prolongable) condition:

D2′. if $z \in L$, then there exists a symbol $a \in \Sigma$ such that $za \in L$ ($az \in L$).

In Xie (1995a) the name "class D of languages" was used for those languages that satisfy the conditions D1 and D2′ (right-prolongable).

Two consequences can be obtained from the condition D1 directly.

Since ε, the empty string, is a substring of each string, it follows that every dynamical language contains the empty word ε.

Lemma 2.1.2 If L is a dynamical language, then ε, the empty string, belongs to L.

Considering the relationship between the condition D1 and the complement language, namely, the language obtained from a given language by complementation, it follows the second consequence.

Lemma 2.1.3 If L is a language $L \subseteq \Sigma^*$ that satisfies the condition D1 and its complement language is denoted by $L' = \Sigma^* - L$, then

$$z \in L' \Longrightarrow xzy \in L' \text{ for all } x, y \in \Sigma^*$$

This fact is expressed in de Luca and Varricchio (1990) by

$$L \text{ is factorial } \Longleftrightarrow L' \text{ is a two-sided ideal of } \Sigma^*$$

It may also be written as

$$L' = \Sigma^* L' \Sigma^*$$

2.2 Distinct Excluded Blocks

The notion of forbidden blocks (or forbidden words) has appeared frequently in works concerned mainly with symbolic flows (see, e.g., Adler 1991 and Section 2.3). In Wolfram (1984) it appeared under the name of distinct excluded blocks, and was used to analyze the complexity of cellular automata. D'Alessandro and Politi (1990), Xu, Liu and Liu (1994) have developed some complexity measures from this notion for some systems.

In this section we give a general discussion about this notion under the name given by Wolfram. The result obtained here will be used in Parts II and III as a basic tool in the study of complexity. The material of this section is largely developed in Xie (1995a). Its abbreviation DEB will be used in the sequel.

2.2.1 Definition and Properties

Definition 2.2.1 A string x is a *distinct excluded block* (DEB) of a language L, if $x \in L'$, the complement language of L, but each proper substring of x belongs to L. The set of all DEB of language L is denoted by L''.

Associated with the notion of DEB are the following notions (see, e.g., Wolfram 1984).

Definition 2.2.2 A language L is called *finite (infinite) complement* if its L'' is a finite (infinite) set.

It is easy to show the existence of DEB for a given language $L \subsetneq \Sigma^*$. For instance, if $L' \neq \emptyset$, then the shortest string of L' is a DEB of L.

Although we may obtain DEB's for every nonempty language, but in general the set L'' found thus far cannot characterize L completely. It is easy to find two different languages whose set of all DEB's are the same. For the dynamical languages, however, the situation is quite different.

Lemma 2.2.3 If a nonempty language $L \subseteq \Sigma^*$ satisfies the condition D1, then

$$L = \Sigma^* - \Sigma^* L'' \Sigma^*$$

Proof. By Lemma 2.1.2 the empty string $\varepsilon \in L$, hence each string of L' contains at least one nonempty substring as its DEB. Thus it follows that

$$L' \subseteq \Sigma^* L'' \Sigma^*$$

On the other hand, from Lemma 2.1.3 we obtain

$$\Sigma^* L'' \Sigma^* \subseteq L'$$

and finish the proof. ∎

In order to clarify the relationship between L and L'' further, we need to introduce some operations on languages as follows (Hopcroft and Ullman 1979):

$$\begin{aligned} MIN(L) &= \{x \in L \mid \text{ no proper prefix of } x \text{ is in } L\}, \\ MIN'(L) &= \{x \in L \mid \text{ no proper suffix of } x \text{ is in } L\}, \\ R(L) &= \{x \mid x^R, x \text{ is written backward, belongs to } L\}. \end{aligned}$$

Here x^R is called the *mirror* of a string x, and $R(L)$ the *mirror* of a language L (see Subsection *A.4.2*).

It is straightforward to verify that

$$MIN'(L) = R \circ MIN \circ R(L)$$

and obtain the following conclusion.

Lemma 2.2.4 If L is a dynamical language, then

$$\begin{aligned} L'' &= MIN \circ MIN'(\Sigma^* - L) \\ &= MIN' \circ MIN(\Sigma^* - L) \\ &= MIN'(\Sigma^* - L) \cap MIN(\Sigma^* - L). \end{aligned}$$

The next question is: which set (of strings) can be the set of all DEB's for a dynamical language? Here we need the operator "π" on nonempty strings defined in Subsection *1.1.5*. (Recall that, for instance, $\pi 10110 = 0110$, and $10110\pi = 1011$.)

Proposition 2.2.5 There exists a dynamical language L for a given set $U \subseteq \Sigma^*$ such that $L'' = U$ if and only if the set U satisfies the following conditions:

1. No string in U is a proper substring of another string in U.
2. If card $S = n$, and $x_1, \cdots, x_n$ are n strings in U with different last symbols, and x_n is the longest one among them, then at least one string among $x_1\pi, \cdots, x_{n-1}\pi$ is not a suffix of $x_n\pi$.
3. If $y_1, \cdots, y_n$ are n strings in U with different first symbols, and y_n is the longest one among them, then at least one string among $\pi y_1, \cdots, \pi y_{n-1}$ is not a prefix of πy_n.

Proof. Since the "only if" part is trivial, only the proof of the "if" part is given. Using the given set U, a language L can be constructed as follows:

$$L = \Sigma^* - \Sigma^* U \Sigma^*$$

We should show first that L is a dynamical language and then $L'' = U$.

Assume the contrary that L does not satisfy the condition D1, then there is a string $x = uvw \in L$ but $v \notin L$. Now we have $v \in L'$, and $v = v_1 v_2 v_3$ with $v_2 \in U$ by the definition of L. Writing $x = uvw = u(v_1 v_2 v_3)w = (uv_1)v_2(v_3 w)$ leads to $x \in L'$ that contradicts to $x \in L$.

Now we show that L is right-prolongable. Assume the contrary that there is a word $z \in L$ with the property that $za \notin L$ for every $a \in L$. Since L satisfy the condition D1 already, we can use Lemma 2.1.3 for $za \in L'$ to obtain the shortest suffix of za, which belongs to U. (It is easy to see this suffix is a DEB of L.) These suffixes provide the strings $x_1, \cdots, x_n$ required. Since each $x_i\pi$ is a suffix of the same string z, it contradicts to the condition 2.

Similarly it can be proved that each word of L is left-prolongable. The proof of $L'' = U$ is easy and omitted. ∎

2.2.2 L and L″ in Chomsky Hierarchy

Formal languages can be classified by their grammatical complexity into different levels in Chomsky hierarchy (see, e.g., Hopcroft and Ullman 1979 and Appendix B). The following inclusion relations in Chomsky hierarchy are well-known:

$$\mathcal{L}(\mathrm{RGL}) \subsetneq \mathcal{L}(\mathrm{CFL}) \subsetneq \mathcal{L}(\mathrm{CSL}) \subsetneq \mathcal{L}(\mathrm{RL}) \subsetneq \mathcal{L}(\mathrm{REL}).$$

Here the abbreviations RGL, CFL, CSL, RL, and REL stand for regular language, context-free language, context-sensitive language, recursive language, and recursively enumerable language. (See Subsection *B.2.1* for details about them.)

Proposition 2.2.6 A dynamical language L is regular if and only if its L'' is regular.

Proof. Since the set $\mathcal{L}(\mathrm{RGL})$ is closed under concatenation, complementation, MIN, and R, it is a direct consequence of Lemmas 2.2.3 and 2.2.4. ∎

Remark. Since each finite language is regular, it follows that each finite complement language is regular.

By the new result about context-sensitive languages due to Immerman (1988) and Szelepcsényi (1987) (see Davis, Sigal and Weyuker 1994), the class $\mathcal{L}(\mathrm{CSL})$ is closed under the operation of complementation. Thus it is easy to obtain the following result.

Proposition 2.2.7 A dynamical language L is context-sensitive if and only if its L'' is context-sensitive.

It is very easy to obtain following two results.

Proposition 2.2.8 A dynamical language L is recursive if and only if its L'' is recursive.

Proposition 2.2.9 If L is a dynamical language, and neither of L and L'' is recursive, then there are three possibilities:

1. L is recursively enumerable, L'' is non-recursively enumerable,
2. L is non-recursively enumerable, L'' is recursively enumerable,
3. both L and L'' are non-recursively enumerable.

An open problem about the notion of DEB is the following conjecture.

Conjecture 1 A dynamical language L is context-free if and only if its L'' is context-free.

The levels of the grammatical complexity of L and L'' in Chomsky hierarchy are shown in Figure 2.1, where the results of Propositions 2.2.6–2.2.9 are represented by "$\longleftrightarrow$", and Conjecture 1 is represented by "$\leftarrow$? $\rightarrow$".

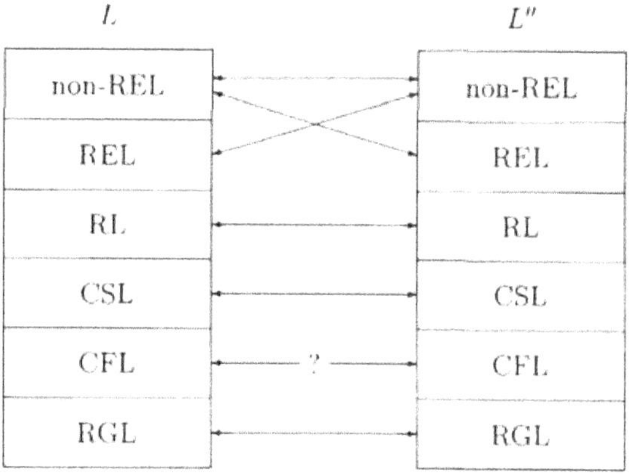

Figure 2.1: The complexity levels of L and L'' in Chomsky hierarchy.

Finally, although we do not know if Conjecture 1 is true, but a dynamical language L may be constructed such that both L and L'' are context-free languages, and neither of them is regular. Its basic idea is contained in Hurd (1990a).

Example 2.2.10 First let a language K be defined by

$$K = \{1^p 0^n 1 0^n 1^q \mid p, n, q \geq 0\},$$

then the language

$$L = \{x \mid x \text{ is a substring of } y \in K\}$$

is obviously dynamical. Using the fact that the language $\{0^n 1 0^n \mid n \geq 0\}$ is context-free (cf. Example B.2.5), we know L is also a context-free language. On the other hand, from

$$L \cap 10^* 10^* 1 = \{10^n 10^n 1 \mid n \geq 0\}$$

and the language of its right-hand side is non-regular, we know L is non-regular.

Now we can calculate the set of all DEB's of L directly:

$$L'' = \{01^m0 \mid m > 1\} \cap \{10^n10^m \mid m > n\} \cap \{0^n10^m1 \mid m < n\} \cap \{010^n10 \mid n > 0\}.$$

Using this expression and the fact that

$$L'' \cap 0^*10^*1 = \{0^n10^m1 \mid m < n\}$$

is non-regular, it follows that L'' is a context-free language, but not a regular language.

2.2.3 A Natural Equivalence Relation

Since the idea contained in the Myhill-Nerode Theorem is indispensable to our presentation in the book, we include its proof completely in Appendix A (see Subsection *A.3.3*) for readers' reference. Its content can be summarized as follows.

If a language $L \subseteq \Sigma^*$ is given, then it introduces a natural equivalence relation R_L on Σ^*: for $x, y \in \Sigma^*$, xR_Ly if and only if $xz \in L \Leftrightarrow yz \in L$ for all $z \in \Sigma^*$. The language L is regular if and only if the index of R_L, that is, the number of equivalence classes of Σ^* with respect to R_L, is finite.

In this subsection it is shown that if L is dynamical then the index of R_L can be calculated from L'', the set of all DEB's of L. This idea will used later in Parts II and III, respectively.

The first result in this aspect is a consequence of Lemma 2.1.3

Proposition 2.2.11 If $L \subseteq \Sigma^*$ is a dynamical language, then its complement $L' = \Sigma^* - L$, seen as a subset of Σ^*, is exactly an equivalence class of R_L.

Proposition 2.2.12 Let L be dynamical and L'' be its set of all DEB's. If a set V is defined by

$$V = \{v \mid v \text{ is a proper prefix of some } y \in L''\},$$

then for each word $x \in L$ there exists a string $v \in V$ such that

$$xR_Lv.$$

Proof. Let a string v be the longest suffix of x which belongs to V. That is, v is the longest prefix-suffix of x with respect to the set V (cf. Definition 1.2.6). We will show that xR_Lv. Here $v = \varepsilon$ is allowed. (Since $\varepsilon \in L$ from Lemma 2.1.2.)

Let z be an arbitrary string over Σ. Consider the strings xz and vz. If $vz \notin L$, then since vz is a suffix of xz, the condition D1 implies $xz \notin L$.

On the other hand, consider the case of $xz \notin L$. If $z \notin L$, then $vz \notin L$ immediately. Otherwise we may suppose $z \in L$, and, using Lemma 2.2.3, obtain decompositions of

$$x = x_1x_2, \quad z = z_1z_2,$$

such that $x_2 z_1$, a substring of $xz = x_1 x_2 z_1 z_2$, is a DEB. From $x, z \in L$ we know that $x_2 \neq \varepsilon$, and $z_1 \neq \varepsilon$. These imply that the substring x_2 is a proper prefix of a DEB, and so $x_2 \in V$. Comparing x_2 with v leads to the conclusion that x_2 is a suffix of v. Thus we see vz, which contains a DEB as its substring, does not belong to L. From

$$xz \notin L \iff vz \notin L \text{ for all } z \in \Sigma^*$$

we obtain xR_Lv and complete the proof. ∎

These two Lemmas enable us to calculate the index of R_L from L''. In order to do so we need only to work on the set V, if L'' is known already. The following fact is convenient to this calculation.

Proposition 2.2.13 For two given strings $u, v \in V$, uR_Lv if and only if for each $z \in \Sigma^*$, uz contains a DEB as its suffix $\iff vz \in L'$ and vice versa.

Two examples are given to show how to calculate index of R_L from L'' and construct the minDFA that accepts language L (by the idea in the proof of Myhill-Nerode Theorem).

Example 2.2.14 If a language $L \subset (0+1)^*$ is given by its $L'' = \{0100, 0010, 001100\}$, then the minDFA for L may be obtained as follows.

First calculate the set V in Proposition 2.2.12:

$$V = \{\varepsilon, 0, 00, 001, 0011, 00110, 01, 010\}.$$

Using the rule of Proposition 2.2.13, we may establish

$$01\ R_L\ 0011, \text{ and } 001\ R_L\ 010\ R_L\ 00110.$$

Combining these equivalence relations with Propositions 2.2.11 and 2.2.12, the index of R_L is 6 and the six equivalence classes are

$$[\varepsilon], [0], [00], [001], [0011], \text{ and } L',$$

where $[x]$ is the class containing string x. Now it is routine to obtain the required minDFA M as shown in Figure 2.2 (see Subsection *A.3.3*). Here the filled circles are used to represent accepting states. The unique nonaccepting state is represented by a circle labeled with L'. As a matter of fact, this language L is the set of configurations after one time step of elementary cellular automaton of rule 222 (Wolfram 1984, 1986).

Example 2.2.15 This is also a language of elementary cellular automaton (of rule 18) after one time step. The set of DEB's is given by the regular expression (see Subsection *A.2.2*):

$$L'' = 11(0^*10^+1)^*1.$$

Although here the set V is infinite, we can still use the foregoing method to determine all equivalence classes as follows:

$$[\varepsilon], [1], [11], [110], [1101], \text{ and } L'$$

The minDFA accepting L is illustrated in Figure 2.3.

Remark. Readers who are interested in the complexity of cellular automata can see, e.g., Golze (1976), Wolfram (1986), Hurd (1988, 1990a, 1990b), Gutowitz (1990), Jackson (vol 2. 1990), Hurd, Kari and Čulik (1992), Kari (1992).

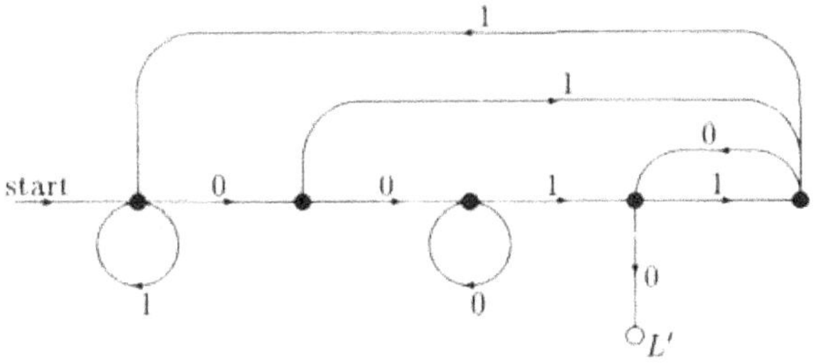

Figure 2.2: The minDFA for Example 2.2.14.

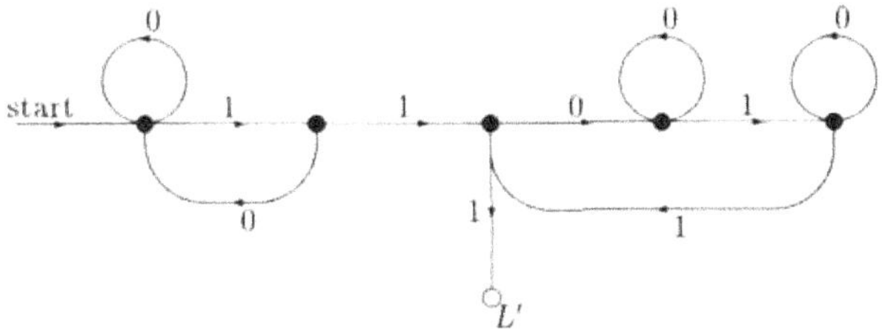

Figure 2.3: The minDFA for Example 2.2.15.

2.3 Symbolic Flows

The notion of symbolic flows, or shift systems, has been used and analyzed in dynamical systems as well as in ergodic theory (see, e.g., Smale 1967 and references therein). Here we only give the definition involved and discuss the relation between symbolic flows and dynamical languages.

2.3.1 Symbolic Flows and Dynamical Languages

Let Σ be a finite alphabet, and $X = \Sigma^Z$ be the set of all two-sided infinite symbolic sequences over Σ:

$$X = \{x = (\cdots x_{-1}x_0x_1\cdots) \mid \text{each } x_i \in \Sigma\}.$$

The set Σ is given the discrete topology and X the induced product topology. An operation shift σ on X is defined by

$$(\sigma(x))_i = x_{i+1} \quad \text{for } i \in Z.$$

The pair (X, σ) is said to be the *full shift (system)* over Σ, and a restriction of it is defined as a symbolic flow (subshift system) as follows.

Definition 2.3.1 If Y is a nonempty, closed, σ-invariant subset of X, then the pair (Y, σ) (or simply Y) is called a *symbolic flow* over Σ.

For every subset $Y \subseteq X = \Sigma^Z$ we can define a natural language associated with it.

Definition 2.3.2 Let Y be a subset of $X = \Sigma^Z$, the language L_Y, associated with Y, is defined by

$$L_Y = \{z \in \Sigma^* \mid z \text{ is a substring of some element } x \in Y\},$$

and called the *associate language* of Y.

It is trivial to verify that for every $Y \neq \emptyset$ its associate language L_Y must satisfy the conditions D1 and D2 in Definition 2.1.1.

On the other hand, for a given language $L \subseteq \Sigma^*$, we may define a subset of $X = \Sigma^Z$ as follows.

Definition 2.3.3 Let $L \subseteq \Sigma^*$ be a given language, the subset Y_L, associated with L, is defined by

$$Y_L = \{x \in \Sigma^Z \mid \text{each finite substring of } x \text{ is a word of } L\}.$$

It is, however, possible that for some language L the set Y_L may be empty. The relations between symbolic flows and languages are clarified by following Theorems. They were mentioned or appeared in references, e.g., Weiss (1973), Hurd (1988), and Fischer (1975a, 1975b).

Theorem 2.3.4 A language $L \subseteq \Sigma^*$ is dynamical if and only if the set Y_L is, with the shift operator σ, a symbolic flow and $L = L_{Y_L}$.

Proof. The "if" part is trivial, since $L = L_{Y_L}$ guarantees the language L satisfying the conditions D1 and D2 already.

In order to prove the "only if" part, first we show that the set Y_L is nonempty. Since L is nonempty by Definition 2.1.1, we can take a word $u \in L$. Using the prolongable property D2 inductively, we obtain a biinfinite symbolic sequence $x \in \Sigma^Z$. Combining this construction with condition D1, we see that each substring v of x is a word of L. Thus we have $x \in Y_L$.

From Definition of Y_L it is obviously σ-invariant. Next we have to show that the set $Y_L \subset \Sigma^Z$ is closed under the product topology of Σ^Z. If $x \in \overline{Y}_L$, the closure of Y_L, and v is a finite substring of x, then we have $x = uvw$, where u, w are infinite prefix and suffix of x. From the product topology of Σ^Z there exists an element $y \in Y_L$ such that both y and x have the same substring v at the same place. This implies that $v \in L$. Since each finite substring of x is a word of L, if follows that $x \in Y_L$ and $\overline{Y_L} = Y_L$. Thus Y_L is a symbolic flow.

Compare L with L_{Y_L}. If there is a word $u \in L$, then we already find an infinite sequence $x \in Y_L$ which contains u as its substring as above. It implies $u \in L_{Y_L}$ and $L \subseteq L_{Y_L}$.

On the other hand, if $u \in L_{Y_L}$, then there exists an element $x \in Y_L$ such that u is a substring of x. From the definition of Y_L, each substring of x, including u, belongs to L, and we obtain $L_{Y_L} \subseteq L$ that completes the proof. ∎

Remark. The conditions of Y_L being symbolic flow and $L = L_{Y_L}$ are independent to each other. It is easy to construct an example, in which Y_L is a symbolic flow, but $L_{Y_L} \subsetneq L$.

Theorem 2.3.5 A subset $Y \subseteq \Sigma^Z$ is a symbolic flow if and only if its associate language L_Y is dynamical and $Y = Y_{L_Y}$.

Proof. The "if" part is easy. Since a dynamical language is nonempty, the hypothesis implies that $L_Y \neq \emptyset$, and hence the set $Y = Y_{L_Y}$ is nonempty. Then using Theorem 2.3.4 is enough.

In order to prove the "only if" part, we need only to establish the equality $Y = Y_{L_Y}$. If $x \in Y$, then each finite substring of x belongs to L_Y by definition, thus we have $x \in Y_{L_Y}$ and $Y \subseteq Y_{L_Y}$.

On the other hand, if $x \in Y_{L_Y}$, then each finite substring of x belongs to L_Y. Considering each substring p_x of x and the definition of L_Y, there exists an element $y \in Y$ such that y has p_x as its substring. Let $y = up_xv$, where u, v are infinite, then from the property of σ-invariant of Y, we can suppose the substring p_x occupies the same place in both x and y. Since this p_x may have any length, it leads to the conclusion that $x \in \overline{Y}$. Since Y is closed under the product topology, we obtain the required $x \in Y$. It means that $Y_{L_Y} \subseteq Y$ and finishes the proof. ∎

Remark. If Y is nonempty and σ-invariant, but not closed, then the language L_Y is still dynamical, but $Y \subsetneq Y_{L_Y}$.

For simplicity in this subsection we only discuss two-sided infinite sequences, but the Definitions and Theorems involved can be established similarly for one-sided infinite sequences. In applications, for example, of one-dimensional dynamical systems, it is more natural to use conditions D1 and D2′ to define dynamical languages that associate with the one-sided infinite sequences. Note that it is possible that the languages associated with one-sided infinite sequences may still satisfy the conditions D1 and D2.

2.3.2 Subshifts of Finite Type

This subclass of symbolic flows was first introduced by Parry (1964) under the name of *intrinsic Markov chains*. Subshifts of finite type occur naturally in the work of the Smale school on Axiom A diffeomorphisms (Smale 1967, Bowen 1970a, 1970b, Alekseev and Yakobsen 1981). The abbreviation SFT is used in the sequel.

There are several ways to define the subshift of finite type. One of the simplest way is: selecting a finite set of strings $w_1, \cdots, w_n$, and declare $y \in Y$ if and only if no w_i occur in y. This idea is exactly that of DEB discussed in Section 2.2. It is easy to see that in order to obtain a symbolic flow Y, the strings $w_1, \cdots, w_n$ have to satisfy the conditions claimed in Proposition 2.2.5. Thus from the viewpoint of languages, we may formally give its definition as follows.

Definition 2.3.6 A symbolic flow Y is a *subshift of finite type* if its associate language L_Y is finite complement, that is, L''_Y, the set of all DEB's of L_Y, is finite.

We can also define the subshift of finite type by describing which blocks of some fixed length are allowed to appear. This leads to the second definition.

Definition 2.3.7 A subshift (symbolic flow) Y is of the finite type if there is a positive integer k and a set of k-blocks C such that

$$Y = \{x = \cdots x_{-1}x_0x_1 \cdots \mid x_{i+1} \cdots x_{i+k} \in C \text{ for every } i \in Z\}.$$

A change in the state space allows us to assume that, in Definition 2.3.6, all w_i have length 2, or, in Definition 2.3.7, the integer $k = 2$. Thus we have the third definition of subshift of finite types.

Definition 2.3.8 Let a transition matrix $A = (t_{ij})$, where $t_{ij} \in \{0, 1\}$, and no row or column consisting entirely of zeros, and let

$$Y_A = \{x = x_1 \cdots x_n \cdots \mid A(x_i, x_{i+1}) = 1 \text{ for all } i \geq 0\},$$

where A is indexed by Σ. A symbolic flow Y is a subshift of finite type if it is isomorphic to some Y_A.

Here an isomorphism is a continuous one-to-one map h from Y to Y_A such that $h \circ \sigma = \sigma \circ h$. If the alphabet $\Sigma = \{1, 2, \ldots, m\}$ is used, then the transition matrix A is an $m \times m$ matrix, and $t_{ij} = 0$ if and only if ij is a DEB.

The name of (topological or intrinsic) Markov chain (or Markovian system) for the subshift of finite type comes from the fourth definition, in which the possible sequences are determined by a chain rule (Parry 1964, Fischer 1975a).

Definition 2.3.9 A symbolic flow Y, for which there exists an nonnegative integer m such that for $n \geq m$ the implication

$$a_1 \cdots a_n \in L_Y \text{ and } a_{n-m+1} \cdots a_{n+1} \in L_Y \Longrightarrow a_1 \cdots a_{n+1} \in L_Y$$

holds, is called a subshift of finite type. The minimal number m is called the *order* of Y.

It is easy to see that, if Y in not the full shift and the longest DEB of L_Y is of length l, then the order of Y is $m = l - 1$.

Example 2.3.10 The language of Example A.1.6, which is shown in Figure A.3 and discussed in Proposition A.1.7, is both regular and dynamical. Let $\Sigma = \{0, 1\}$, the subshift of finite type that corresponds the language can be defined in several ways by Definitions above. (Here we use one-sided infinite sequences as elements of flow.)

First we have

$$Y = \{x = x_1x_2 \cdots \mid x_{i+1}x_{i+2}x_{i+3} \neq 100 \text{ for each } i \geq 0\}$$

from Proposition A.1.7.

Let a set of 3-blocks C be defined by

$$C = \{000, 001, 010, 011, 101, 110, 111\}.$$

We may obtain Y by

$$Y = \{x = x_1x_2\cdots \mid x_{i+1}x_{i+2}x_{i+3} \in C \text{ for each } i \geq 0\}.$$

By a simple mapping

$$\phi(x_1x_2x_3\cdots) = (x_1x_2)(x_2x_3)\cdots,$$

we can take 00, 01, 10, 11 as the new states and then obtain the transition matrix A by

$$\begin{pmatrix} 1 & 1 & 0 & 0 \\ 0 & 0 & 1 & 1 \\ 0 & 1 & 0 & 0 \\ 0 & 0 & 1 & 1 \end{pmatrix}$$

For instance, since 100 is not allowed in $x \in Y$, hence the succession of (10)(00) is impossible. From the definition of ϕ there is also no (10)(10) and (10)(11) appearing in the symbolic flow Y_A. Thus we obtain 0100, the third row of the matrix A. The mapping ϕ provides the isomorphism between Y and Y_A.

It is easy to see that Y is a topological Markov chain of order 2.

2.3.3 *Sofic Systems*

The notion of sofic systems was introduced by Weiss (1973) by the method of finite semigroup. The name "sofic" comes from the Hebrew word for "finite". It has been proved that the sofic systems are the class of symbolic flows which are homomorphic images of subshifts of finite type. Here a homomorphism is a continuous mapping between symbolic flows and commutes with the shift operator. It was mentioned in Weiss (1973) that the sofic systems are essentially those that can be defined by finite automata. From the point of view of language theory, it was also pointed out by Goodwyn that a dynamical language L is regular if and only if the symbolic flow Y_L is a sofic system (Coven and Paul 1977, Lind 1984, Boyle, Kitchens and Marcus 1983). This statement was proved later as Theorem 5.3.1 in Hurd (1988). The name of sofic language is often used in references.

Here we use this fact as the definition of sofic systems.

Definition 2.3.11 A symbolic flow Y is a *sofic system* if its associate language L_Y is a regular language.

It is clear that the study of sofic systems is equivalent to that of regular dynamical languages. For example, if we call a sofic system which is not a subshift of finite type *strictly sofic*, then we have the following fact (see Definition 2.2.2 for the name of infinite complement language).

Theorem 2.3.12 A symbolic flow Y is a *strictly sofic system* if and only if its associate language L_Y is an infinite complement regular language.

By this Theorem we can understand that the so-called sofic languages are exactly the infinite complement regular dynamical languages.

The following Theorem in Weiss (1973) is a direct consequence of the Myhill-Nerode Theorem.

Let Y be a symbolic flow, and L_Y its associate language. For $u \in L_Y$ define

$$F(u) = \{v \in L_Y \mid uv \in L_Y\},$$

namely, the set of allowable followers of word u in L_Y.

Theorem 2.3.13 A symbolic flow Y is a sofic system if and only if the set $\{F(u) \mid u \in L_Y\}$ is finite.

Proof. Using Theorem 2.3.5 we know L_Y is a dynamical language. Consider the natural equivalence relation R_L introduced in Subsections *2.2.3* (and *A.3.2*). From Proposition 2.2.11 the complement language L'_Y of L_Y, as a subset of Σ^*, is an equivalence class of R_L. It is easy to establish that

$$uR_L w \iff F(u) = F(w)$$

by Definitions of $F(\cdot)$ and R_L, and hence a one-to-one bijection between the set of $F(\cdot)$ and the set of equivalence classes of R_L. Thus the hypothesis of Theorem is equivalent to say that the language L_Y can be divided into finite equivalent classes of R_L, and hence the index of R_L is finite. Using the Myhill-Nerode Theorem now is sufficient to complete the proof. ∎

2.4 Graphs and Dynamical Languages

An important tool in study of complexity is to use graphs, including transition diagram of automata. In the sequel an automaton is not distinguished from its transition diagram. See Appendix A, and Weisbuch (1991).

In this section it is shown that if a dynamical language L is regular, then the minimum states deterministic automaton accepting L is exactly a graph generating L. The abbreviation minDFA is used in the sequel.

Furthermore, the minDFA for a given language need not be the minimal description for it. Sometimes a smaller graph may be found, in the sense that the number of vertices is strictly smaller than the number of states of the minDFA. But the theory of minimal graph has not been established completely. In this section we present the result due to Fischer (1975a, 1975b), which give the answer of finding the minimal graphs in the case of transitive regular dynamical languages, that is, transitive sofic systems.

2.4.1 Graphs and Shannon-Graphs

A graph used in this book is a *directed graph*, denoted $G = (V, A)$, consists of a finite set of *vertices* V and a set of ordered pairs of vertices A called *arcs*. An arc starts from vertex v to vertex w can be denoted by (v, w) or $v \to w$. We also assume that G is *nondegenerate*: each vertex has at least one arc leading from it and at least one arc terminating at it.

If a graph G is given, and each arc of G is labeled by a symbol from an alphabet Σ, then each finite path corresponds a string over Σ. Taking all finite paths in G, we obtain a language generated by graph G, denoted by $L(G)$. Since G is nondegenerate, it is easy to see that languages generated by those graphs are dynamical.

The relation between graphs and dynamical languages is clarified by the following Theorem.

Theorem 2.4.1 A dynamical language is regular if and only if it is generated by a graph.

Proof. The "only if" part. Let L be a regular dynamical language. Since L is regular, we may find a minDFA $M = (Q, \Sigma, \delta, q_0, F)$ to accept L (see Section A.3 for details). If $L = \Sigma^*$, then we have $Q = F = \{q_0\}$ and $\delta(q_0, a) = a$ for each $a \in \Sigma$. So M is already the graph desired. Otherwise, from Proposition 2.2.11 we know that there exists only one non-accepting state, denoted q_{na}, in Q, and $Q = F \cup \{q_{na}\}$. Moreover, it is a dead state, that is to say, each arc starting from q_{na} will return to it immediately:

$$\delta(q_{na}, a) = q_{na} \text{ for each } a \in \Sigma.$$

Thus we can obtain a smaller finite automaton to accept L by deleting the unique non-accepting state and all arcs associated with it from M (cf. Wolfram 1984, Hurd 1988). The conditions D1 and D2 imply also that it is not necessary to have a particular starting state among all states of M. As a matter of fact, each finite path, which is through accepting states, corresponding a word of the language L. In this sense, each accepting state may be used as a starting state. Therefore, we obtain a graph $G = (V, A)$, where

$$V = F, \text{ and } A = \{(v, w) \mid \delta(v, a) = w \text{ for some } a \in \Sigma\},$$

and each $(v, w) \in A$ is labeled by a if $\delta(v, a) = w$. Since L is dynamical, it is easy to verify that this graph G is nondegenerate and $L = L(G)$.

The "if" part. If a language $L \subset \Sigma^*$ is generated by a graph $G = (V, A)$, then the idea of Subsection A.1.2 can be used to construct a nondeterministic finite automaton (NFA) with ε-move as follows. Let $M = (Q, \Sigma, \delta, q_0, F)$, where $Q = V \cup \{q_0\}$, $F = V$, $\delta(v, a) = w$ for each arc $(v, w) \in A$ labeled by $a \in \Sigma$, and $\delta(q_0, \varepsilon) = v$ for each $v \in V$. It is easy to verify that L is accepted by M and, consequently, is regular. ∎

Remark. From Theorem 2.3.4 we already know that each dynamical language corresponds a unique symbolic flow. It is easy to see that each element of this flow, an infinite symbolic sequence, corresponds at least one infinite path through the graph discussed above. Thus Theorem 2.4.1 is equivalent to say: a symbolic flow is a sofic system if and only if it is generated by a graph.

The graphs generating languages are called *admissible finite automata* in Hurd (1988).

In a graph, generally speaking, there may exist multiple arcs starting from one vertex of the graph, labeled by the same symbol. Removing this uncertain nature we obtain deterministic graphs, which will be called Shannon-graphs as in Fischer (1975a,1975b), since C. Shannon considered symbolic sequences generated by such graphs, and used them to study certain discrete noiseless channels (Shannon 1949).

Definition 2.4.2 A *Shannon-graph* G is a graph which has the property that different arcs starting from the same vertex are differently labeled.

In a Shannon-graph a path is full determined by its starting vertex and the symbolic string (or sequence) generated by it.

The following result due to Csizar and Komlos (1968) is proved here by using the idea of minDFA.

Theorem 2.4.3 If a dynamical language is generated by a graph, then it is also generated by a Shannon-graph.

Proof. If L is generated by a graph, then L is a regular language. The procedure of constructing a minDFA M for L and then modifying it in the proof of Theorem 2.4.1 provides exactly a Shannon-graph. ∎

It is possible that the graph, which is obtained by modifying the minDFA as described above, may not be a minimal description, namely, a graph having minimal vertices, for a given dynamical language. An example will be given in the next subsection.

The concept of minimal Shannon-graphs is defined similarly as the notion of minDFA in Section A.3: a minimal Shannon-graph of a dynamical language L is a graph that has minimal number of vertices among all Shannon-graph generating L.

The problems about minimal Shannon-graph are: 1. for a given regular dynamical language L, is its minimal Shannon-graph unique in some sense? 2. How to decide a given Shannon-graph is minimal or not. 3. How to obtain a minimal Shannon-graph for a given language?

As we know these problems are still open, but for transitive regular dynamical languages the answer has been obtained by Fischer (1975a,1975b). This is the content of the next subsection. Before that we will present two properties of minimal Shannon-graphs as follows. Both of them are due to Fischer.

Some notations used below are

$$F_L(w) = \{u \in L \mid wu \in L\}, \text{ and}$$
$$F_G(\alpha) = \{u \in \Sigma^* \mid u \text{ is generated by a path leading from vertex } \alpha \text{ of } G\}.$$

The first lemma gives a necessary condition for a Shannon-graph G being minimal.

Lemma 2.4.4 If G is a minimal Shannon-graph generated language L, then the mapping $\alpha \mapsto F_G(\alpha)$ is injective.

Proof. Suppose $\alpha \neq \beta$ and $F_G(\alpha) = F_G(\beta)$. Remove the vertex β and all arcs starting from it, and let those arcs terminating at β end in the vertex α. Thus we obtain a Shannon-graph with a smaller number of vertices which also generates L, a contradiction. ∎

The second Lemma is indeed a wonderful finding.

Lemma 2.4.5 Let G be a Shannon-graph generating language L. If the mapping $\alpha \mapsto F_G(\alpha)$ is injective then there exists for each $w \in L$ a word $v \in F_L(w)$ such that all paths generating the word wv have the same end-vertex.

Proof. Assume that a string w is generated by several paths ending in $\alpha_1, \ldots, \alpha_k$, say. Let v_1 be a word which does belong to at least one but not all of the sets $F_G(\alpha_i)$, $1 \leq i \leq k$. Then wv_1 is generated by paths having at most $k-1$ different end-vertices. By iterating this argument the statement follows. ∎

2.4.2 *Transitive Languages*

Transitive dynamical languages are an important subclass of dynamical languages. For example, it is easy to show that every dynamical language generated from cellular automaton after finite time steps is transitive (Wolfram 1984, Hurd 1988).

First of all, two Definitions are needed for our discussion.

Definition 2.4.6 A language L is *transitive* if for each pair of words $u, v \in L$ there exists a string w such that $uwv \in L$.

For convenience, in this Subsection a transitive regular dynamical language is simply referred to as a transitive language.

Definition 2.4.7 A graph G is *transitive* if for each pair (α, β) of vertices, there exists a path leading from α to β.

Let us call two graphs *isomorphic*, if there exists a (graph theoretic) isomorphism, such that each arc and its image are labeled by the same symbol.

The main result of this Subsection is the following Theorem.

Theorem 2.4.8 (i) If L is a transitive regular dynamical language, then all minimal Shannon-graph generating L are isomorphic.

(ii) A transitive Shannon-graph G generating L is minimal if and only if the mapping $\alpha \mapsto F_G(\alpha)$ is injective.

Let a language L be generated by a graph G. It is easy to see that if G is transitive, then L is transitive, but its converse is not true. If G is a minimal Shannon-graph, then the converse claim is true as shown by following Lemma.

Lemma 2.4.9 If G is a minimal Shannon-graph, and language $L = L(G)$, generated from G, is transitive, then G is also transitive.

Proof. Since G is minimal, there is no vertex of G is superfluous, for each vertex α there exists a word $w(\alpha) \in L$ such that each path generating $w(\alpha)$ touches α. Now for two vertices α, β and their corresponding words $w(\alpha), w(\beta)$ there exists a word $x \in L$ which makes the relation $w(\alpha)xw(\beta) \in L$ true. Now each path that generates $w(\alpha)xw(\beta)$ must contain a subpath which starts from α and ends in β. ∎

Lemma 2.4.10 Let G and G' be transitive Shannon-graphs generating language L such that the mapping $\alpha \mapsto F_G(\alpha)$ and $\alpha' \mapsto F_{G'}(\alpha')$ both are injective. Then G and G' are isomorphic.

Proof. Applying Lemma 2.4.5 we obtain the existence of a word $w \in L$ such that all paths generating w in G have the same end-vertex β and all paths generating w in G' have the same end-vertex β'. Therefore $F_G(\beta) = F_L(w) = F_{G'}(\beta')$. For each vertex α of G we choose a path from β to α. If ν is the word generated by this path, then $F_G(\alpha) = F_L(w\nu)$. Using the word ν and its corresponding path in G' we can find a vertex α' of G' such that $F_G(\alpha') = F_L(w\nu)$. Obviously the mapping $\alpha \mapsto \alpha'$, where $F_G(\alpha) = F_{G'}(\alpha')$, is a bijection of the vertex sets and can be extended to an isomorphism of the graphs. ∎

Remark. In this proof we see that each element of $\{F_G(\alpha) \mid \alpha \text{ is a vertex of } G\}$ belongs to the set $\{F_L(u) \mid u \in L\}$, but in general these two sets are not coincide (cf. Fischer 1975b). It is possible that some $F_L(u)$ do not correspond to any $F_G(\alpha)$ even though G is a minimal transitive Shannon-graph. This point will be clarified through Example 2.3.10 later.

Proof of Theorem 2.4.8. (i) If G and G' are two minimal Shannon-graph generating L, then, by using Lemmas 2.4.4 and 2.4.9, both graphs are transitive, and both mappings $\alpha \mapsto F_G(\alpha)$ and $\alpha' \mapsto F_{G'}(\alpha')$ are injective. Using Lemma 2.4.10 leads directly to the conclusion.

(ii) Let G be a transitive Shannon-graph generating L. If G is minimal, then the mapping $\alpha \mapsto F_G(\alpha)$ is injective by Lemma 2.4.4. On the other hand, if this mapping is injective, then using Lemma 2.4.10 to G and the minimal Shannon-graph G' as in (i), we see that G is isomorphic to G', and, consequently, G itself is minimal also. ∎

Now consider how to find a minimal Shannon-graph for a given transitive language L. First of all, using the procedure described in proof of Theorem 2.4.1 we can obtain a Shannon-graph G. An alternative way is to use the elements of set $\{F_L(u) \mid u \in L\}$ as the vertices of graph (Fischer 1975b).

If the graph thus found is transitive, then it is the minimal Shannon-graph generating L.

Proposition 2.4.11 Let L be a regular dynamical language and M its minDFA. If the graph G, which is obtained from M by deleting its unique non-accepting state and arcs involved, is transitive, then G is the minimal Shannon-graph generating L.

Proof. By the conclusion of Myhill-Nerode Theorem the vertex set V of graph G consists of all equivalence classes, that is, $V = \{[u] \mid u \in L\}$, where $[u]$ is the equivalence class containing u. From Definition of equivalence relation R_L (of Subsection 2.2.3) we see that, if $[u] \neq [v]$, that is, $uR_L v$ does not hold, then $\{w \in L \mid uw \in L\} \neq \{w \in L \mid vw \in L\}$. It means that the mapping $\alpha \mapsto F_G(\alpha)$ is injective, and our conclusion follows from Theorem 2.4.8(ii). ∎

For instance, it is easy to see from Figure 2.2 that, using the minDFA shown there, a transitive graph is obtained. Thus it is the minimal Shannon-graph generating L, the language L of Example 2.2.14.

If G is not transitive, namely, if there exist two vertices α and β such that there is a path from α to β, but there is no path from β to α, then we can reduce the graph G by deleting all those vertices, which have a path to α, and arcs involved, and the reduced graph G' can still generate L. As a matter of fact, if a word $u \in L$ is generated by a finite path which touches the vertices and arcs in $G \setminus G'$, then by the property of transitivity of L, the word u can also be generated by a path starting from the vertex β. Since there is no path from α to β, so this path which starting from β will not touch any vertex in $G \setminus G'$. It implies that the reduced graph G' can also generate word u, and $L(G) = L(G') = L$. Proceed this way, we obtain a transitive Shannon-graph generating L.

In one word, when L is transitive, we may delete all transient subgraph from G and obtain a reduce transitive graph, which still generate the same language L.

The next step is to see if the mapping $\alpha \mapsto F_G(\alpha)$ is injective. If it is not, then the procedure described in proof of Lemma 2.4.4 can be used. Here the following Lemma is helpful. It provides an efficient method to decide if a given transitive Shannon-graph is minimal or not. The notation L_n is the set of all words in L, which are of length n.

Lemma 2.4.12 A transitive Shannon-graph G generating L with n vertices is minimal if and only if the mapping $\alpha \mapsto F_G(\alpha) \cap L_n$ is injective.

Proof. The "if" part is trivial. Suppose G is minimal Shannon-graph generating L. What we need to prove is

$$\operatorname{card}\{F_G(\alpha) \cap L_n \mid \alpha \text{ is a vertex of } G\} = n.$$

Assuming the contrary, we would have some i, $1 \leq i < n$ such that $f_i = f_{i+1}$, where

$$f_i = \operatorname{card}\{F_G(\alpha) \cap L_i \mid \alpha \in G\}$$

is nondecreasing. Since there are only n vertices in G, it implies that for some vertices α and β, the following equalities hold

$$F_G(\alpha) \cap L_i = F_G(\beta) \cap L_i,$$
$$F_G(\alpha) \cap L_{i+1} = F_G(\beta) \cap L_{i+1}.$$

It is easy to see that, in order to ensure $f_i = f_{i+1}$, the first equality of them must imply the second one.

Consider the sets $F_G(\alpha) \cap L_{i+2}$ and $F_G(\beta) \cap L_{i+2}$. If a word $a_1 \cdots a_{i+2} \in F_G(\alpha) \cap L_{i+2}$, then $a_1 \in F_G(\alpha) \cap L_1$. Denote a_1 by $a \in \Sigma$. Since $F_G(\alpha) \cap L_1 = F_G(\beta) \cap L_1$, if the arc starting from vertex α and labeled by a ends at vertex α_a, then we may assume the arc starting from β and labeled by a ends at β_a. Write

$$F_G(\alpha) \cap L_{i+1} = \cup_a a(F_G(\alpha_a) \cap L_i),$$
$$F_G(\beta) \cap L_{i+1} = \cup_a a(F_G(\beta_a) \cap L_i).$$

If both left-hand sides are equal, then for each possible a we have $F_G(\alpha_a) \cap L_i = F_G(\beta_a) \cap L_i$. It implies that all these equalities still hold if the subscript i is replaced by $i+1$ and, consequently,

$$F_G(\alpha) \cap L_{i+2} = F_G(\beta) \cap L_{i+2}.$$

Proceed inductively this way, we see

$$F_G(\alpha) \cap L_j = F_G(\beta) \cap L_j \text{ for each } j \geq i,$$

which leads to $F_G(\alpha) = F_G(\beta)$, a contradiction. ∎

Now we give a language and calculate its graphs and sets involved in this Section. This is the language in Example 2.2.15. It is easy to see that it is a transitive language.

Example 2.4.13 In Figure 2.4(a) we see a Shannon-graph generating the language

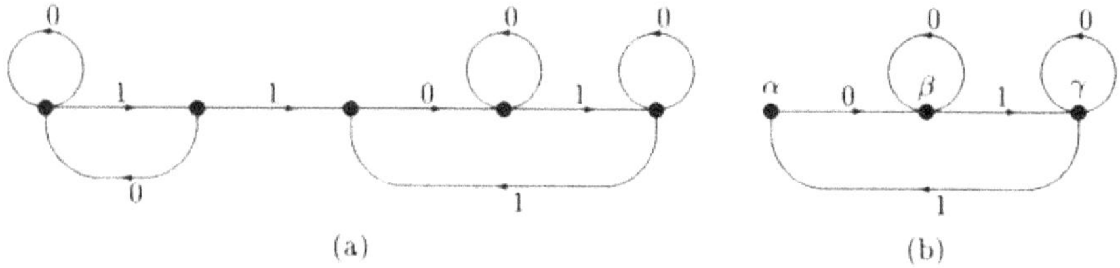

Figure 2.4: The Shannon-graphs generating L of Example 2.2.15

L of Example 2.2.15. It is a modification of the minDFA given in Figure 2.3 by deleting the unique non-accepting state and arcs involved. But it is not a minimal description of L by graph. The minimal Shannon-graph generating L is shown in Figure 2.4(b). It has only three vertices, denoted by α, β and γ. Using Lemma 2.4.5 it is easy to obtain results listed below:

$$F_G(\alpha) = F_L(011), F_G(\beta) = F_L(0110), F_G(\gamma) = F_L(01101).$$

But there are another two elements in $\{F_L(u) \mid u \in L\}$:

$$F_L(0) = F_G(\beta) \cup F_G(\gamma), F_L(1) = F_G(\alpha) \cup F_G(\gamma) = L.$$

We also list the calculation in Lemma 2.4.12:

$$F_G(\alpha) \cap L_3 = \{000, 001, 010, 011\},$$
$$F_G(\beta) \cap L_3 = \{000, 001, 010, 011, 100, 101, 110\} = L_3,$$
$$F_G(\gamma) \cap L_3 = \{000, 001, 010, 100, 101\}.$$

Starting from a minimal transitive Shannon-graph the following result of Fischer can be used to decide if its language is finite complement, that is, if the associate sofic system is a subshift of finite type.

Theorem 2.4.14 If G is a minimal transitive Shannon-graph, then the language $L = L(G)$ is finite complement if and only if there exists a nonnegative integer m such that for each $u \in L_m$ all paths generating u have the same end-vertex.

Proof. The "if" part. It suffices to show that, if x is a DEB of L, then $|x| \leq m+1$. Assume the contrary that there exists a DEB $x = a_1 \cdots a_l$ with its length $l \geq m+2$. Since both $x\pi$ and πx belong to L, they can be generated by paths below:

$$v_1 \xrightarrow{a_1} v_2 \xrightarrow{a_2} \cdots \xrightarrow{a_{l-2}} v_{l-1} \xrightarrow{a_{l-1}} v_l, \quad w_1 \xrightarrow{a_2} w_2 \xrightarrow{a_3} \cdots \xrightarrow{a_{l-1}} w_{l-1} \xrightarrow{a_l} w_l.$$

Since $|a_2 \cdots a_{l-1}| = l-2 \geq m$, we conclude that $v_l = w_{l-1}$. It implies that the path

$$v_1 \xrightarrow{a_1} v_2 \xrightarrow{a_2} \cdots \xrightarrow{a_{l-1}} v_l \xrightarrow{a_l} w_l$$

generates the word x, which contradicts the fact of x being a DEB.

The "only if" part. It is trivial if the complement language L' is empty, since then the graph G has only one vertex. Otherwise, let the longest DEB of L is of length l. We claim that $m = l-1$ is the required integer. Assume the contrary that there exists a word $w \in L_m$, which is generated by two paths: from α_1 to α_2, and from β_1 to β_2, and $\alpha_2 \neq \beta_2$. Since G is a Shannon-graph it implies $\alpha_1 \neq \beta_1$. Using Lemma 2.4.5 and the fact that G being transitive, there exist words u, v such that all paths generating u end at α_1, and all paths generating v end at β_1. Thus all paths generating uw end at α_2, and all paths generating vw end at β_2. Since $F_G(\alpha_2) \neq F_G(\beta_2)$, we have $F_L(uw) \neq F_L(vw)$. This means there exists a nonempty string $a \in \Sigma^*$ such that $uwa \in L$, but $vwa \notin L$ (or vice versa). So the string vwa must contain a DEB as its suffix, and its length is strictly longer than $|wa| \geq m+1 = l$, a contradiction. ∎

2.5 Topological Entropy

In this section the notion of topological entropy is introduced from the viewpoint of languages as in Parry (1964). As a complexity measure of languages, we will discuss the topological entropy for unimodal maps in Part II (Chapter 8) and for circle homeomorphisms in Part III (Chapter 9). Its use as a complexity measure for general dynamical systems has a long history (see, e.g., Brudno 1983, Keller 1991, Li 1991).

Definition 2.5.1 Let Y be a symbolic flow, $L = L_Y$ be its language. The quantity

$$h(Y) = \lim_{n\to\infty} \frac{1}{n} \log N_n,$$

where $N_n = \mathrm{card}\{x \in L \mid |x| = n\}$, is called the *topological entropy* of Y.

Since there exists a bijection between symbolic flows and dynamical languages (by Theorems 2.3.4 and 2.3.5), we can also use this quantity to define the topological entropy for a dynamical language directly and denote it $h(L)$.

Theorem 2.5.2 For a symbolic flow Y or a dynamical language L the limit

$$h(Y) = h(L) = \lim_{n\to\infty} \frac{1}{n} \log N_n$$

always exists.

Proof. From the property D1 of the dynamical languages (by Definition 2.1.1) we have

$$z = uv \in L, \ |u| = m, \ |v| = n \implies u, v \in L,$$

and

$$N_{n+m} \leq N_n \cdot N_m.$$

The following procedure is standard in analysis (Pólya and Szegö 1972). Let $\alpha = \inf(1/n)\log N_n$, it is obvious that

$$\alpha \leq \liminf_{n\to\infty} \frac{1}{n} \log N_n.$$

For every $\varepsilon > 0$, choose N such that

$$\frac{1}{N} \log N_N < \alpha + \varepsilon.$$

For this number N, every $n > 0$ can be written $n = mN + k$ with $0 \leq k < N$. Then

$$\frac{1}{n}\log N_n \leq \frac{m}{n}\log N_N + \frac{1}{n}\log N_k \leq \frac{mN}{n}(\alpha + \varepsilon) + \frac{1}{n}\log N_k.$$

Since $mN/n \to 1$ as $n \to \infty$, we have $\limsup (1/n) \log N_n \leq \alpha + \varepsilon$. Since ε is arbitrary, we have proved

$$\limsup_{n\to\infty} \frac{1}{n} \log N_n \leq \alpha.$$

Thus $\lim (1/n) \log N_n$ exists and equals α. ∎

If μ is a σ-invariant normalized Borel-measure (simply admissible measure) on Y, then the limit

$$h(Y,\mu) = \lim_{n\to\infty} \left(-\frac{1}{n} \sum_{z\in L_n} \mu(z)\log\mu(z)\right)$$

always exists and is called the *measure-theoretic entropy* of Y with respect to μ (Walters 1982). Since the sum $-\sum_{z\in N_n} \mu(z)\log\mu(z)$ is maximized by the equal distribution to each z of length n for which $\mu(z) > 0$, we see that

$$h(Y,\mu) \leq h(Y).$$

It is proved by Fischer (1975a) that for transitive sofic systems there exists one and only one admissible measure μ_0, such that

$$h(Y, \mu_0) = h(Y),$$

and μ_0 is ergodic with respect to σ. This result extends Parry's result for transitive subshift of finite type in Parry (1964). A symbolic flow having this property is called intrinsically ergodic (Weiss 1973).

If G is a Shannon-graph for Y or L, then a connection matrix $A = (a_{ij})$ can be calculated from G, where a_{ij} is the number of arcs starting from the i-th and ending in the j-th vertex (with arbitrary enumerating of the vertices). It is obvious that A is a nonnegative matrix. The following Theorem is a folklore, which was proved in Parry (1964) and Walters (1982) for the case of transitive subshifts of finite type.

Theorem 2.5.3 If G is a Shannon-graph for the symbolic flow Y, and A is the connection matrix associated with G, then the topological entropy of y is

$$h(Y) = \log \beta,$$

where β is the largest positive eigenvalue of A.

Proof. Assume the number of vertices of G is k. It is easy to see that a_{ij}^n, the (i, j) element of A^n, is the number of paths starting from the i-th and ending in the j-th vertex and of length n. Since G is a Shannon-graph, a_{ij}^n is also the number of words of L_n, which can be generated from the i-th vertex to the j-th vertex. Thus the sum $\sum_{j=1}^k a_{ij}^n$ is the number of words generated from the i-th vertex and of length n. Therefore, we obtain

$$\sum_{j=1}^{k} a_{ij}^n \le N_n = \mathrm{card} L_n \text{ for each } i, 1 \le i \le k$$

and

$$N_n = \mathrm{card} L_n \le \sum_{i=1}^{k} \sum_{j=1}^{k} a_{ij}^n.$$

Denote a norm of $k \times k$ matrix by $\|(b_{i,j})\| = \sum_{i,j=1}^k |b_{i,j}|$. An estimation of N_n is given by

$$\frac{1}{k}\|A^n\| \le N_n \le \|A^n\|.$$

From Person-Frobenius Theorem (Gantmacher 1959) for nonnegative matrices, the spectral radius of matrix A is β. Using the spectral radius theorem, we have

$$\|A^n\|^{1/n} \to \beta$$

and complete the proof. ∎

Part II

GRAMMATICAL COMPLEXITY OF UNIMODAL MAPS

CHAPTER 3
LANGUAGES OF UNIMODAL MAPS

The study of interval maps from the point of view of dynamical systems has a long history. As a matter of fact, it began with the appearance of electronic computer (Ulam and von Neumann 1947, Stein and Ulam 1964). Lorenz's pioneer work (1963) about chaos pointed out the importance of one-dimensional iterated systems in study of general dynamical systems. There are many papers and books that contributed completely or partially to this study by different approaches. The Part II, including Chapters 3–8, of this book is devoted to the study of the grammatical complexity of unimodal maps, which are the typical and simplest maps among others. The content of Chapter 3 is a preparation for the study later on in following Chapters.

In Section 3.1 the definition of unimodal maps is given, and some terminology of dynamical systems is explained. The logistic map and the tent map are presented with the bifurcation diagrams of them. Some standard references about unimodal maps are listed in Subsection *3.1.3*. Readers who are familiar with these materials can skip to Section 3.3.

In Section 3.2 the method of coarse-grained description is used to obtain symbolic sequences, that is, itineraries, from orbits of unimodal maps. An ordering relation is introduced to reflect the order of points on real line in a weaker sense. A brief history of symbolic dynamics is mentioned in Subsection *3.2.1*.

Section 3.3 gives the definition of the languages of unimodal maps, which are completely determined by a special itinerary, namely, the kneading sequence. The notation KS is used to denote a given kneading sequence of unimodal map throughout Part II.

For a given kneading sequence some basic facts concerned string manipulations are collected in Section 3.4. These results are used frequently below.

Finally, in Section 3.5 the connection between periodic sequences and periodic orbits is discussed first, and then the grammatical complexity of the set of all periodic sequences is analyzed. It is proved that in Chomsky hierarchy this set is strictly a context-sensitive language.

3.1 Unimodal Maps

3.1.1 Definition and Terminology

A map f from the interval $I = [0, 1]$ to itself is said to be a *unimodal map* on I if the following conditions are satisfied:

1. f is continuous.
2. f reaches its maximal value $f(c)$ at some interior point c of interval I, and is strictly increasing on $[0, c)$, and strictly decreasing on $(c, 1]$.
3. $f(0) = f(1) = 0$.

The point $c \in (0, 1)$ is called the *critical point* of f. Note that the selection of interval $I = [0, 1]$ and the condition 3 are not essential, other choices are possible.

Introduce some notations. Starting from a given point $x \in I$, we can iterate it to obtain a sequence $x, f(x), f(f(x)), \ldots$. Using the notations $f^2(x) = f(f(x))$, and $f^n(x) = f(f^{n-1}(x))$ generally, we can denote this sequence by $\{f^n(x)\}_{n \geq 0}$, where the convention $f^0(x) = x$ is taken. It can also be written as

$$f^{\cdot}(x) = (x, f(x), f^2(x), \cdots, f^n(x), \cdots),$$

and called the *orbit* of iterated map f starting from the point x.

Some terminology of dynamical systems are as follows.

A point x is a *fixed point* of f if $f(x) = x$. A point x is a *periodic point of period* n if $f^n(x) = x$. The least positive n for which $f^n(x) = x$ is called the *primitive period* of x. A fixed point is a periodic point with the primitive period 1.

A fixed point p is *stable* if there is an open neighborhood U of p so that if $x \in U$ then

$$\lim_{n \to \infty} f^n(x) = p.$$

A periodic orbit $\{p, f(p), \ldots, f^{n-1}(p)\}$ of period n is stable if there is an open neighborhood U of p so that if $x \in U$ then

$$\lim_{n \to \infty} |f^n(x) - f^n(p)| = 0.$$

For a differentiable f it is easy to show that if p is a fixed point and $|f'(p)| < 1$ then p is stable. Similarly, if $\{p, f(p), \ldots, f^{n-1}(p)\}$ is a periodic orbit with period n and

$$|(f^n)'(p)| = |f'(f^{n-1}(p)) \cdots f'(f(p)) f'(p)| < 1,$$

then this periodic orbit is stable, where $(f^n)'(p)$ is called the *multiplier* of a periodic orbit. A *superstable periodic orbit* is a periodic orbit with zero multiplier.

A basic fact for unimodal maps is as follows.

Proposition 3.1.1 Let f be an unimodal map on $I = [0, 1]$. For each $n \geq 1$, there exists a partition of I by $0 = x_0 < x_1 < \cdots < x_N = 1$, such that f^n is injective, that is, strictly monotone, on each subinterval $[x_{i-1}, x_i]$ for i, $1 \leq i \leq N$. Each interior point x_i, $1 \leq i < N$, is a local maximum or minimum point of f^n, and a preimage of point c, namely, $f^k(x_i) = c$ for some $0 \leq k < n$.

Proof. From the conditions about a unimodal map the set $\{x \mid f^k(x) = c, 0 \le k < n\}$ is finite. Denote the points of this set by $x_1 < \cdots < x_{N-1}$ and $x_0 = 1$ and $x_N = 1$. Taking all these points as dividing points of $I = [0, 1]$, we obtain a partition of I. We will show that this partition is the required one.

Consider a subinterval $[x_{i-1}, x_i]$, where $1 \le i < N$. By the construction above and the continuity of f, we see that for each k, $0 \le k < n$, $f^k([x_{i-1}, x_i])$ cannot have the point c as its interior point. Therefore, either

$$f^k([x_{i-1}, x_i]) \subseteq [0, c] \text{ or } f^k([x_{i-1}, x_i]) \subseteq [c, 1]$$

holds for each k with $0 \le k < n$. Since f is injective on either $[0, c]$ or $[c, 1]$, it follows that each of $f, f^2, \ldots, f^n$ is injective on $[x_{i-1}, x_i]$.

Now consider an interior dividing point x_i, $1 \le i < N$. Let k be the first integer on which $f^k(x_i) = c$. If $k = n - 1$, then it is obvious that $f^n(x_i) = f(c)$ and x_i is a maximum point of f^n Otherwise f^{k+1} will have x_i as its maximum point and map $O(x_i)$, a neighborhood of x_i, into a subinterval $[f(c) - \delta, f(c)]$ where δ is a positive number. Let δ be sufficiently small such that each $f, f^2, \ldots, f^{n-k+1}$ is injective on $[f(c) - \delta, f(c)]$, then it follows that each of $f^{k+1}, \ldots, f^n$ has x_i as either its maximum point or minimum point. ∎

Corollary 3.1.2 Let f be a unimodal map on $I = [0, 1]$ and x a point of I. (1) For a given integer $n > 0$, there exists $\varepsilon > 0$ such that f^n is injective on $[x - \varepsilon, x]$ and $[x, x + \varepsilon]$ respectively. (2) If x satisfies conditions $f^k(x) \ne c$ for all k, $0 \le k < n$, then there exists a neighborhood $O(x)$ of x such that f^n is injective on $O(x)$.

3.1.2 Logistic Map and Tent Map

The most famous family of unimodal maps is the *logistic map* (or the *quadratic map*):

$$f(x) = bx(1 - x), \quad 0 \le x \le 1, \tag{3.1}$$

where b is a parameter between 0 and 4. There are other equivalent forms of logistic map used in references, e.g., $1 - ax^2$, $a - x^2$ etc..

It is interesting to note that the logistic map is known by the name of *Verhulst law* in biology and other sciences (Cramer 1993, Çambel 1993).

Another family of unimodal maps, the *tent map*, is also important in many aspects (see, e.g., Milnor and Thurston 1988).

$$f(x) = \begin{cases} sx, & \text{when } 0 \le x \le 0.5, \\ s(1 - x), & \text{when } 0.5 < x \le 1, \end{cases} \tag{3.2}$$

where s is a parameter between 0 and 2.

3.1.3 *Simple Models with Complicated Dynamics*

It is fascinating to see the *bifurcation diagram* for one-parameter families of unimodal maps generated by personnel computers (May 1976, Collet and Eckmann 1980).

The bifurcation diagrams for the logistic map and the tent map are shown in Figures 3.1 and 3.2. It seems that for the former the interesting region of the parameter b is between 3 and 4, and for the latter the corresponding region of s is between 1 and 2. As a matter of fact, the surjective case of (3.1), for which the parameter $b = 4$, was studied very early by Ulam and von Neumann (1947).

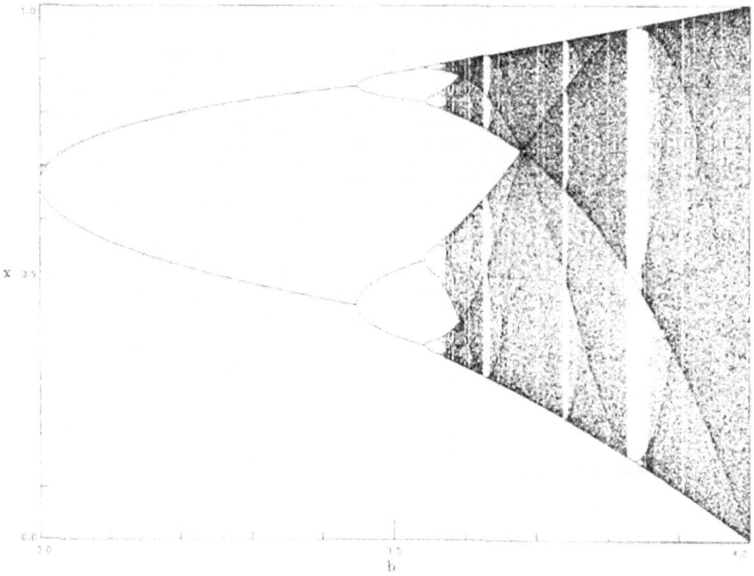

Figure 3.1 The bifurcation diagram for logistic map (3.1).

Thus although the function f of a unimodal map can be very simple, but if an arbitrary point $x_0 \in I$ is iterated indefinitely, then we obtain a one-dimensional dynamical system which can have very complicated behaviors for some ranges of parameters. The article of May (1976), whose title is "Simple mathematical models with very complicated dynamics", has played an important role in stimulating research in this field.

Here the difficulty arises from the fact that for most situations we cannot have useful analytic expressions for orbits obtained through iterations. In order to know the dynamical behaviors of iterated maps such as one-dimensional dynamical systems we have to use new methods created in recent decades. The papers of Sarkovskii (1964),

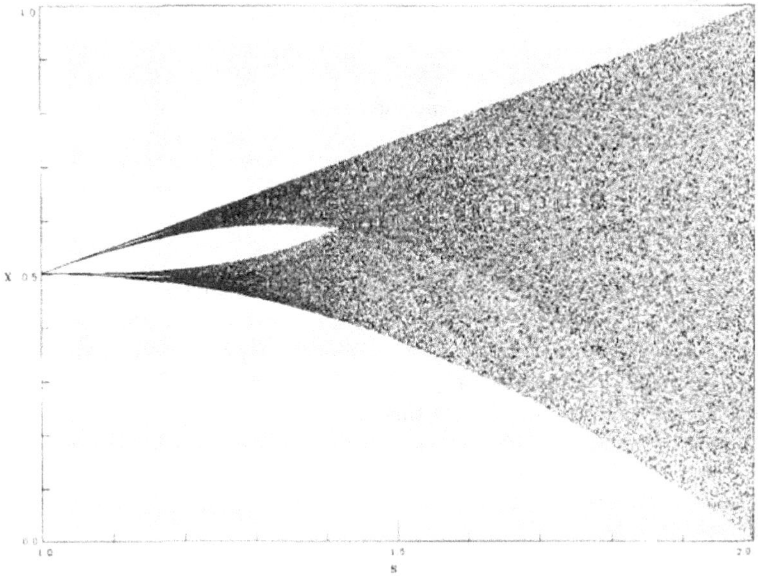

Figure 3.2 The bifurcation diagram for tent map (3.2).

Parry (1966), Li and Yorke (1975), Milnor and Thurston (1988) are very important for research in this field. For further references see the following books and references therein: Collet and Eckmann (1980), Preston (1983, 1988), Mira (1987), Devaney (1989), Hao (1989), Block and Coppel (1992), de Melo and van Strien (1993), Alsedà, Llibre, and Misiurewicz (1993).

3.2 Symbolic Dynamics

3.2.1 A Brief History

Here a brief history about symbolic dynamics is appropriate. Systematic studies of symbolic sequences started with the work of Thue at the turn of the century (Thue 1906, 1912), but his results were not well-known before 60' (Hedlund 1967). On the other hand, at about the same time symbolic analysis for geodesic flows was introduced by Hadamard (1898). In a series of papers (Morse 1921a, 1921b, Morse and Hedlund 1938) this tool is used to prove the existence of geodesics with different properties. It was explicitly proposed to separate the symbolic aspects from the differential aspects and to develop the symbolic theory on its own account (Morse and Hedlund 1938, Gottschalk and Hedlund 1955). An article of Alekseev and Yakobson (1981) summarize the development of symbolic dynamics in studies of dynamical systems.

It was published in *Physics Reports* with a foreword by Ford, who commended the editors for their daring foresight to publish this highly abstract, mathematical article in a physical journal. The symbolic dynamics as a practical tool in study of chaos is developed by Hao (1989, 1991) and his colleagues. For the content of symbolic flow, including subshifts of finite type and sofic systems, see Section 2.3.

3.2.2 *Itinerary and Kneading Sequence*

The method of symbolic dynamics is essentially a coarse-grained description of dynamical systems. Here the meaning of *coarse-grained description* is to describe dynamical behaviors of some systems by symbolic sequences, that is, the itineraries, instead of the orbits themselves. this method has been successfully applied to many problems, and is the starting point of this book.

For a unimodal map f it suffices to use three symbols $\{0, 1, c\}$ to describe all orbits $f^{\bullet}(x)$. If we denote by Σ' the alphabet $\{0, c, 1\}$, which is an extension of $\Sigma = \{0, 1\}$, then for each orbit $f^{\bullet}(x)$ we can obtain a one-sided infinite symbolic sequence over Σ' as follows. First let

$$A(x) = \begin{cases} 0, & \text{when } x \in [0, c), \\ c, & \text{when } x = c, \\ 1, & \text{when } x \in (c, 1], \end{cases}$$

and then define

$$A(f^{\bullet}(x)) = (A(x), A(f(x)), A(f^2(x)), \ldots, A(f^n(x)), \ldots).$$

We call this infinite sequence the *itinerary* of point x, and denote it by $I(x)$.

Since $f(c)$ is the maximum value of f over the interval $I = [0, 1]$, we will show that for a given unimodal map f, the most important one, among all its itineraries, is the itinerary of point $f(c)$, which is referred to as *the kneading sequence* of f. In this book we use the notation

$$KS = I(f(c)) = e_1 e_2 \cdots e_n \cdots$$

exclusively to denote a given kneading sequence of unimodal map. Sometimes we also use the notation $KS(f)$ to emphasize the relation of kneading sequence with f. We will see that for a given unimodal map f nearly all possible itineraries are determined by the kneading sequence of f.

Remark. From now on the notation c has twofold meanings in our discussion. It can be either the critical point of a unimodal map f in the interval $I = [0, 1]$ or a symbol used in our coarse-grained description. Its concrete meaning, however, is clear in each use from context.

3.2.3 An Order Relation of Sequences

It is easy to see that for a given unimodal map f, not every infinite sequence over $\Sigma' = \{0, c, 1\}$ can be a real itinerary $I(x)$ for some point $x \in [0, 1]$. What is the condition for an infinite sequence being an itinerary for a given f?

First of all we need to discuss the appearance of c in an itinerary. Because the symbol $c \in \Sigma'$ corresponds the unique critical point $c \in (0, 1)$, hence if a symbol c appears in an itinerary $I(x)$, then the infinite suffix of $I(x)$ after c can only be the kneading sequence $KS(f)$. Similarly, if c appears in the kneading sequence itself, then this kneading sequence must be of the form

$$KS = (e_1 e_2 \cdots e_{m-1} c)^\infty,$$

where $e_i \neq c$ for $i = 1, \ldots, m-1$. It has a clear dynamical meaning that in this case the critical point c is a periodic point of f with a primitive period n, and, if f is differentiable, then the periodic orbit $\{c, f(c), \cdots, f^{n-1}(c)\}$ is superstable, since its multiplier

$$(f^n)'(c) = f'(f^{n-1}(c)) \cdots f'(f(c)) f'(c) = 0.$$

We assume that each infinite sequence over $\Sigma' = \{0, c, 1\}$ discussed below satisfies this necessary condition about the appearance of symbol c in itineraries.

Of course, this condition is just a necessary requirement. In order to establish the sufficient condition for an infinite sequence being an itinerary we have to introduce the shift operator σ mentioned in Subsection *1.1.4* and define an order relation between strings which are essentially the same one which is presented in Section 1.2 as an additional order structure for the free monoid $\{0, 1\}^*$.

Let s be an infinite sequence over $\Sigma' = \{0, c, 1\}$:

$$s = s_1 s_2 \cdots s_n \cdots,$$

which satisfies the condition about the appearance of symbol c. A shift operator σ is defined on all such sequences that

$$(\sigma(s))_i = s_{i+1} \text{ for all } i \geq 1.$$

If s is indeed an itinerary of a point x, namely, $s = I(x)$ for some point $x \in I$, then it is easy to verify the commutation rule that

$$\sigma(I(x)) = I(f(x)),$$

or, simply, $\sigma \circ I = I \circ f$.

Now we introduce the order relation between those infinite sequences over Σ'. Here we use notations $<$, $>$, $\leq$, $\geq$, and $=$ to denote the relation. First we define an order relation between symbols of Σ':

$$0 < c < 1.$$

Next we define the notion of even string and odd string as follows: a nonempty finite string $s_1 \cdots s_n$ consisting entirely of symbols 0 and 1 is called *even* (*odd*) if the number of 1's appearing in $s_1 \cdots s_n$ is even (odd) (cf. Section 1.2).

Let two infinite sequences over Σ' be

$$s = s_1 \cdots s_n \cdots \text{ and } t = t_1 \cdots t_n \cdots$$

, and both of them satisfy the necessary condition about the appearance of symbol c.

Compare s and t. If $s_i = t_i$ for all $i \geq 1$, then we define $s = t$. Otherwise we have $s \neq t$. Compare s_i and t_i from $i = 1$. If $s_1 \neq t_1$, then define $s < t$ (or $s > t$) if and only if $s_1 < t_1$ (or $s_1 > t_1$). In the general case, if $s_i = t_i$ for $i = 1, \ldots, n$ but $s_{n+1} \neq t_{n+1}$, then the string $s_1 \cdots s_n (= t_1 \cdots t_n)$ is referred to as the (longest) *common prefix* of s and t. Since both s and t satisfy the necessary condition about the appearance of symbol c, and $s \neq t$, this common prefix must consist entirely of symbols 0 and 1. We define

$$s < t\,(s > t) \iff s_{n+1} < t_{n+1}\,(s_{n+1} > t_{n+1})$$

if the common prefix $s_1 \cdots s_n (= t_1 \cdots t_n)$ is even; and

$$s < t\,(s > t) \iff s_{n+1} > t_{n+1}\,(s_{n+1} < t_{n+1})$$

if the common prefix $s_1 \cdots s_n (= t_1 \cdots t_n)$ is odd. It can be seen from the following Proposition that the order relation defined thus far is compatible with the order of real numbers.

Proposition 3.2.1 If $x, y \in I$ and $I(x), I(y)$ are their itineraries, then

$$x < y \Longrightarrow I(x) \leq I(y).$$

Proof. Let $x < y$ and $I(x) \neq I(y)$. Denote $I(x) = s_1 \cdots s_n \cdots$ and $I(y) = t_1 \cdots t_n \cdots$. We know that $s_1 = A(x)$ and $t_1 = A(y)$. If $s_1 \neq t_1$ then it is obvious that $x < y \Rightarrow s_1 < t_1$ and the proof is done. Now suppose that the common prefix of $I(x)$ and $I(y)$ is $s_1 \cdots s_n = t_1 \cdots t_n$, which does not contain symbol c at all. This implies that both $f^i(x)$ and $f^i(y)$ are at the same side of the critical point c for $i = 0, \ldots, n-1$. Since f is order-preserving on $[0, c)$ and order-reversing on $(c, 1]$, if the common prefix of $I(x)$ and $I(y)$ is even, then f^n is order-preserving on the interval $[x, y]$ and so $x < y \Rightarrow f^n(x) < f^n(y)$. This leads to $s_{n+1} < t_{n+1}$ since $s_{n+1} \neq t_{n+1}$. Otherwise we have $x < y \Rightarrow f^n(x) > f^n(y)$ and $s_{n+1} > t_{n+1}$. In all cases the claim is proved. ∎

A consequence of this Proposition is that

$$I(x) < I(y) \Longrightarrow x < y.$$

Here some simple facts are: 0^∞ is the smallest sequence, and 10^∞ is the largest sequence, that is to say, for every infinite sequence s over $\Sigma' = \{0, c, 1\}$, we have

$$0^\infty \leq s \leq 10^\infty$$

Remark. It is easy to see that the selection of order relation between itineraries is not arbitrary. If we require that this order relation reflects (in a weak sense) the natural order of points in the interval $I = [0, 1]$, then there is no other choice but the one defined above.

We need the following notations in the sequel for the case of periodic kneading sequence. If a kneading sequence containing c is written as $KS = (e_1 \cdots e_{m-1}c)^\infty$ where m is the first index for which $e_m = c$, then two new strings are defined by

$$(e_1 \cdots e_{m-1}c)_- = \begin{cases} e_1 \cdots e_{m-1}0, & \text{if } e_1 \cdots e_{m-1} \text{ is even,} \\ e_1 \cdots e_{m-1}1, & \text{if } e_1 \cdots e_{m-1} \text{ is odd.} \end{cases} \tag{3.3}$$

and

$$(e_1 \cdots e_{m-1}c)_+ = \begin{cases} e_1 \cdots e_{m-1}0, & \text{if } e_1 \cdots e_{m-1} \text{ is odd,} \\ e_1 \cdots e_{m-1}1, & \text{if } e_1 \cdots e_{m-1} \text{ is even.} \end{cases} \tag{3.4}$$

Evidently, we always have

$$(e_1 \cdots e_{m-1}c)_- < e_1 \cdots e_{m-1}c < (e_1 \cdots e_{m-1}c)_+.$$

3.2.4 Conditions of being Itinerary

The following Proposition gives a necessary condition for an infinite sequence over Σ' being an itinerary of a point of I.

Proposition 3.2.2 Let KS be the kneading sequence of a unimodal map f. If $s = I(x)$ is an itinerary of a point $x \in I$, then

$$\sigma^i(s) \leq KS \text{ for all } i \geq 1.$$

Proof. From the relation $\sigma \circ I = I \circ f$ we have $\sigma^i(I(x)) = I(f^i(x))$. Since $f^i(x) \leq f(c)$ for each $x \in I$ and $i \geq 1$, using Proposition 3.2.1 is enough. ∎

A consequence of this Proposition is that

$$\sigma^i(KS) = I(f^{i+1}(c)) \leq KS.$$

Recalling that (in Definition 1.2.1) an infinite sequence s satisfying $\sigma^i(s) \leq s$ is called a maximal (or shift-maximal) sequence, thus each kneading sequence is maximal. Its converse is also true. This proof can be found in Milnor and Thurston (1988), de Melo and van Strien (1993).

It is easy to construct an example to show that the necessary condition above is not sufficient to ensure an infinite sequence being an itinerary of a point of I. For instance, in Figure 3.3 there are two examples of unimodal maps having the same kneading sequence $KS = 10^\infty$. The itinerary $I(x)$ of the point x in Figure 3.3 (a) is 010^∞. (It is also easy to find itineraries of form 0^n10^∞ for each $n > 0$ in Figure 3.3 (a).) But there exists no such itinerary for the unimodal map in Figure 3.3 (b). As a matter of fact, since the fixed point 0 there is unstable, if a point y has 010^∞ as

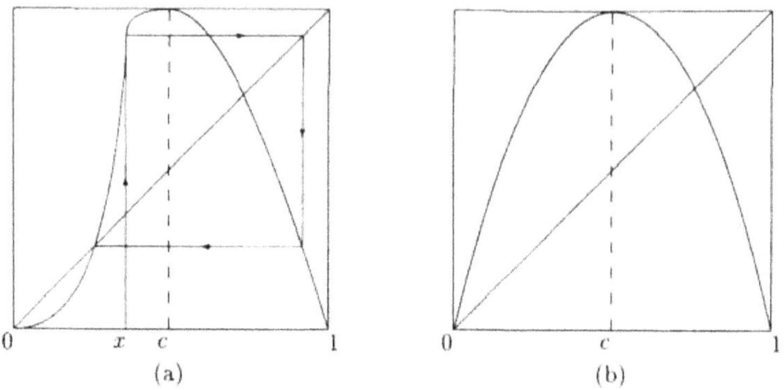

Figure 3.3: Two examples of surjective unimodal map.

its itinerary, then it must hold that $f^2(y) = f^3(y) = \cdots = 0$ which leads to $f(y) = 1$ *and* $y = c$, *a contradiction.*

It is clear from these examples that a necessary and sufficient condition for a sequence being an itinerary must depend on the function f itself and is inconvenient for our study of symbolic dynamics. Instead we will give a sufficient condition below, which serves our purpose completely.

Proposition 3.2.3 Let $KS = e_1 \cdots e_n \cdots$ be the kneading sequence of a unimodal map f, and t be an infinite sequence over $\Sigma' = \{0, c, 1\}$. If KS does not contain the symbol c, and $t = t_1 \cdots t_n \cdots$ satisfies inequalities

$$\sigma^i(t) < KS \text{ for all } i \geq 1,$$

then there exists a point $x \in I$ such that $t = I(x)$.

Proof. Since $I(1) = 10^\infty$ and $I(0) = 0^\infty$ we need only to discuss the case of $0^\infty < t < 10^\infty$. Let two subsets of $I = [0, 1]$ be defined by

$$L_t = \{x \in I \mid I(x) < t\} \text{ and } R_t = \{x \in I \mid I(x) > t\}.$$

It is obvious that both of them are nonempty as $0 \in L_t$ and $1 \in R_t$ hold. We will prove that both subsets L_t and R_t are open in the interval $I = [0, 1]$. Since then $I = L_t \cup R_t$ cannot hold, it leads to the conclusion directly. Because the proofs for L_t and R_t is quite similar, we only give the proof for L_t.

Suppose a point $y \in L_t$ and $I(y) = s = s_1 \cdots s_n \cdots$. Since $s < t$ we have a finite string $s_1 \cdots s_{n-1} = t_1 \cdots t_{n-1}$ as their common prefix and $s_n \neq t_n$. Since $s \neq t$ there is no symbol c in $s_1 \cdots s_{n-1}$. We assume that $s_1 \cdots s_{n-1}$ being even. (The discussion for the case of odd common prefix is quite the same and omitted.) In this case $s < t$ implies that $s_n < t_n$. There are two possibilities to be discussed: $s_n = 0$ and $s_n = c$.

If $s_n = 0$ then by the continuity of f we can find a neighborhood $O(y)$ of the point y such that for each $z \in O(y)$ the itinerary $I(z)$ have the same prefix of length n with $I(y)$, and $I(z) < t$ holds. Thus we obtain $O(y) \subseteq L_t$ (cf. Corollary 3.1.2).

Otherwise, we have $s_n = c$, $t_n = 1$, and

$$s = t_1 \cdots t_{n-1} c e_1 e_2 \cdots, \text{ and } t = t_1 \cdots t_{n-1} 1 \, t_{n+1} \cdots$$

Using the condition of $\sigma^i(t) < KS$ for $i \geq 1$ we have $t_{n+1} \cdots < KS = e_1 \cdots e_n \cdots$. Assume the longest common prefix of them being $e_1 \cdots e_m = t_{n+1} \cdots t_{n+m}$ and $e_{m+1} \neq t_{n+m+1}$. By the assumption that KS containing no symbol c we have $e_{m+1} \neq c$. Again by the continuity of f we can find a neighborhood $O(y)$ of y such that for each $z \in O(y)$ its itinerary is either

$$I(z) = s_1 \cdots s_{n-1} 0 e_1 \cdots e_{m+1} \cdots$$

or

$$I(z) = s_1 \cdots s_{n-1} 1 e_1 \cdots e_{m+1} \cdots$$

Since

$$t = t_1 \cdots t_{n-1} 1 \, t_{n+1} \cdots t_{n+m} t_{n+m+1} \cdots = s_1 \cdots s_{n-1} 1 e_1 \cdots e_m t_{n+m+1} \cdots,$$

we obtain $I(z) < t$ for both cases of $I(z)$. As a matter of fact, if $I(z) = s_1 \cdots s_{n-1} 0 \cdots$ then it is obviously true since $s_1 \cdots s_{n-1}$ is even. Otherwise, the prefix $s_1 \cdots s_{n-1} 1$ is odd, and $t_{n+1} \cdots t_{n+m+1} < e_1 \cdots e_{m+1}$ holds. Thus the proof is completed. ∎

This Proposition gives a sufficient condition for an infinite sequence being itinerary, but it is not a necessary condition. It is possible that for an itinerary $s = I(x)$ the equality $\sigma^i(s) = KS$ holds for some i. For instance, in Figure 3.3 (a) we have an itinerary $s = 01^\infty$ and a kneading sequence $KS = 10^\infty$, and the equality $\sigma(s) = KS$ holds.

For the case of kneading sequences containing c, a similar sufficient condition can be obtained as follows. (Here the notation (3.3) is used.)

Proposition 3.2.4 Let a kneading sequence be $KS = (e_1 \cdots e_{m-1} c)^\infty$ where m is the first index for which $e_m = c$. If an infinite sequence t over $\Sigma = \{0, 1\}$ satisfies conditions

$$\sigma^i(s) < ((e_1 \cdots e_{m-1} c)_-)^\infty,$$

then there exists a point x such that $I(x) = t$.

Remark. The proof of this Proposition can be found in Collet and Eckmann (1980). Here we only point out that it is possible there exists no point x for which $I(x) = ((e_1 \cdots e_{m-1} c)_-)^\infty$. For example, if the parameter s in the family of tent map (3.2) takes the value

$$s = (\sqrt{5} + 1)/2 = 1.618\ldots,$$

then we have $KS = (10c)^\infty$. But there exists no x such that $I(x) = (101)^\infty$. Assume the contrary, then we find that f^i is injective on $[x, 1/2]$ for all $i \geq 0$. This leads to

$$|f^i([x, 1/2])| = (s)^i \times |x - 1/2|.$$

Since $s > 1$ it is impossible.

3.3 Definition of Languages

A natural way to define a language generated by a given unimodal map f is to consider all itineraries of f, and then define the language L to be the set of all finite blocks appearing in itineraries. But the discussion in the previous Section shows that this set of itineraries can be different for unimodal maps having the same kneading sequence. Thus at the beginning we cannot take this natural way as granted.

We will start our work by introducing a new concept of admissible sequences, and after some discussion we will prove that the natural way of defining languages is indeed correct (by Corollary 3.3.5).

3.3.1 Admissible Sequences

Definition 3.3.1 Let KS be a given kneading sequence for a unimodal map f. an infinite sequence consisting of symbols 0 and 1 is called an *admissible sequence* with respect to KS if it satisfies conditions:

$$\sigma^i(s) \leq KS \text{ for each } i \geq 0.$$

This Definition implies the following consequences.

1. The requirement $s \leq KS$ means that we will not consider itineraries $I(x)$ for $x \in (f(c), 1]$ (if $f(c) \neq 1$) in the sequel. Because

$$f((f(c), 1]) \subsetneq [0, f(c)] \text{ and } f([0, f(c)]) \subseteq [0, f(c)]$$

always hold for unimodal maps, any interesting dynamical behaviors must happen in the subinterval $[0, f(c)]$, hence this restriction is acceptable. Note that if $f([0, f(c)]) \subsetneq [0, f(c)]$ happens then we have $f(c) < c$, which corresponds a trivial case.

2. Since the symbol c is not allowed to appear in admissible sequences, an itinerary containing c is not considered admissible. From our definition of unimodal maps we know that the critical point c can have at most two direct preimages (or inverse images), and hence the set of all preimages $f^{-n}(c)$ $(n \geq 0)$ of c,

$$\{x \in I \mid f^n(x) = c \text{ for some } n \geq 0\},$$

is at most a countable infinite set. Thus the itineraries excluded is at most countable. On the other hand, if a point x has an itinerary $I(x)$ containing c as $I(x) = s_1 \cdots s_{n-1} c s_{n+1} \cdots$, where n is the first index for which $s_n = c$, then by the continuity of f the finite block $s_1 \cdots s_{n-1}$ must appear in other itineraries which containing no symbol c at all. As to the remaining part $s_{n+1} \cdots$, the suffix of $I(x)$ after c, we know that it must equal $KS = I(f(c))$, the kneading sequence. This discussion is effective to the kneading sequence itself. If KS contains symbol c, then it can be written as

$$KS = (e_1 \cdots e_{n-1} c)^{\infty},$$

where n is the first index for which $e_n = c$.

Thus we see if a point $x \in [0, f(c)]$ is not a premage of the critical point c, then its itinerary $I(x)$ is an admissible sequence with respect to $KS = I(f(c))$.

3. From the discussion about Figure 3.3 we know that It is possible that not every admissible sequence is an itinerary for some point in I. It means that an admissible sequence may not be a coarse-grained description of a real orbit of a given unimodal map. An example is $s = 010^\infty$ for Figure 3.3 (b).

3.3.2 The Languages $\mathcal{L}(KS)$

In the following definition of languages generated from unimodal maps we use Σ to denote the alphabet $\{0, 1\}$ and Σ^* the set of all finite strings over Σ as in Chapter 1.

In the discussion about strings we also need the order relation either between finite strings or between a finite string and an infinite sequence. It is easy to obtain this new order relation from the order relation defined for infinite sequences above. What we need to note is: if $s \preceq t$, $s \not\prec t$, and s, t are not of the same length, then the shorter string among s, t is the proper prefix of the longer one.

Definition 3.3.2 Let KS be the kneading sequence of a unimodal map f. The *language L generated by f* is the set

$$L = \{x \in \Sigma^* \mid x \text{ is a prefix of an admissible sequence } s \text{ with respect to } KS\}.$$

Remark. It is obvious that if s is an admissible sequence, then each shift of s, $\sigma^i(s)$ for each $i > 0$, is also admissible. Thus an equivalent definition of L is

$$L = \{x \in \Sigma^* \mid x \text{ is a substring of an admissible sequence } s \text{ with respect to } KS\}.$$

From this definition we see that L is determined completely by KS, so instead of $L(f)$ we will use the notation $L = \mathcal{L}(KS)$ to denote a language generated by a given kneading sequence KS in the sequel of the book, where $\mathcal{L}$ may be understood as an operator on KS: first determine all admissible sequences with respect to a given kneading sequence KS, and then take all prefixes (or substrings) of them.

3.3.3 Each Word is "Real"

We will show that for the languages $L = \mathcal{L}(KS)$ there is no such difficulty as shown in Figure 3.3. Although not every admissible sequence is a coarse-grained description of a real orbit, but each finite block appearing in admissible sequences is "real" in the following sense.

Theorem 3.3.3 If $KS = e_1e_2\cdots$ is the kneading sequence of a unimodal map f, and $L = \mathcal{L}(KS)$ is the language of f, then for each word $t \in L$, there exists a point $x \in [0, f(c)]$ such that t is a finite substring of $I(x)$.

First we need a Lemma from Collet and Eckmann (1980) (lemma II.3.7).

Lemma 3.3.4 Let $KS = (e_1 \cdots e_{m-1}c)^\infty$, where m is the first index for which $e_m = c$, be the kneading sequence of a unimodal map f. Given $k \geq 1$, there exists a neighborhood $O(c)$ of the critical point c of f such that for $x \in O(c)$ and $x \neq c$, the itinerary of $f(x)$ has the form

$$I(f(x)) = ((e_1 \cdots e_{m-1}c)_-)^k \cdots$$

Proof. By Corollary 3.1.2 there exists a subinterval $(w, f(c)]$ on which f^{m-1} is injective. From the condition of $KS = I(f(c)) = e_1 \cdots e_{m-1}c \cdots$, each itinerary $I(y)$ of $y \in (w, f(c))$ begins from $(e_1 \cdots e_{m-1})_-$. Since $c \in I$ is the maximal point of f, there exists a neighborhood $O(c)$ of c such that $f(O(c)) \subseteq (w, f(c)]$. This finishes the proof for $k = 1$. For the general case of $k > 1$, we can find a smaller subinterval $(w_1, f(c)] \subseteq (w, f(c)]$ such that $f^{ik}(w_1, f(c)]) \subseteq (w, f(c)]$ for $i = 0, 1, \ldots, m$. Selecting a $O(c)$ for which $f(O(c)) \subseteq (w_1, f(c)]$ is enough for our purpose. ∎

Proof of Theorem 3.3.3. Assume the contrary that there exists a word $t \in L$ which does not appear in any itinerary $I(x)$ of $x \in [0, f(c)]$. This means that there exist only two possibilities between t and any $I(x)$ of $x \in [0, f(c)]$: either $t < I(x)$ or $t > I(x)$.

Define two subsets of $[0, f(c)]$ (as in the proof of Proposition 3.2.3):

$$L_t = \{x \in [0, f(c)] \mid I(x) < t\} \text{ and } R_t = \{x \in [0, f(c)] \mid I(x) > t\}.$$

Note that here t is finite while $I(x)$ is infinite.

We will prove that both of them are nonempty and open in the interval $[0, f(c)]$, which leads to $[0, f(c)] = L_t \cup R_t$ and a contradiction. Thus there must exist a point $x \in [0, f(c)]$ such that the itinerary $I(x)$ has t as its prefix.

Since $I(0) = 0^\infty$ and $I(f(c)) = KS$, so t is neither a 0-block nor a prefix of KS, and we have $0 \in L_t$ and $f(c) \in R_t$. Hence both L_t, R_t are nonempty.

Next we prove that L_t is open.

Let $t = t_1 \cdots t_l$ where $l = |t|$ is the length of t. Take a point $y \in L_t$ and denote $I(y) = s = s_1 s_2 \cdots$. We will find a neighborhood $O(y)$ of y such that $O(y) \subseteq L_t$.

Here there are two cases to be treated separately.

1. First discuss the case of KS containing no symbol c. Since $I(y) = s < t$, there exists an integer n, $1 \leq n \leq l$, such that $s_1 \cdots s_{n-1} = t_1 \cdots t_{n-1}$ and $s_n \neq t_n$. Assume $t_1 \cdots t_{n-1}$ is an even string, then $t_n = 1$. (The discussion of $t_1 \cdots t_{n-1}$ being odd is similar and thus omitted.) Then there are two possibilities for s_n: $s_n = 0$ or $s_n = c$.

If $s_n = 0$, then by the continuity of f there exists a neighborhood $O(y)$ of point y such that each $I(z)$ of $z \in O(y)$ has the same prefix of length n with $I(y) = s$ and thus $I(z) < t$, which leads to $O(y) \subseteq L_t$.

If $s_n = c$, then the first step is to show that $n < l$. Otherwise if $n = l$, then we can find a neighborhood $U(y)$ of point y on which f^{n-1} is injective. From $s_n = c$ we have $f^{n-1}(c) = c$, and on $U(y)$ the image of f^{n-1} can reach both sides of point c. Thus we can find a point x such that $I(x)$ having t as its prefix, a contradiction.

Starting from $n < l$, and using the fact that the suffix of s after c must be KS, we have

$$I(y) = s = s_1 \cdots s_{n-1} c e_1 e_2 \cdots,$$
$$t = s_1 \cdots s_{n-1} 1 t_{n+1} \cdots t_l.$$

Since $t \in L$ implies $t_{n+1} \cdots t_l \in L$ and $t_{n+1} \cdots t_l \leq KS = e_1 e_2 \cdots$, we have

$$t_{n+1} \cdots t_l \leq s_{n+1} \cdots s_l = e_1 \cdots e_{l-n}.$$

By assumption there is no the symbol c appearing in KS, it is easy to find a neighborhood $O(y)$ such that for each $z \in O(y)$, $z \neq y$, we have either

$$I(z) = s_1 \cdots s_{n-1} 0 s_{n+1} \cdots s_l \cdots$$

or

$$I(z) = s_1 \cdots s_{n-1} 1 s_{n+1} \cdots s_l \cdots,$$

and hence $I(z) < t$ and $O(y) \subseteq L_t$.

2. Assume $KS = (e_1 \cdots e_{m-1} c)^\infty$, where m is the first index for which $e_m = c$, and continue the discussion above from the point of

$$I(y) = s = s_1 \cdots s_{n-1} c (e_1 \cdots e_{m-1} c)^\infty,$$
$$t = s_1 \cdots s_{n-1} 1 t_{n+1} \cdots t_l.$$

Assuming that $e_1 \cdots e_{m-1}$ is even, then by using Lemma 3.3.4 we can find a neighborhood $O(y)$ of y such that each $I(z)$ in which $z \in O(y)$ and $z \neq y$ is either

$$I(z) = s_1 \cdots s_{n-1} 0 (e_1 \cdots e_{m-1} 0)^k \cdots$$

or

$$I(z) = s_1 \cdots s_{n-1} 1 (e_1 \cdots e_{m-1} 0)^k \cdots,$$

where k is a positive integer such that $(e_1 \cdots e_{m-1} 0)^k$ is longer than $t_{n+1} \cdots t_l$.

If $I(z) = s_1 \cdots s_{n-1} 0 \cdots$ then we have $I(z) < t$ already. Otherwise, since $t \in L$ implies $t_{n+1} \cdots t_l \leq KS$ and $t_{n+1} \cdots t_l \leq (e_1 \cdots e_{m-1} 0)^\infty$, we also have

$$t \geq I(z) = s_1 \cdots s_{n-1} 1 (e_1 \cdots e_{m-1} 0)^k \cdots,$$

as $s_1 \cdots s_{n-1} 1 = t_1 \cdots t_n$ being odd. By assumption made from beginning, t cannot be a prefix of $s_1 \cdots s_{n-1} 1 (e_1 \cdots e_{m-1} 0)^k$, thus it must be $I(z) < t$ and $O(y) \subseteq L_t$.

The discussion of $e_1 \cdots e_{m-1}$ being odd is similar and omitted.

This finishes our discussion of L_t being open. The proof for R_t being open is similar and thus omitted. ∎

Corollary 3.3.5 The language $L = \mathcal{L}(KS)$ can be equivalently defined by

$$L = \{x \in \Sigma^* \mid x \text{ is a prefix of an itinerary}\}.$$

Remark. The proof of this Theorem is similar to that of Proposition 3.2.3. Their conclusions, however, are different. Both sets L_t and R_t in Proposition 3.2.3 are proved to be nonempty and open. But in Theorem 3.3.3 we have established that

$$\{x \in [0, f(c)] \mid I(x) \text{ has } t \text{ as its prefix}\} \neq \emptyset$$

by *reductio ad absurdum*, and since this set is obviously open, hence both L_t and R_t are closed.

3.3.4 $\mathcal{L}(KS)$ are Dynamical Languages

Recalling Definition 2.1.1 of dynamical languages. It is easy to have the following result.

Theorem 3.3.6 If KS is the kneading sequence of a unimodal map f, then, $\mathcal{L}(KS)$, the language generated from f, is a dynamical language.

Proof. For a given word $t \in L = \mathcal{L}(KS)$, by Definition 3.3.2 there exists an admissible sequence s with respect to KS such that t is a prefix of s. Hence each substring of t, say y, is also a substring of s, and hence $y \in L$ holds. It is also obvious that there exists a symbol $a \in \Sigma = \{0, 1\}$ such that $ta \in L$. Thus L satisfies the right-prolongable condition.

Next we show that L is also left-prolongable. By Theorem 3.3.3 there exists a point $x \in [0, f(c)]$ such that t is a prefix of the itinerary $I(x)$. In order to prove the word t is left-prolongable, we need only to show that there exists a preimage $y \in I$ such that $f(y) = x$ and $y \neq c$. Since $x = 0$ is a fixed point of f, there is no difficulty for $x = 0$. Then suppose $x \neq 0$ below. If $0 < x < f(c)$, then, using the fact that f is surjective on the interval $[0, f(c)]$, there are two preimages $y_1 < c < y_2$ such that $f(y_1) = f(y_2) = x$. Finally, if $x = f(c)$, then by the continuity of f, we can find a point y near x such that $I(y)$ has also the string t as its prefix and proceed as before. ∎

From Theorem 2.3.4 we know that for each dynamical language L there exists a unique symbolic flow Y_L such that $L = L_{Y_L}$. It is easy to see that for the language $L = \mathcal{L}(KS)$ the corresponding symbolic flow is exactly the set of all admissible sequences. This justify our foregoing step in defining $\mathcal{L}(KS)$ by taking all admissible sequences.

Let Σ be $\{0, 1\}$ and $X = \Sigma^N$ the set of all one-sided infinite sequences over Σ. Introduce a product topology into X by the discrete topology provided for Σ. If KS is the kneading sequence of a unimodal map f, then two subsets of X are defined as follows:

$$Y = \{s \in \Sigma^N \mid s \text{ is admissible with respec to } KS\}$$

and

$$Y_1 = \{s \in \Sigma^N \mid s = I(x) \text{ for } x \in [0, f(c)]\}.$$

Theorem 3.3.7 If KS is the kneading sequence of a unimodal map f and $L = \mathcal{L}(KS)$ the language generated from f, then Y, the set of all admissible sequences with respect to KS, is the symbolic flow corresponding L with the shift operator σ, and is the closure of Y_1, the set of all itineraries $I(x)$ for $x \in [0, f(c)]$.

Proof. First of all we prove that Y is a closed set. Assume the contrary that there exists an element $s = s_1 \cdots s_n \cdots \in X$ such that s is in the closure of Y, but it is not an admissible sequence with respect to KS. From Definition 3.3.1 of admissible sequences there exists an integer i such that

$$\sigma^i(s) > KS$$

This implies that there exists another integer $j > i$ such that

$$s_i s_{i+1} \cdots s_j > e_1 e_2 \cdots e_{j-i+1}.$$

Since s is in the closure of Y there exist an admissible sequence $t = t_1 \cdots t_n \cdots \in Y$ such that

$$t_1 \cdots t_j = s_1 \cdots s_j.$$

But then we also have $\sigma^i(t) > KS$ and a contradiction. Thus we have proved that $Y = \overline{Y}$, and Y is closed.

By the same argument and using the conclusion of Theorem 3.3.3 it is clear that $Y = \overline{Y_1}$.

Since $0^\infty \in Y$ holds for each kneading sequence, Y is nonempty. Thus $Y \subset X$ is a nonempty, closed, σ-invariant subset of X, and Y is a symbolic flow. From Theorem 2.3.5 we know that the language $L = \mathcal{L}(KS)$ is just the associate language of Y, and $Y = Y_{L_Y}$. ∎

3.4 Some Facts of Strings for a Given Kneading Sequence

3.4.1 Prefix-Suffixes with respect to Kneading Sequence

First we introduce a useful notion of prefix-suffix with respect to a given kneading sequence. As a fact, this is an extension of prefix-suffix (to itself) in Definition 1.2.6. The same idea has also been used in Proposition 2.2.12.

Definition 3.4.1 Let $z \in \Sigma^*$. A *prefix-suffix* (PS) of z with respect to a given KS is a substring z' of z which satisfies two conditions at the same time: (1) z' is a suffix of z, and (2) z' is a prefix of KS.

We also need the notion of the *longest prefix-suffix* of a string with respect to a kneading sequence and use the abbreviation LPS to denote it. Note that, both of PS and LPS can be ε, but if there is a nonempty PS for a string, then its LPS is certainly nonempty.

3.4.2 Conditions for $x \in \mathcal{L}(KS)$

If a kneading sequence KS is given, then It is easy to decide whether a finite string x over Σ is a word of the language $L = \mathcal{L}(KS)$.

Proposition 3.4.2 If KS is the kneading sequence of a unimodal map f and x a finite string over $\Sigma = \{0, 1\}$, then $x \in L = \mathcal{L}(KS)$ if and only if its each suffix, including x itself, satisfies

$$\sigma^i(x) \leq KS \text{ for each } i, 0 \leq i \leq |x| - 1.$$

Proof. The "only if" part is simple. If $x \in L = \mathcal{L}(KS)$, then by Definition of L there exists an admissible sequence s such that x is a prefix of s. From $\sigma^i(s) \leq KS$ for each $i \geq 0$ it follows that $\sigma^i(x) \leq KS$ for every i, $0 \leq i \leq |x| - 1$.

The "if" part. If KS is 0^∞ or c^∞, then there is only one admissible sequence $s = 0^\infty$. Thus it is simple to obtain the language

$$\mathcal{L}(KS) = \{0^n \mid n \leq 0\}$$

and the claim is obviously true.

Otherwise, then KS must begin from symbol 1. If x is a 0-block, that is, a string consisting of 0 only, then since 0^∞ is less than KS, the claim is true. For the other cases, the proof depends on finding x's nonempty prefix-suffix with respect to KS. Consider the string x. If its last symbol is 1, then there is no problem. If x's last symbol is 0, then since x is not a 0-block, it must have a suffix of the form 10^m for some $m > 0$. The condition $10^m \leq KS$ implies that 10^m must be a prefix of KS. Thus 10^m is a nonempty PS of x with respect to KS.

Using the notion of the longest prefix-suffix (LPS), we can have a decomposition of x in the form of

$$x = y(e_1 \cdots e_k) \text{ for some } k > 0,$$

where $e_1 \cdots e_k$ is the LPS of x with respect to $KS = e_1 \cdots e_n \cdots$.

Now we will show that the sequence

$$s = y(KS) = ye_1 \cdots e_n \cdots$$

is admissible. If this is true, then since s has x as its prefix, it follows that $x \in L$ immediately.

It is obvious that

$$\sigma^i(s) \leq KS \text{ for all } i \geq |y|$$

hold since KS itself is maximal. If $0 \leq i < |y|$ then we have

$$\sigma^i(s) = (\sigma^i(y)e_1 \cdots e_k)e_{k+1} \cdots$$

Since the string $\sigma^i(y)e_1 \cdots e_k = \sigma^i(x)$ is a suffix of x we have

$$\sigma^i(y)e_1 \cdots e_k \leq KS$$

by the hypothesis of Proposition. From the definition of LPS we see that this suffix of x, $\sigma^i(y)e_1\cdots e_k$, cannot be a prefix of KS, thus we obtain

$$\sigma^i(y)e_1\cdots e_k < KS$$

and the required relation $\sigma^i(s) < KS$. ∎

Remark. If we change the conditions in Proposition by using suffix instead of prefix, that is, all prefixes of string x, including x itself, satisfy the similar condition, then the claim is not true. For instance, let

$$KS = 1011011010110\cdots \text{ and } x = 10110110100,$$

then each prefix of x, say y, satisfies $y \leq KS$, but from $\sigma^8(x) = 100 > KS$ we know that $x \notin \mathcal{L}(KS)$.

3.4.3 $\mathcal{L}(KS)$ when KS contains c

A kneading sequence KS containing c can be written as

$$KS = (e_1\cdots e_{m-1}c)^\infty,$$

where m is the first index for which $e_m = c$.

The following fact is just a simple application of Proposition 3.4.2.

Proposition 3.4.3 If $KS = (e_1\cdots e_{m-1}c)^\infty$ be the kneading sequence of a unimodal map f, where m is the first index for which $e_m = c$, then

$$\mathcal{L}(KS) = \mathcal{L}(KS_1),$$

where $KS_1 = ((e_1\cdots e_{m-1}c)_-)^\infty$.

Remark. This Proposition shows that in discussion of complexity of languages generated by kneading sequences we may restrict our consideration to those kneading sequences which contain no symbol c in the rest of the book.

3.4.4 Two Lemmas about Dual Strings

If x is a word of the language $L = \mathcal{L}(KS)$ generated by a given KS, then its dual string $\overline{x}$ introduced in Subsection *1.1.5* is a useful concept in Part II.

The two simple Lemmas below are due to Wang (Wang and Xie 1994). (cf. Lemmas 1.2.8 and 1.2.9.)

Lemma 3.4.4 If $x \in L = \mathcal{L}(KS)$, then $\overline{x}$ has no even PS with respect to KS.

Proof. First of all, $\overline{x}$ itself cannot be a prefix of KS. Otherwise we would have $x > KS = \overline{x}\cdots$ which contradicts $x \in L$ by Proposition 3.4.2.

Now if y is a proper even PS of $\overline{x}$ with respect to KS, then x has $\overline{y}$ as its suffix and KS has y as its prefix. Then we have $\overline{y} > y$, a contradiction with $x \in L$ again. ∎

Lemma 3.4.5 If x is an odd prefix of a given KS and has no even PS with respect to KS, then $\overline{x} \in L = \mathcal{L}(KS)$.

Proof. It suffices to examine the condition of Proposition 3.4.2 to $\overline{x}$.

From the conditions x satisfies, we see that x has the properties (P1) and (P2) in Definition 1.2.7. Using Lemma 1.2.9, the dual string $\overline{x}$ also satisfies (P1) (and (P2)), that is, every suffix of $\overline{x}$, say y, satisfies the inequality $y \leq \overline{x}$. Since x is odd, we obtain

$$y \leq \overline{x} < x$$

as required by Proposition 3.4.2. ∎

Remark. If $KS = x^\infty$ for some x, then the notion of PS of x (to itself) and that of PS of x with respect to KS coincide.

3.5 Periodic Sequences and Periodic Orbits

There are many results about periodic orbits of one-dimensional dynamical systems. The most beautiful one among them is the Sarkovskii Theorem (Sarkovskii 1964). For unimodal maps it can be proved by combinatorial argument as shown, e.g., in Collet and Eckmann (1980).

In this section two problems about periodic sequences are discussed. In the first subsection we clarify the relation between periodic sequences that are admissible for a given kneading sequence, and periodic orbits that the corresponding unimodal map may have. In the second subsection is solved a new question, that is, how complicated is the set of all periodic sequences for a given kneading sequence from the viewpoint of languages.

3.5.1 Periodic Sequences which are Itineraries

From Subsection *3.3.3* we see that it is important to understand the connection between symbolic strings (or sequences) and real orbits of dynamical systems. Theorem 3.3.3 is a basic result in this connection. Here we will discuss the same problem for periodic sequences and periodic orbits.

A basic result in this aspect is that if a periodic sequence s is admissible, then it must be an itinerary, that is, $s = I(x)$ for some point $x \in I$ (cf. Collet and Eckmann 1980, Milnor and Thurston 1988, Devaney 1989). More precisely, we have the following Theorem.

Theorem 3.5.1 Let KS be a kneading sequence which contains no symbol c. If $s = B^\infty$ is a periodic admissible sequence with respect to KS, and $n = |B|$ is the primitive period of s, then there exists a periodic point $x \in I$ with primitive period n such that $s = I(x)$.

Proof By assumption the critical point c is not periodic, hence s contains no symbol c. Without loss of generality, we can assume that $s = B^\infty$ is already a maximal sequence (in Definition 1.2.1), that is,

$$\sigma^i(s) \leq s \text{ for each } i \geq 0.$$

Otherwise we can use

$$\max\{s, \sigma(s), \ldots, \sigma^{n-1}(s)\}$$

to replace s.

First we show that there exists a point $x \in I$ such that $s = I(x)$. If $KS = s$, then let $x = f(c)$. Otherwise we have $s < KS$ and also $\sigma^i(s) < KS$ for each $i > 0$. By Proposition 3.2.3 there exists a point $x \in I$ such that $s = I(x)$.

If $f^n(x) = x$, denote the primitive period of x by m. From $f^n(x) = x$ we have m dividing n. From $I(x) = s$ and n is the primitive period of s it follows that $n = m$.

Otherwise, if B is even, we consider the points x and $f^n(x)$. Without loss of generality assume $x < f^n(x)$. Since each point $z \in [x, f^n(x)]$ has $I(x)$ as its itinerary, from Proposition 3.1.1 we see that f^n is strictly increasing on $[x, f^n(x)]$. Thus we have $f^n(x) < f^{2n}(x)$. Proceeding inductively this way we see that the sequence $\{f^{kn}(x)\}$ is strictly increasing as k increasing, and hence the limit

$$\lim_{k \to \infty} f^{kn}(x) = y$$

exists and $f^n(y) = y$. Since the critical point c is not periodic, the itinerary $I(y)$ contains no symbol c. By the continuity of f we have $I(y) = I(x) = s$. Arguing as before the primitive period of point y is exactly n.

If B is odd, then we can consider the points x and $f^{2n}(x)$. Similarly we obtain a periodic point y such that $f^{2n}(y) = y$ and $I(y) = I(x) = s$. The primitive period of y can be either n or $2n$. If it is $2n$ that $f^n(y) \neq y = f^{2n}(y)$ then by the continuity of f there exists a periodic point z between y and $f^n(y)$ with period n and $I(z) = s$. ∎

Corollary 3.5.2 Assume the hypothesis of Theorem 3.5.1 holds. If a point $x \in I$ is periodic with the primitive period m and $I(x) = s = B^\infty$, then the only possibilities are: if B is even then $m = n$, and if B is odd then either $m = n$ or $m = 2n$.

Proof As a matter of fact, from $f^m(x) = x$ and $I(x) = s$ we have n dividing m. If $m > n$, in the case of B being even we consider the points x and $f^n(x)$. Without loss of generality, assume $x < f^n(x)$. Since f^n is strictly increasing on $[x, f^n(x)]$, we would have $x < f^n(x) < f^{2n}(x) < \cdots$ and contradicts $x = f^m(x)$. If B is odd, we consider the points x and $f^{2n}(x)$. If $x \neq f^{2n}(x)$ it also leads to a contradiction. ∎

Remark. If $KS = (e_1 \cdots e_{m-1}c)^\infty$, then the conclusion of Theorem 3.5.1 and its Corollary need not be true. For instance, it is possible that for some unimodal maps f there exists no point x such that $I(x) = ((e_1 \cdots e_{m-1}c)_-)^\infty$. With the exception of this sequence, however, a similar result as Theorem 3.5.1 can be obtained.

3.5.2 Complexity of Windows

A family of unimodal maps $f = f_\mu(x)$ with $\mu_0 \le \mu \le \mu_1$ is said to be a *full family* if the following conditions are satisfied.

1. $f_{\mu_0}(x) = 0$ for all $x \in I$.
2. $KS(f_{\mu_1}) = 10^\infty$
3. The map $\mu \to f_\mu$ is a continuous curve in the space of C^1-unimodal maps.

The name of full family comes from the following result, whose proof can be found, e.g., in Collet and Eckmann (1980), de Melo and van Strien (1993).

Proposition 3.5.3 Let f_μ be a full family of unimodal maps. For every maximal infinite sequence s, there exists a parameter $\mu \in [\mu_0, \mu_1]$ such that the map f_μ has s as its kneading sequence, that is, $KS(f_\mu) = s$.

A typical example of full family is the family of logistic map 3.1, while an example of non-full family is the family of tentmap 3.2.

If for some μ' the kneading sequence $KS(f_{\mu'})$ contains the symbol c, then the map $f_{\mu'}$ has c as its superstable periodic point. From the continuity of f_μ for both x and μ there is a neighborhood (μ_a, μ_b) of μ' such that for each $\mu \in (\mu_a, \mu_b)$ the kneading sequence $KS(\mu)$ is periodic. If each f_μ satisfies some more analytical condition such as *negative Schwarzian derivative*, then we will see the *periodic windows* on the bifurcation diagram of f_μ. A typical example is the Figure 3.1 for the logistic map (3.1).

The order of appearance of periodic windows has been discussed by Metropolis, Stein, and Stein (1973). Its improvement is obtained by Zheng (1989c).

Here we consider the set of all periodic kneading sequences and discuss its grammatical complexity in the Chomsky hierarchy (see Subsection *B.2.1*). It is referred to as the complexity of windows in the sequel. it is obvious that we need only to consider the maximal finite strings (see Subsection *1.2.2*) as the representatives of windows.

Theorem 3.5.4 If $L = \{x \in \Sigma^* \mid x \text{ is maximal }\}$, then L is a context-sensitive language (CSL), but not a context-free language (CFL).

Proof. From Definition 1.2.1 of maximal strings, it is easy to construct a linear bounded automaton (LBA) to accept L and hence to know that L is context-sensitive (see Subsection *B.1.4*).

In order to prove that L is not context-free, we have to use the The Ogden Lemma. (Its content and an example is in Appendix B.) It says that for each context-free language L there is an integer $k(L)$ such that for each word $z \in L$, if any k or more distinct positions in z are designated as distinguished, then there is a decomposition of $z = uvwxy$ such that:

(i) $uv^iwx^iy \in L$ for each $i \geq 0$.
(ii) w contains at least one of the distinguished positions.
(iii) Either u and v both contain distinguished positions, or x and y both contain distinguished positions.
(iv) vwx contains at most k distinguished positions.

The fact we will use below is simple: if a string x is of the form $10^i1\cdots0^j\cdots$ with $i < j$ then it is impossible to be a maximal string. A substring consisting of symbol 0 entirely is called a *0-block*.

Assume the contrary that L is context-free and then use the Ogden Lemma to L to obtain the integer $k(L)$ asserted in Lemma. Take a string $z = 10^k110^k110^k \in L$, and designate all positions of the substring 10^k in middle of z as the distinguished positions:

$$z = 10^k1\overbrace{100\cdots0}^{k+1}110^k$$

Consider any decomposition of $z = uvwxy$ which satisfies the properties (i)–(iv) in Lemma. There are several cases to be considered.

(1) Both u and v contain distinguished positions. Since w contains at least one of the distinguished positions, we see that v must be a 0-block, and u has 10^k11 as its prefix. This implies that if the exponent i in uv^iwx^iy is big enough ($i > k$) then uv^iwx^iy cannot be maximal.

(2) Otherwise, v cannot be a 0-block which contains distinguished positions. But from the condition (iii) of the Lemma, each of w, x, y contains distinguished positions, so that x must a 0-block.

(2.1) v is not a 0-block (including $v = \varepsilon$), then the strings uv^iwx^iy is not maximal for sufficiently large i.
(2.2) v is a 0-block and contains no distinguished positions at all, then the string uwy has a prefix of 10^j1 with some $j < k$, and y has 0^k as its suffix. It implies that uwy cannot be maximal.

Thus, any decomposition of $z = uvwxy$ which satisfies the conditions of (i)–(iv) of Lemma will lead to contradiction, and hence L is not a context-free language. ∎

CHAPTER 4
REGULAR LANGUAGES OF UNIMODAL MAPS

In this chapter we study the regular languages generated from unimodal maps, which are on the lowest level of the Chomsky hierarchy (see Appendices A and B). The main result is that the language $\mathcal{L}(KS)$ is regular if and only if the kneading sequence KS is either periodic or eventually periodic. A useful approach to characterize the complexity of these languages is to use the theory of finite automata. We will explicitly give the structures of the minimal deterministic finite automata (minDFA) accepting those $\mathcal{L}(KS)$.

The first section is used to discuss some simple examples of unimodal maps from the viewpoint of languages. For these examples those results that can be obtained from entropy and Lyapunov exponent are presented for comparison.

In Section 4.2 the main Theorem 4.2.1 is presented and proved by different approaches.

Sections 4.3 and 4.4 give the minDFA's for the two classes of regular languages generated from unimodal maps. Here it is emphasized that the minDFA is a complete description of regular languages and the number of states is the most important complexity measure for these languages.

The final Section 4.5 is devoted to discuss the $*$-composition law (Derrida, Gervois and Pomeau 1978) and the generalized composition law (Zheng 1989b), which are generated by the two classes of regular languages discussed above.

4.1 Some Examples

4.1.1 The Surjective Unimodal Map

The first example of unimodal maps we will discuss is the surjective case, which corresponds the case of $b = 4$ for the logistic map (3.1), and the case of $s = 2$ for the tent map (3.2). Both of them have the kneading sequence $KS = 10^\infty$ Two examples of surjective unimodal maps have been shown in Figure 3.3.

Example 4.1.1 Discuss the language $L = \mathcal{L}(KS)$ for $KS = 10^\infty$

From this kneading sequence the language $\mathcal{L}(KS)$ is easily to be found. Since the sequence 10^∞ is the largest sequence according the order relation defined in Subsection *3.2.3*, any infinite sequence over $\Sigma = \{0, 1\}$ is admissible. Therefore, every finite string over Σ^* is a word of the language $L = \mathcal{L}(10^\infty)$. Thus we obtain $L = \Sigma^*$, the largest language over Σ in the sense of inclusion.

Therefore, if $KS = 10^\infty$, then $L = \mathcal{L}(10^\infty) = \Sigma^*$, and $L' = L'' = \emptyset$.

The minDFA accepting L is the simplest one and is covered in Appendix A. (See Example A.1.2 and Figure A.2 (a).) In Tables A.1 and A.2 we give its grammar and regular expression. For reader's convenience we reproduce them here in Figure 4.1.

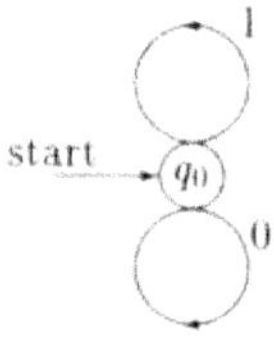

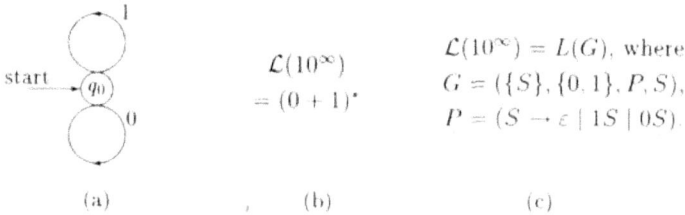

$$\mathcal{L}(10^\infty) = (0+1)^*$$

$\mathcal{L}(10^\infty) = L(G)$, where
$G = (\{S\}, \{0,1\}, P, S)$,
$P = (S \to \varepsilon \mid 1S \mid 0S)$.

(a) , (b) (c)

Figure 4.1: (a) minDFA, (b) regular expression, (c) right linear grammar for $\mathcal{L}(10^\infty)$

From (a) we see that the minDFA can be written as

$$M = (Q, \{0,1\}, \delta, q_0, F),$$

where both Q, the set of states, and F, the set of accepting states, are $\{q_0\}$, and δ, the transition function, is defined by $\delta(q_0, 0) = \delta(q_0, 1) = q_0$.

By the discussion in Subsection 3.2.4 around Figure 3.3 (a) we know that not every admissible sequence is an itinerary. But from Theorem 3.3.3 it follows that every finite symbolic behavior (over $\{0,1\}$) can be found in a surjective unimodal map if an appropriate initial condition is taken. Furthermore, since now each periodic sequence is admissible, by Theorem 3.5.1 there exist periodic orbits with any given periodic itinerary.

In the rest of this subsection we will compare the result above with the results obtained by other methods.

In Section 2.5 the concept of topological entropy is defined for symbolic flows and dynamical languages. It appeared in Parry (1964) by the name of absolute entropy. In general setting the topological entropy was introduced by Adler, Konheim, and McAndrew (1965). From Misiurewicz and Szlenk (1977, 1980) we know that the topological entropy for unimodal maps is exactly the same quantity evaluated for languages $\mathcal{L}(KS)$ by the Definition given in Section 2.5 (also see Alsedà, Llibre and Misiurewicz (1993)).

For the case of $KS = 10^\infty$ the calculation of topological entropy is trivial. Using L_n to denote the set of all words of $L = \mathcal{L}(KS)$ which are of length n, then since $L = \Sigma^*$ we have $N_n = \mathrm{card} L_n = 2^n$. Hence we have

$$h(L) = \lim_{n\to\infty} \frac{\log N_n}{n} = \log 2.$$

From this calculation we see that the topological entropy here is exactly the measure of multitude of orbits of a unimodal map. For the surjective unimodal map every symbolic string is a word of language $\mathcal{L}(10^\infty)$, hence its topological entropy is the largest one that a unimodal map can reach.

The concept of measure-theoretic entropy was introduced by Kolmogorov and Sinai (Walters 1982). We have also introduced this concept in Section 2.5 for symbolic flows on which a σ-invariant normalized Borel measure is defined. It is easy to understand that the measure-theoretic entropy is also a measure of multitude of dynamical behaviors, but with different weights for different orbits. Among those measures the m-absolutely continuous invariant probability measure (a.c.i.p.m.) is the most important one (here m is the Lebesgue measure).

For the logistic map (3.1) with parameter $b = 4$, the density function of μ, the a.c.i.p.m., is

$$\rho(x) = \frac{1}{\pi\sqrt{x(1-x)}},$$

and the measure μ for the tentmap (3.2) with parameter $s = 2$ is simply the Lebesgue measure. For both surjective maps we have $h_\mu = h(L) = \log 2$.

The concept of *Lyapunov exponent* is used widely in study of dynamical systems. It is a measure of the average asymptotic divergence rate of nearby trajectories. If the Lyapunov exponent of a unimodal map is positive, then it shows explicitly the sensitive dependence of orbits on initial conditions. The Lyapunov exponent at a point $x \in I$ is defined by

$$\lambda(x) = \lim_{n\to\infty} \frac{1}{n} \log |(f^n)'(x)| = \lim_{n\to\infty} \frac{1}{n} \sum_{k=0}^{n-1} \log |f'(f^k(x))|$$

if this limit exists. Otherwise, it is defined by the corresponding lim sup and denoted by $\bar{\lambda}(x)$. Usually, we suppose here that the function f is continuously differentiable. But for the tentmap (3.2) it is easy to see that except the preimages of critical point 1/2, we have $\lambda(x) = \log s$ everywhere. For the logistic map $f(x) = bx(1-x)$ $(3 \le b \le 4)$ the discussion of its Lyapunov exponent is much more difficult (Collet and Eckmann 1980, Keller 1991). For the surjective logistic map, however, we can calculate it directly by the formula

$$\lambda(x) = \int \log |f'| \rho(x)\, dx,$$

where the density function $\rho(x)$ is given above, and obtain the value $\lambda(x) = \log 2$ for m-$a.e.\, x \in I$.

Here it is clear that even for surjective case it is not easy to calculate the measure-theoretic entropy h_μ for a given measure μ and the Lyapunov exponent $\lambda(x)$ if f is continuously differentiable. But in order to calculate the topological entropy it is only a problem of counting the number of words of the same length, and hence a combinatorial calculation. Since $h = \sup\{h_\mu\}$ for all Borel measures, this also provides an estimation for measure-theoretic entropy.

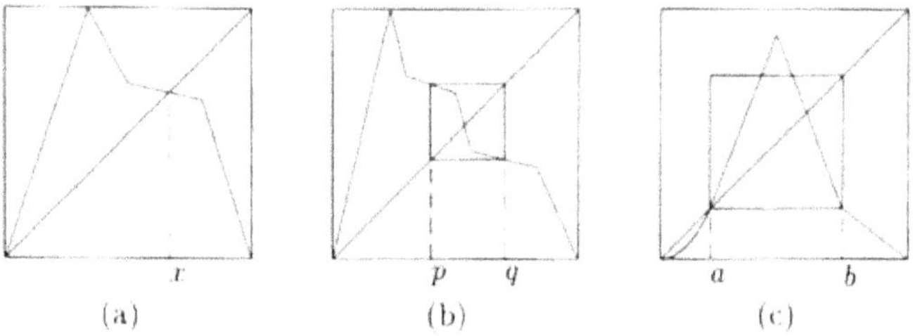

Figure 4.2. Unimodal maps with $KS=10^\infty$ but different topological behaviors.

Finally, a unimodal map having the kneading sequence 10^∞ need not be surjective. In Figure 4.2 are shown three examples of unimodal maps. The first map in (a) has a stable fixed point x. The second map in (b) has a stable periodic orbit $\{p, q\}$ with period 2. The third map in (c) has a Cantor set between a and b together with the fixed point 0 as its non-wandering set. Hence they have different topological behaviors, but from the combinatorial point of view they are quite the same. They have the same kneading sequence $KS = 10^\infty$, and thus the same topological entropy $h = \log 2$.

4.1.2 More Examples

By the order relation between kneading sequences we have

$$0^\infty < 1^\infty < (10)^\infty < \cdots < 10^\infty$$

In this subsection we discuss the left end of kneading sequences, that is, 0^∞ and 1^∞.

Example 4.1.2 Discuss the language $L = \mathcal{L}(KS)$ for $KS = 0^\infty$. For the logistic map (3.1) when $0 < b < 2$ and for the tentmap (3.2) when $0 < s < 1$ the kneading is sequence 0^∞.

The language $L = \mathcal{L}(0^\infty)$ is easy to determined. Since 0^∞ itself is the smallest sequence, it is also the unique admissible sequence. Therefore we obtain

$$L = \mathcal{L}(0^\infty) = \{0^n \mid n \geq 0\}.$$

This is a simple regular language. Its regular expression is 0^* (see A.2.2). In Figure 4.3 (a) we give the minDFA for this language. Here we have the finite automaton

$$M = (Q, \{0, 1\}, \delta, q_0, F),$$

where $Q = \{q_0, q_1\}$, $F = \{q_0\}$. The state q_1 is the unique non-accepting state. From Lemma 2.2.11 we know that if L is a dynamical language and $L' \neq \emptyset$, then in its minDFA there is always only one non-accepting state. Moreover, each arc starting from this non-accepting state will return to it immediately. Hence we call this state a dead state.

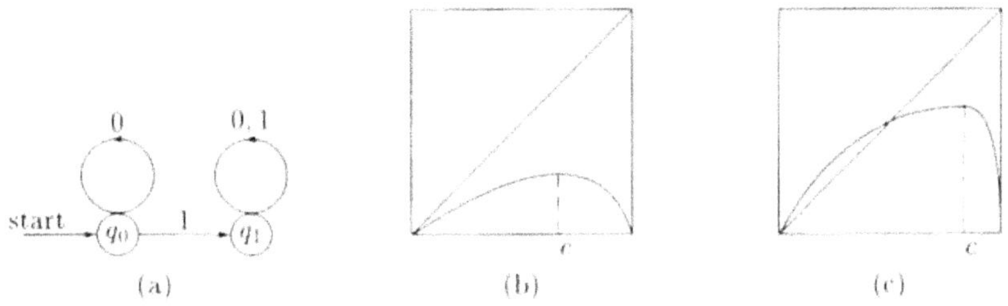

Figure 4.3: (a) minDFA for, and (b), (c) two unimodal maps with $KS=0^\infty$

The transition function δ is very simple that

$$\delta(q_0, 0) = q_0, \quad \delta(q_0, 1) = q_1, \quad \delta(q_1, a) = q_1 \text{ for } a = 0, 1.$$

In Figure 4.3 (b), (c) we give two examples with the same $KS = 0^\infty$ but having different behavior. The map of (b) has only one fixed point 0, which is globally stable. The map of (c) has two fixed points 0 and p, where 0 is unstable, and p is stable and attracts any point except 0. By Corollary 3.5.2 it is impossible for a unimodal map having $KS = 0^\infty$ that there exists any periodic orbit with period greater than 1. But such a map can have many fixed point with different attracting properties. An extreme case is the tentmap 3.2 with the parameter value $s = 1$. Then $KS = c^\infty$ and its language $\mathcal{L}(KS)$ is the same with above. Here we have a continuum of fixed point $[0, \frac{1}{2}]$.

Remark. This is the unique case for unimodal maps that the inequality $f(c) < c$ happens and $f([0, f(c)]) \subsetneq [0, f(c)]$, which has been mentioned in Subsection *3.3.1*.

Example 4.1.3 Discuss the language $L = \mathcal{L}(KS)$ for $KS = 1^\infty$. For (3.1), the logistic map, this case happens when $2 < b < 1 + \sqrt{5} = 3.236\ldots$. But for the tentmap (3.2) the kneading sequence 1^∞ does not exist. The proof is as follows (cf the Remark at the end of Section 3.2).

Since $I(f(c)) = 1^\infty$, the critical point c is not periodic, and hence $c \neq f(c)$. Since f^i is injective on the subinterval $[c, f(c)]$ for each $i > 0$, the length of $f^i([c, f(c)])$ equals to $(s)^i \times |c - f(c)|$. It is impossible if $s > 1$. If $s \leq 1$, however, then its kneading sequence cannot begin from the symbol 1.

From the order relation between admissible sequences it is easy to find that the only possible admissible sequences in this case are

$$0^\infty, \text{ and } 0^m 1^\infty \text{ for all } m \geq 0.$$

Thus we can determine the language

$$\mathcal{L}(1^\infty) = \{0^m 1^n \mid n \geq 0, m \geq 0\}.$$

This language is already discussed in Appendix A in Example A.1.4. Its minDFA is given in Figure A.2 (b), its grammar in Table A.1, and its regular expression in Table A.2. Here we note that by Corollary 3.5.2 there may exist fixed point and periodic orbit with period 2, and both of them has the itinerary 1^∞. As a matter of fact, for the logistic map with $2 < b < 3$ there exists no periodic orbit with primitive period 2, but if $3 < b < 1 + \sqrt{5}$ then there exists periodic orbit with period 2. The parameter value 3 is called the first *period-doubling point* of logistic map as seen from Figure 3.1.

4.2 Decide Regularity from Kneading Sequence

A symbolic sequence s is said to be *eventually periodic* if some shift of s, that is, $\sigma^i(s)$ for some $i \geq 0$, is periodic. Of course, a periodic sequence is eventually periodic, but the converse is not true.

The main result of this chapter is the following Theorem.

Theorem 4.2.1 Let KS be the kneading sequence of a unimodal map f. The language $\mathcal{L}(KS)$ is regular *if and only if the kneading sequence* $KS(f)$ is *eventually periodic*.

The "if" part of this Theorem is well-known in references (e.g., Grassberger 1986, 1988a, 1988b, Crutchfield and Young 1990, Crutchfield 1994, Auerbach 1990, Auerbach and Procaccia 1990, Friedmann 1991, Moore 1991a, 1991b, Hao 1991).

Its converse, the "only if" part of this Theorem, appeared in Xie (1993a, 1993b), Wang and Xie (1994).

There are several proofs of this Theorem available for both its "only if" part and "if" part. For convenience we will give them in the form of several Propositions below.

4.2.1 Calculations of Equivalence Classes

Since the discussion in the sequel relies heavily on the notions of the equivalence relation R_l and equivalence classes, we will give three simple examples to show how to do calculation involved. For reader's convenience the content of the Myhill-Nerode Theorem and its proof are included in Subsection *A.3.3* of Appendix A for reference.

If the language $L = \mathcal{L}(KS)$ is regular, then there exists a minimum states deterministic finite automaton (minDFA) to accept L (see Section A.3 for concepts and notations involved). Let this minDFA be denoted by

$$M = (Q, \Sigma, \delta, q_0, F)$$

The quantity $N = \text{card}\, Q$ is referred to as the so-called *regular complexity* of L in Wolfram (1984). It is the basis of many other measure of complexity developed later (Grassberger 1986, Peliti and Vulpiani 1987, Lindgren and Nordahl 1988, Wackerbauer, Witt, Atmanspacher, Kurths, and Scheingraber 1994).

From the Myhill-Nerode Theorem, we know that N is the index of equivalence relation R_L, that is, the number of equivalence classes of R_L. In this subsection we give some simple examples of calculations of N for unimodal maps.

Proposition 4.2.2 For unimodal maps the regular complexity $N = 1$ if and only if $KS = 10^\infty$.

Proof. The "if" part is simple and given in Subsection *A.3.4*. If $N = 1$, then it implies L', the complement language of L, is empty. Therefore each string 10^m belongs to L for $m \geq 0$. Since $10^m \leq KS$ holds for every $m \geq 0$, we obtain $KS = 10^\infty$. ∎

Proposition 4.2.3 For unimodal maps the regular complexity $N = 2$ if and only if $KS = 0^\infty$.

Proof. The "if" part is simple. From Example 4.1.2 we already have $L = \{0^m \mid m \geq 0\}$. Any string containing symbol 1 does not belong to L and vice versa. Thus we obtain two equivalence classes L and L'.

Conversely, If $N = 2$ then from Proposition 4.2.2 $L' \neq \emptyset$ holds. Thus we have at least two equivalence classes $[\varepsilon]$ and L' by Lemma 2.1.2 and Lemma 2.2.11. Since $N = 2$ we have $L = [\varepsilon]$. Consider the elements of $[\varepsilon]$. It is easy to know that $0^m R_L \varepsilon$ for each $m \geq 0$. As a matter of fact, KS is either 0^∞ or of the form $1\cdots$. By Proposition 3.4.2 in both cases we have $\varepsilon z = z \in L \to 0^m z \in L$, and the other half is trivial. Thus we have $\{0^m \mid m \geq 0\} \subset [\varepsilon]$.

If there exist in $[\varepsilon]$ other words than 0-block, then $1 \in L$ holds. From $0^m R_L 1$ and take $z = 0^n$, we find that $10^n \in L$ for each $n \geq 0$. Hence we must have $KS = 10^\infty$. Using Proposition 4.2.2 leads to a contradiction with $N = 2$. Therefore we see that $L = [\varepsilon] = \{0^m \mid m \geq 0\}$ and $KS = 0^\infty$. ∎

Proposition 4.2.4 For unimodal maps the regular complexity $N = 3$ if and only if $KS = 1^\infty$.

Proof. The "if" part. From Example 4.1.3 and by the calculations in Subsection *A.3.4* we have that $L = \mathcal{L}(1^\infty) = \{0^m 1^n \mid m \geq 0, n \geq 0\}$ and $N = 3$. Here the equivalence classes are $[\varepsilon]$, $[1]$, and L'.

The "only if" part. Using Proposition 4.2.2 and 4.2.3 we know that the kneading sequence KS begins from 1 and is not 10^∞. Assume the contrary that $KS \neq 1^\infty$, then $KS = 10^k 1\cdots$ for some $k \geq 0$. Arguing as in Proposition 4.2.3 we have $[\varepsilon] = \{0^m \mid m \geq 0\}$. Thus from $N = 3$ we have $[\varepsilon]$, $[1]$, and L'. If $k > 0$ then we have $10 \in L$ and $10 \in [1]$. But from $10 R_L 1$ and $0^m \in L$ for each $m \geq 0$ we have $10^n \in L$ for each $n > 0$, which implies $KS = 10^\infty$, a contradiction. ∎

Remark. From these examples we see that if a kneading sequence KS begins from 10 then the index of R_L is greater than 3, and at least three classes are clearly known: $[\varepsilon] = \{0^m \mid m \geq 0\}$, $[1]$, and L'.

4.2.2 The Role of Prefixes of Kneading Sequence

It turns out that in the calculation of the index N the prefixes of kneading sequence play a special role, and the notions of PS and LPS introduced in Definition 3.4.1 are essential for reaching conclusions in the sequel (cf. the set V in Proposition 2.2.12).

Lemma 4.2.5 If two words $x, y \in L = \mathcal{L}(KS)$ have the same LPS with respect to the given KS, then $xR_L y$ holds.

Proof. Write $x = uv$ and $y = u'v$ where v is the common LPS of them with respect to the KS. Assume that $xz \in L$ for $z \in \Sigma^*$. Recall Proposition 3.4.2 that every suffix of xz, say w, satisfies $w \le KS$. Consider any suffix, say w', of $yz = u'vz$. If its length is less than or equal to $|vz|$, the length of string vz, then from $xz = uvz \in L$, we have $w' \le KS$. Otherwise we can write this suffix of yz as $w' = y'z$, where $|y'| > |v|$. Since v is the LPS of y, the string y', as a suffix of y, cannot be a prefix of KS. But as a subword of $y \in L$, we also have $y' \le KS$. Combining these facts we have $y' < KS$, and consequently $w' = y'z < KS$. Again using Proposition 3.4.2, we deduce that $yz \in L$. In the similar way we can also establish that $yz \in L \Rightarrow xz \in L$, and obtain $xR_L y$. ∎

Recall that N, the index of R_L, is the number of the equivalence classes of R_L. Let N' be the number of those classes which contains at least one prefix of KS.

Lemma 4.2.6 Every equivalence class of R_L, except L' (the complement of L), contains a prefix of KS. If $KS \neq 10^\infty$, then $N = N' + 1$.

Proof. If $KS = 10^\infty$, then there is only one equivalence class L and the first claim is trivial. If $KS = 0^\infty$, then from Proposition 4.2.3 we know that the only class, except L', is $L = \{0^m \mid m \ge 0\}$, and the claim is also trivial.

For general kneading sequences, as shown by Proposition 4.2.2, the class $[\varepsilon] = \{0^m \mid m \ge 0\}$ contains ε, and all 0 blocks as its elements. For other classes, say $[x]$, its element x must contain symbol 1, hence we can have decomposition $x = uv$ such that v is the nonempty LPS of x with respect to KS as in the proof of Proposition 3.4.2. By Lemma 4.2.5 we have $xR_L v$ and hence $v \in [x]$.

The formula of $N = N' + 1$ if $KS \neq 10^\infty$ is obviously true. ∎

By this Lemma we need only to work on all prefixes of a given kneading sequence KS to decide that if the language $L = \mathcal{L}(KS)$, generated from KS, is regular. Furthermore, if L is regular, we can use the formula $N = N' + 1$ (for $KS \neq 10^\infty$) to calculate the index N of R_L and obtain the minDFA that accepts L.

4.2.3 The Necessary Part of Theorem 4.2.1

Assume that $KS = e_1 \cdots e_n \cdots$ and denote the n-th prefix of KS for $n \in \mathbf{N}$ by $e^{(n)} = e_1 \cdots e_n$, and $e^{(0)} = \varepsilon$.

Lemma 4.2.7 The language L is regular if and only if there exist two distinct prefixes of KS, $e^{(i)}$ and $e^{(j)}$, $i \neq j$, such that

$$e^{(i)} R_L e^{(j)}.$$

Proof. The "only if" part is obvious. If L is regular, then, from the Myhill-Nerode Theorem, the index N of R_L is finite. Thus we can find some two distinct $e^{(i)}$ and $e^{(j)}$ belonging to one class.

Conversely, suppose that $e^{(i)} R_L e^{(j)}$ for some $0 \leq i < j$. We claim that each prefix $e^{(j+l)}$ for $l \geq 1$ belongs to an equivalence class $[e^{(k)}]$ for some k with $0 \leq k < j$.

Proceed inductively on l.

For $l = 1$, if $e_{j+1} = e_{i+1}$, then it is trivial that $e^{(j+1)} R_L e^{(i+1)}$ and $k = i + 1 \leq j$. If $e_{j+1} = \bar{e}_{i+1}$ ($\bar{0} = 1$, $\bar{1} = 0$), then $e^{(j+1)} R_L e^{(i)} \bar{e}_{i+1}$. Since the string $e^{(i)} \bar{e}_{i+1}$ is not a prefix of KS, we can find its LPS denoted by $e^{(i')}$ for some $i' \leq i$. From Lemma 4.2.5 we obtain $e^{(i)} \bar{e}_{i+1} R_L \bar{e}^{(i')}$ and $e^{(j+1)} R_L e^{(i')}$.

Now assume that the statement is true for l and consider the case of $l + 1$. From the inductive hypothesis we have $e^{(j+l)} R_L e^{(k)}$ for some k, $0 \leq k < j$. If $e_{j+l+1} = e_{k+1}$, then $e^{(j+l+1)} R_L e^{(k+1)}$. If $e_{j+l+1} = \bar{e}_{k+1}$, then $e^{(j+l+1)} R_L e^{(k)} \bar{e}_{k+1}$, we can find $e^{(k)} \bar{e}_{k+1}$'s LPS and argue as before. ∎

Remark. If L is a regular language, using $N = N' + 1$, and the fact that $N' \leq j$ from the foregoing discussion, we obtain the estimation $N \leq j + 1$.

Definition 4.2.8 Let z be a prefix of a given kneading sequence KS. z is called a *prefix of the first kind*, if both strings $z0$ and $z1$ belong to $L = \mathcal{L}(KS)$. Otherwise z is called a *prefix of the second kind*.

Some simple facts about these notions are listed in the following Lemma without proof.

Lemma 4.2.9 (1) If $e^{(i)} R_L e^{(j)}$, $i \neq j$, then either both or neither of the prefixes $e^{(i)}$ and $e^{(j)}$ is of the first kind.

(2) If $e^{(i)} R_L e^{(j)}$, $i \neq j$, and $e^{(i)}$, $e^{(j)}$ are both of the second kind, then $e_{i+1} = e_{j+1}$. (It is easy to see from the following facts: $e^{(i)} e_{i+1} R_L e^{(j)} e_{i+1}$, $e^{(i)} \bar{e}_{i+1} R_L e^{(j)} \bar{e}_{i+1}$ and $e^{(j)} e_{j+1} \in L, e^{(j)} \bar{e}_{j+1} \notin L$.)

(3) If $e^{(i)}$ is a prefix of the first kind, then $e^{(i)} e_{i+1} = e^{(i+1)} > e^{(i)} \bar{e}_{i+1}$.

(4) If $e^{(i+1)}$ is an even string, then $e^{(i)}$ must be a prefix of second kind, but the converse is wrong.

The following Proposition is the most important step towards our goal.

Proposition 4.2.10 If two prefixes $e^{(i)} R_L e^{(j)}$ for some $i < j$, and are of the same parity, then

$$KS = e_1 \cdots e_i (e_{i+1} \cdots e_j)^{\infty}$$

Proof. Observe first that the string $e_{i+1} \cdots e_j$ is even.

We claim that for each integer $l \geq 1$, the following statement are true.

$$e_{i+l} = e_{j+l}, \quad e^{(i+l)} R_L e^{(j+l)}, \quad \text{and } e_{i+l+1} \cdots e_{j+l} \text{ is even .} \tag{4.1}$$

Proceed inductively on l.

For $l = 1$, if both $e^{(i)}$, $e^{(j)}$ are of the second kind, then we have $e_{i+1} = e_{j+1}$ by Proposition 4.2.9 (2). It is clear that $e^{(i+1)} R_L e^{(j+1)}$ and $e_{i+2} \cdots e_{j+1}$ is even. If both $e^{(i)}$, $e^{(j)}$ are of the first kind, then from Proposition 4.2.9 (3) we have the following inequalities:

$$e^{(i)} e_{i+1} > e^{(i)} \bar{e}_{i+1}, \text{ and}$$

$$e^{(j)} e_{i+1} = e^{(i)} e_{i+1} \cdots e_j e_{i+1} > e^{(i)} e_{i+1} \cdots e_j \bar{e}_{i+1}.$$

Hence we have $e_{j+1} = e_{i+1}$ and the statement (4.1) holds again.

Now assume that (4.1) holds for l and consider the case of $l+1$. If both $e^{(i+l)}$, $e^{(j+l)}$ are of the second kind, then (4.1) holds for $l+1$ as before. Otherwise, from the inductive hypothesis, the string $e_{i+l+1} \cdots e_{j+l}$ is even. Using Proposition 4.2.9 (3) again, and observing the inequalities,

$$e^{(i+l)} e_{i+l+1} > e^{(i+l)} \bar{e}_{i+l+1},$$

$$e^{(i+l)} e_{i+l+1} \cdots e_{j+l} e_{i+l+1} > e^{(i+l)} e_{i+l+1} \cdots e_{j+l} \bar{e}_{i+l+1},$$

we have $e_{i+l+1} = e_{j+l+1}$ and (4.1) holds too.

From (4.1) we see that the block $e_{i+1} \cdots e_j$ will repeat infinitely many times in the kneading sequence $e_1 \cdots e_n \cdots$, and we obtain

$$KS = e_1 \cdots e_i (e_{i+1} \cdots e_j)^\infty$$

as required. ∎

Remark. If all prefixes $e^{(l)}$ for l greater than some integer n are of the second kind, then there are two of them, say $e^{(i)}$ and $e^{(j)}$, satisfying the condition $e^{(i)} R_L e^{(j)}$, and the conclusion of Proposition 4.2.10 holds. Here the requirement about the parity of $e^{(i)}$ and $e^{(j)}$ is not necessary.

Finally, we give the proof that if the language $L = \mathcal{L}(KS)$ is a regular language, then the kneading sequence KS is eventually periodic.

Proposition 4.2.11 If $L = \mathcal{L}(KS)$ is regular, then KS is eventually periodic.

Proof. By the Myhill-Nerode Theorem the equivalence relation R_L is of finite index. Except L', each equivalence class contains prefixes of KS. Since there are infinitely many prefixes of KS but only finite equivalence classes of R_L, there exist two distinct prefixes $e^{(i)}$ and $e^{(j)}$ belonging to one class and of the same parity. Using Lemma 4.2.10 completes the proof. ∎

4.2.4 The Maximal Path in minDFA

The first proof of Proposition 4.2.11 appeared in Chen, Lu and Xie (1993), where the main idea is to find a maximal path in the deterministic finite automaton (DFA) accepting L. This proof was also included in the book (Xie 1994).

In this Subsection we only give a sketch of this proof to explain its idea.

Let $L = \mathcal{L}(KS)$ be a regular language generated by KS. Since L is regular there exists a DFA

$$M = (Q, \Sigma, \delta, q_0, F)$$

accepting L. Here this M can be supposed to be a minDFA.

From Lemma 2.2.11 we know that if $KS \neq 10^\infty$, then there is only one non-accepting state, that is, card $Q \setminus F = 1$. Since 10^∞ is eventually periodic and its discussion is trivial, we assume that $KS \neq 10^\infty$ holds in the sequel of this subsection.

Deleting the unique non-accepting state U' and all arcs ending at it, we obtain a graph, which is also denoted by F below. (see Section 2.4 for the concept of graph.) The main idea here is to find a special infinite path in F, which is called a maximal path in our discussion below.

Beginning from q_0, the starting state of M, an infinite path in F consists of infinite arcs from states to states, and hence corresponds an infinite sequence

$$t = t_1 t_2 \cdots t_n \cdots$$

Since each prefix of t, say $t_1 \cdots t_n$, satisfies the condition

$$t_1 \cdots t_n \leq KS = e_1 \cdots e_n \cdots,$$

we see that t is an admissible sequence given in Definition 3.3.1.

Now determine a special maximal path as follows. Proceed inductively on n. Starting from q_0, if there are two arcs leaving q_0 and ending at states of F, then let $t_1 = 1$, otherwise let $t_1 = 0$. It is easy to see that except the case of $KS = 0^\infty$ we always have $t_1 = 1$ (see Figure 4.3 (a)).

Assume that we have obtained a finite path of length n in F, namely, a finite string $t_1 \cdots t_n$, and the state arrived by this path is $p \in F$. In order to determine the next arc leaving p, that is, the next symbol t_{n+1}, we have two situations to be discussed.

1. There are two arcs leaving from p and ending at states of F, then we select t_{n+1} such that

$$t_1 \cdots t_n t_{n+1} = \max\{t_1 \cdots t_n 0, t_1 \cdots t_n 1\}.$$

2. There is only one arc leaving from p and ending at a state of F. In this case we select t_{n+1} the symbol by which this arc is labeled.

Since M is a DFA and $L = \mathcal{L}(KS)$ is a dynamical language, each finite path in F can be elongated further. (Recalling the property D2 in Definition 2.1.1.) Thus there is no other situation possible except these two situations, and the inductive proof is completed. We call this path the *maximal path* of M (or F), and the corresponding sequence t the *maximal sequence*.

It is easy to prove that $t = KS$. (Here we suppose that there is no symbol c in KS as implied by Proposition 3.4.3.) As a matter of fact, each finite string $x \in \Sigma^*$ corresponding a finite path in F, and by the construction of t we have $x \leq t$ evidently. Thus we know that each admissible sequence s satisfies the inequality

$$s \leq t.$$

Since KS is the largest admissible sequence, and t itself is admissible, we obtain $t = KS$.

Finally, we prove that this maximal sequence $t = KS$ is eventually periodic. Here we use the idea implied in the proof of Lemma A.1.1.

Write the sequence of states and symbols generating the maximal sequence t by

$$q_0 \xrightarrow{t_1} q_1 \xrightarrow{t_2} \cdots \xrightarrow{t_n} q_n \xrightarrow{t_{n+1}} \cdots,$$

where q_0 is the starting state of the minDFA M, and the other indices of q_n's only reflect the order of generating symbols of t. Of course, each q_n represents a state of M. Since all q_n, $n \geq 0$, belong to F, which is a finite set, there exist infinitely many states q_n which coincide. Then we have to treat two cases separately.

Case 1. There exist infinitely many q_n's which have two arcs ending at states of F. Consider these q_n's and find the one which appears infinitely many times, then we can have integers $j \geq 0$ and $l > 0$ such that

$$q_j = q_{j+l}, \quad t_{j+1} = t_{j+l+1}.$$

From the procedure of generating t we have

$$t_1 \cdots t_j t_{j+1} > t_1 \cdots t_j \bar{t}_{j+1}$$

and

$$t_1 \cdots t_j t_{j+1} \cdots t_{j+l} t_{j+l+1} > t_1 \cdots t_j t_{j+1} \cdots t_{j+l} \bar{t}_{j+l+1}.$$

It is obvious from these inequalities that the string $t_{j+1} \cdots t_{j+l}$ is an even string. Thus it can be proved by mathematical induction that

$$q_{j+k} = q_{j+l+k}, \quad t_{j+1+k} = t_{j+l+k}$$

for each $k \geq 0$, which leads to the conclusion

$$KS = t = t_1 \cdots t_j (t_{j+1} \cdots t_{j+l})^\infty.$$

Case 2. If there exists an integer j such that for each $i > j$, the state q_i have only one arc ending at a state of F. It is easy to find an integer $l > 0$ such that the kneading sequence KS has the same form as above. Therefore, the proof is completed.

4.2.5 The Sufficient Part of Theorem 4.2.1

In this subsection we prove the "if" part of Theorem 4.2.1. There are several proofs available in references for it. The proof given here is from Xie (1993). The main tool in this proof is Lemma 4.2.7.

First we obtain a result which is an improvement of Proposition 3.4.2 for the case of periodic kneading sequences.

Lemma 4.2.12 If $KS = x^\infty$, where x is even and $|x| = n$, then $z \in L = \mathcal{L}(KS)$ if and only if every z', which is a substring of z and of length $|z'| \leq |x|$, satisfies the condition $z' \leq x$.

Proof. From Proposition 3.4.2 the proof of "only if" part is trivial.

Consider the "if" part. Assume the contrary that the claim is wrong, then there exists a string $z \notin L$, but each substring z' of z with $|z'| \leq n$ satisfies $z' \leq x$.

From Proposition 3.4.2 there exists at least one suffix of z, denoted by v, such that $v > KS$. Of course we have $|v| > n$. Let v' be the prefix of v with $|v'| = n$. From $v > KS$ we have $v' \geq x$. But since $|v'| = n$, we also have $v' \leq x$, and hence $v' = x$. Using the condition of x being even, the same argument leads to $v = x^l w$, where $l > 0$ and $|w| < n$. Since x is even, we have

$$v > KS \Longleftrightarrow w > KS \Longleftrightarrow w > x.$$

As w is also a substring of z and of length less than n, we have a contradiction. ∎

Proposition 4.2.13 If $KS = x^\infty$, then the language $L = \mathcal{L}(KS)$ is regular.

Proof. Without loss of generality, we assume that the string x is even. From Lemma 4.2.7 it suffices to show that $xR_L x^2$ holds. Consider xz and x^2z for each $z \in \Sigma^*$. If $x^2z \in L$, then it follows that $xz \in L$. On the other hand, if $xz \in L$, then using Lemma 4.2.12 we know that each substring v of xz and of length $v \leq |x|$ satisfies $v \leq x$. But this leads to the conclusion that x^2z also has this property, and, by Lemma 4.2.12 again, we have $x^2z \in L$. ∎

For the discussion of eventually periodic kneading sequences which are not periodic, we need the notions of primitive strings, cyclic shift of strings, and dual strings which are introduced in Chapter 1 (see Definition 1.1.7, Subsections *1.1.4* and *1.1.5*). First we need an estimation about the length of LPS of a special kind of strings obtained from prefixes of those kneading sequences.

Lemma 4.2.14 Let t be a nonempty prefix of $KS = \rho\lambda^\infty$, and $\bar{t}$ be its dual string. If u is the LPS of $\bar{t}$ with respect to KS, then $|u| < |\rho\lambda|$.

Proof. Without loss of generality, we suppose that λ is primitive. Assume the contrary that $|u| \geq |\rho\lambda|$, then from $|t| > |u|$ we have $t = \rho\lambda^j v$ with some $j > 0$, where v is a nonempty prefix of λ. Thus $0 < |v| \leq |\lambda|$ holds and $v = \lambda$ is allowed. Write $\lambda = vv'$.

Now we have $\bar{t} = \rho\lambda^j\bar{v}$ and u is its LPS. Since u is a prefix of KS, we can write it as $u = \rho\lambda^i u'$, where u' is a proper prefix of λ. Since u is also a suffix of $\bar{t}$ we have

$$\bar{t} = \rho\lambda^j\bar{v} = w\rho\lambda^i u'$$

for some string w. Observe the right-hand side of this equality.

If $u' = \varepsilon$, then its last substring λ is a suffix of $\bar{t}$. But then $v'\bar{v} = \lambda$ holds. Since $v'v$ is a cyclic shift of λ, this is impossible since each cyclic shift of λ has the same parity with λ, and $v'\bar{v}$ cannot have the same parity with λ.

If $u' \neq \varepsilon$, then from the fact of λ being primitive, we would have $\bar{v} = u'$. Since u' is a prefix of λ, but $\bar{v}$ is not, a contradiction again. ∎

The following result reflects the structure of language $\mathcal{L}(KS)$ if KS is eventually periodic.

Lemma 4.2.15 Let $KS = \rho\lambda^\infty$ be not periodic, $k = \lfloor |\rho|/|\lambda| \rfloor + 2$, and λ' a nonempty prefix of λ. If λ is even, then

$$\rho\lambda^k\overline{\lambda'} \notin L \Longleftrightarrow \rho\lambda^j\overline{\lambda'} \notin L \quad \forall j > k;$$

and if λ is odd, then

$$\rho\lambda^k\overline{\lambda'} \notin L \Longleftrightarrow \rho\lambda^j\overline{\lambda'} \notin L \quad \forall j > k \text{ and } j \equiv k \pmod 2.$$

Proof. Without loss of generality we suppose that λ is primitive. Since the proofs for both parity of λ are quite similar, we only give the proof for λ being even.

If $\rho\lambda^k\overline{\lambda'} \notin L$, then we can find a suffix v of this string such that $v > KS$ and $\bar{v}$ is a prefix of KS (we can choose v being a DEB of L). If $|v| \leq |\rho\lambda|$, then by the selection of k, we have $|\lambda^{k-1}| \leq |\rho\lambda| < |\lambda^k|$. Thus v is a suffix of $\lambda^k\overline{\lambda'}$, and also a suffix of $\rho\lambda^j\overline{\lambda'}$ for each $j > k$, which leads to $\rho\lambda^j\overline{\lambda'} \notin L$.

If $|v| > |\rho\lambda|$, then we can write it as $v = \rho\lambda^i\overline{\lambda''}$ for some $i > 0$ and λ'', a nonempty prefix of λ. At the same time we have $\rho\lambda^k\overline{\lambda'} = wv = w\rho\lambda^i\overline{\lambda''}$ for some string w. Using the primitivity of λ we have $\lambda' = \lambda''$. If $i < k$ then ρ is a suffix of $\rho\lambda^{k-i}$. It implies that KS is periodic, a contradiction. If $i = k$ then from $\rho\lambda^k\overline{\lambda'} = v > KS = \bar{v}\cdots$ we know that v is odd. Since λ is even, the string $\rho\lambda^j\overline{\lambda'}$ is odd for each $j > k$ and it follows $\rho\lambda^j\overline{\lambda'} > KS = \rho\lambda^j\lambda'\cdots$ as desired. Therefore we obtain

$$\rho\lambda^k\overline{\lambda'} \notin L \Longrightarrow \rho\lambda^j\overline{\lambda'} \notin L \quad \forall j > k.$$

The converse part can be proved similarly and omitted. ∎

Proposition 4.2.16 If $KS = \rho\lambda^\infty$, then the language $L = \mathcal{L}(KS)$ is regular.

Proof. By Proposition 4.2.13 we need only to discuss those kneading sequence which is eventually periodic but not periodic. Without loss of generality, we can also assume that the string λ is primitive.

Taking $k = [|\rho|/|\lambda|] + 2$ as in Lemma 4.2.15, we claim that if λ is even then

$$\rho\lambda^k R_L \rho\lambda^{k+1}$$

is true and so does our Proposition by using Lemma 4.2.7.

Assume that $\rho\lambda^k z \in L$ for $z \in \Sigma^*$. If z is a prefix of λ^∞, then $\rho\lambda^{k+1}z \in L$ is trivial. Otherwise we can decompose the string z into the form of $z'z''$ such that z' is the shortest prefix of z but not a prefix of λ^∞. This implies that $z' = \lambda^{k'}\overline{\lambda'}$ for some $k' \geq 0$ and λ' being a nonempty prefix of λ.

Since $\rho\lambda^k z \in L$, its substring $\rho\lambda^k z' \in L$ too. Let u be its LPS with respect to $KS = \rho\lambda^\infty$. By Lemma 4.2.14 we have $|u| < |\rho\lambda|$. The selection of k leads to $|\lambda^{k-1}| \leq |\rho\lambda| < |\lambda^k|$. Hence u is a suffix of $\lambda^k z'$. By Lemma 4.2.14 u is also the LPS of $\rho\lambda^{k+1}z'$.

Next we use Lemma 4.2.15 and know that $\rho\lambda^{k+1}z' \in L$ holds. By Lemma 4.2.5 we obtain $\rho\lambda^{k+1}z' R_L \rho\lambda^k z'$. Since R_L is right invariant, $\rho\lambda^{k+1}z R_L \rho\lambda^k z$ is also true. This proves that $\rho\lambda^k z \in L \Rightarrow \rho\lambda^{k+1}z \in L$. The converse part is quite the same and omitted.

The proof for the case of λ being odd is the same as before except that we will establish another relation $\rho\lambda^k R_L \rho\lambda^{k+2}$ instead. ∎

Remark. From Theorem 4.2.1 and Propositions 4.2.13, 4.2.16 we see that there are only two kinds of regular languages generated by kneading sequences, one is from periodic kneading sequence and another from eventually periodic kneading sequence but not periodic.

For the logistic map 3.1 we know that if a kneading sequence is periodic for some value of parameter b, then there exists a unique stable periodic orbits, and almost all orbits converge to it. Here the typical behavior is periodic. On the other hand, for $KS = 10^\infty$ each sequence is admissible. If a prefix of an itinerary $I(x)$ is given, say $I(x) = t_1 \cdots t_n \cdots$, then in this case the next symbol t_{n+1} cannot be predicted by using the information contained in $t_1 \cdots t_n$. In this sense its behavior is purely random. For the general case of eventually periodic kneading sequences, similar conclusion is reached by Hao (1989), and is given the name of "fully developed chaos".

Thus we see that for some unimodal maps these two cases are not very complicated, since both of them can be described by regular languages, the lowest level in Chomsky hierarchy. The same point of view can be found in, e.g., Crutchfield and Young (1990), Crutchfield (1994), Grassberger (1986, 1988a, 1988b), Li (1991) for general dynamical systems.

4.3 Minimum States DFA for Periodic Kneading Sequence

In this section we will determine the minimum state deterministic finite automata for the case of of periodic kneading sequences (Xie 1993b).

Some notions about strings are needed in the sequel.

A string x is said to be *even and minimal* (in length), if x is either even and primitive or $x = yy$ with an odd and maximal y. Recalling the notion of maximal string in Definition 1.2.1, if a maximal string x is even and minimal, then x satisfies the condition (P3) in Definition 1.2.7.

4.3.1 Calculation of the Index of R_L

Let $KS = x^\infty$ where x is even and minimal, that is, x satisfies the property (P3) in Definition 1.2.7. It is well-known from the Myhill-Nerode Theorem that the minDFA accepting L is unique up to an isomorphism, that is, a renaming of the states (see the proof of Theorem A.3.2).

It turns out that the most important information about the minDFA accepting $\mathcal{L}(KS)$ is the index of R_L, that is, the number of states in minDFA. The main result of this section is the following theorem.

Theorem 4.3.1 If $KS = x^\infty$, where x is an even and minimal string, then the index of the equivalence relation R_L, that is, the regular complexity of the language $\mathcal{L}(KS)$, is given by $N = |x| + 1$.

First of all we establish several Lemmas as a preparation.

Lemma 4.3.2 For each pair of two strings $x, y \in \Sigma^*$, $xR_L y$ if and only if $xaR_L ya$ for each $a \in \Sigma$.

Proof. If $xR_L y$, then from R_L's definition, it is obvious that $xaR_L ya$ for each $a \in \Sigma = \{0, 1\}$ by the right invariant property of R_L (see Subsection *A.3.1*).

Now assume that $xaR_L ya$ for each $a \in \Sigma = \{0, 1\}$. For each nonempty string z, we can write z as az' for some $a \in \Sigma$. Since $xaR_L ya$, $xaz' \in L \Leftrightarrow yaz' \in L$. So we obtain $xz \in L \Leftrightarrow yz \in L$ for each $z \in \Sigma^+$. If $z = \varepsilon$, and assume that $x \in L$ but $y \notin L$ (or *vice versa*), then from the definition of language L, x is a prefix of an admissible word s. This means that there is a symbol $a \in \Sigma$ such that $xa \in L$. But since $y \notin L$, so does ya and $xaR_L ya$ is false. By this contradiction our Lemma is proved. ∎

Lemma 4.3.3 If a kneading sequence $KS = x^\infty$, where x is even and minimal, and u is a proper PS of x, then $\overline{u}$, the dual string of u, belongs to L.

Proof. Assume the contrary that $\overline{u} \notin L$. Using Proposition 3.4.2, there exists a suffix of $\overline{u}$, say $\overline{u}'$, such that $\overline{u}' > KS$. Since u' is a suffix of u, we have $u' \le KS < \overline{u}'$, which implies that u' is an even prefix of KS. Hence u' is an even proper PS of x. Since x has the property (P3), by Theorem 1.2.10 u' should be odd, a contradiction. ∎

In the following discussion we will use the notations $x = x_1 \cdots x_n$, $|x| = n$, and $x^{(i)} = x_1 \cdots x_i$. We also use the conventions $x^{(0)} = \varepsilon$, $x^{(n)} = x$ and $x_{n+1} = x_1$.

Lemma 4.3.4 If $KS = x^\infty$ where x is even and minimal, and $x \neq 0$, then there exists a positive integer $n' < n$, such that

$$xR_L x^{(n')}. \tag{4.2}$$

Proof. From the hypothesis we see that KS begins from the symbol 1 and the substring $x_2 \cdots x_n$ also contains 1. Thus it has a nonempty LPS with respect to KS, denoted by $x^{(n')}$. Of course, $x^{(n')}$ is a nonempty proper PS of x. Using Theorem 1.2.10, $x^{(n')}$ is odd. Now consider the following situations.

(1) $x_{n'+1} = 1$. Then it is easy to see that $x^{(n'+1)}$ is the LPS of $x_2 \cdots x_n 1$. By applying Lemmas 4.2.5 and 4.2.12,

$$x1R_L x_2 \cdots x_n 1 \text{ and } x1R_L x^{(n')}1 = x^{(n'+1)}$$

are true. On the other hand, since $x^{(n')}$ is odd, we have $x^{(n')}0 > x^{(n')}1 = x^{(n'+1)}$. This means that $x^{(n')}0 \notin L$ and $x0 \notin L$. So we have

$$x0R_L x^{(n')}0$$

too. Using Lemma 4.3.2 we conclude that $xR_L x^{(n')}$ is true.

(2) $x_{n'+1} = 0$. Now the string $x0$ has $x^{(n')}0 = x^{(n'+1)}$ as its LPS. Using Lemma 4.2.5 we obtain $x0R_L x^{(n')}0 = x^{(n'+1)}$. On the other hand, we claim that in this case the LPS of the string $x_2 \cdots x_n 1$ must be 1. Assume the contrary that it has a PS $u1$ with a nonempty u. But then u is a PS of $x = x_1 \cdots x_n$, and by Theorem 1.2.10 u must be odd. Since $u0 > u1$, we see that $u0 \notin L$ and, consequently, $x0 \notin L$. This contradicts the fact that $x0R_L x^{(n')}0 = x^{(n'+1)} \in L$. Now the words $x1$ and $x^{(n')}1$ both have 1 as their LPS, we conclude that

$$x1R_L x^{(n')}1$$

is true. Again using Lemma 4.3.2, our proof is finished. ∎

Proof of Theorem 4.3.1. From the relation (4.2) and Remark of Lemma 4.2.7 we already obtain the estimation for N, the index of R_L, by

$$N \leq |x| + 1.$$

We claim that neither of two distinct prefixes among $x^{(0)}, x^{(1)}, \cdots, x^{(n-1)}$ can belong to one equivalence class of R_L. If this is true, then the number N' of those classes containing prefixes of KS is exactly $n-1$, and the proof is completed (cf. Lemma 4.2.6). We will prove this claim by several steps.

(1) $x^{(n-1)}R_L x^{(n'-1)}$ is false. Here $x^{(n')}$ is the proper PS of x in the relation (4.2). Since $x = x^{(n-1)}x_n$ is even, we have $x^{(n-1)}\bar{x}_n > x$ and $x^{(n-1)}\bar{x}_n \notin L$. Since $u = x^{(n')} = x^{(n'-1)}x_n \in L$, and by Lemma 4.3.3 we also have $\bar{u} \in L$. Thus $x^{(n-1)}R_L x^{(n'-1)}$ is impossible.

(2) $x^{(n-1)}R_L x^{(i)}$ is false for each $0 \leq i \leq n-2$. Assume the contrary, $x^{(n-1)}R_L x^{(i)}$ for some i, $0 \leq i \leq n-2$. From (1) we know that $i \neq n'-1$. Since R_L is right invariant and both prefixes $x^{(n-1)}$ and $x^{(i)}$ are of the second kind (see Lemma 4.2.9), $xR_L x^{(i+1)}$ holds too. Now if $x^{(i+1)}$ is even, we can use Proposition 4.2.10 and obtain

$$KS = x_1 \cdots x_{i+1}(x_{i+2} \cdots x_n)^\infty = x^\infty,$$

and $x_{i+2}\cdots x_n$ is even. This contradicts the condition that x is even and minimal. There remains the case that $x^{(i+1)}$ is odd. Here $xR_L x^{(i+1)}$ and $xR_L x^{(n')}(of\ (4.2)) \Rightarrow x^{(i+1)}R_L x^{(n')}$. Since $i+1 \neq n'$, and $x^{(n')}$ is odd, we can use Proposition 4.2.10 again and hence the claim of (2) is true.

(3) Assume that there are two distinct prefix $x^{(i)}$, $x^{(j)}$, $0 \leq i < j \leq n-2$, such that $x^{(i)}R_L x^{(j)}$. Since R_L is right invariant,

$$x^{(i)}x_{j+1}\cdots x_{n-1}R_L x^{(n-1)}$$

holds too. The claim (2) means that the string of the left-hand side, $x^{(i)}x_{j+1}\cdots x_{n-1}$, is not a prefix of x. But using its LPS, denoted by $x^{(k)}$, and Lemma 4.2.5, we can conclude that

$$x^{(k)}R_L x^{(n-1)}$$

for some k, $0 \leq k < n-1$. This contradicts the established claim (2) and finishes our proof. ∎

Remark. The result of Theorem 4.3.1 coincides with the study in Grassberger (1988a p.674).

4.3.2 Construction of minDFA for $KS = x^\infty$

Using Theorem 4.3.1 and the Myhill-Nerode Theorem we can explicitly construct the minDFA accepting $\mathcal{L}(x^\infty)$ as follows.

Let the kneading sequence be given that

$$KS = (x_1\cdots x_n)^\infty,$$

where the string $x = x_1\cdots x_n$ is even and minimal. Since the cases of $KS = 0^\infty$ and 1^∞ are shown in Examples 4.1.2 and 4.1.3 (see Figure A.2 (b)), we suppose that $n > 2$ in the following discussion.

The minDFA for the language L, generated from $KS = x^\infty$ is a 5-tuple $M = (Q, \Sigma, \delta, q_0, F)$, where Q is a finite set of states given by

$$Q = \{q_0, q_1, \cdots, q_{n-1}, q_d\},$$

in which q_d is the unique non-accepting state (see Lemma 2.2.11). Σ is the input alphabet $\{0,1\}$, $q_0 \in Q$ is the initial state, $F = Q\backslash\{q_d\}$ is the set of accepting states, and δ is the transition function mapping $Q\times\Sigma$ to Q, defined as follows:

(1) $\delta(q_i, x_{i+1}) = q_{i+1}$ for $0 \leq i \leq n-2$.

(2) $\delta(q_0, 0) = q_0$. For $1 \leq i \leq n-2$, if $x^{(i)}\bar{x}_{i+1} \in L$, then define $\delta(q_i, \bar{x}_{i+1}) = q_{f(i)}$, where $f(i)$ is the length of the LPS of $x^{(i)}\bar{x}_{i+1}$. Otherwise we define $\delta(q_i, \bar{x}_{i+1}) = q_d$.

(3) $\delta(q_{n-1}, x_n) = q_{n'}$, where n' is given in the proof of (4.2).

(4) $\delta(q_{n-1}, \bar{x}_n) = q_d$,

(5) $\delta(q_d, a) = q_d\ \forall a \in \Sigma$

From Theorem 4.3.1 and the Myhill-Nerode Theorem it is clear that this M is the minDFA accepting L: $L = L(M)$.

4.3.3 Some Examples

In Figures 4.4 and 4.5 we present the minDFA's for every $\mathcal{L}(x^\infty)$ with $|x|$ from 1 to 6. A few of them have been presented in Figure A.2 (b) (for $KS = 0^\infty$), 4.3 (a) (for $KS = 1^\infty$), and A.3 (a) (for $KS = (101)^\infty$). Note that in the expression $\mathcal{L}(x^\infty)$ the string x is even and minimal, hence for example if $KS = (10)^\infty$ then $x = 1010$. For simplicity the unique non-accepting state and arcs involved is omitted. Using the idea in the proof of Myhill-Nerode Theorem, the key information in these minDFA's is the relation (4.2).

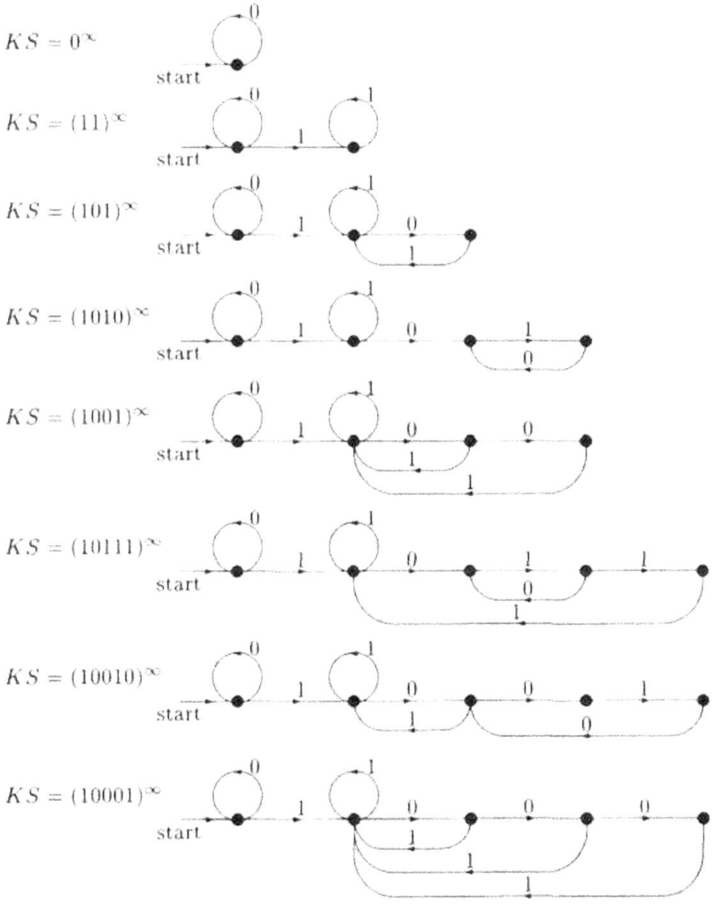

Figure 4.4. The minDFA's for $KS = x^\infty$ with $|x|$ from 1 to 5.

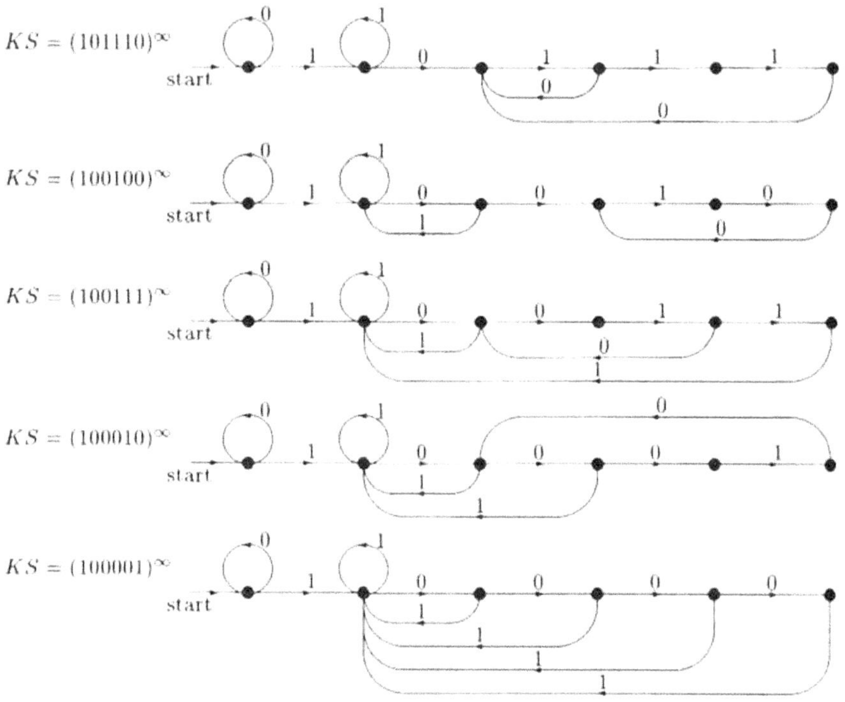

Figure 4.5 The minDFA's for $KS = x^\infty$ with $|x|$=6.

It is easy to give a general discussion for the case of $KS = y^\infty$ in which y is an odd maximal primitive string. Here $x = yy$ is even and minimal. We also denote $|x| = n$ as above. At first we see that the string $x^{(n')}$ in (4.2) is just the string y and thus $n' = |y| = n/2$. As a matter of fact, from the result of Corollary 1.2.11, $x = yy$ cannot have any proper PS longer than $|y|$, otherwise the primitive and maximal string y would have an even PS. Thus we obtain $x^{(n')} = y$. In this case the relation (4.2) becomes $yyR_{\iota}y$. Secondly, we will have $\delta(q_i, x_{i+1}) = q_d$ for each $n' \le i \le n-1$, which implies that each prefix $x^{(i)}$ for $n' \le i \le n-1$ is of second kind. If $x^{(i+1)}$ is even, then $x^{(i)}\bar{x}_{i+1} \notin L$. Otherwise $x_{n'+1}\cdots x_{i+1}$ (a prefix of y) is even, and $x_{n'+1}\cdots x_i\bar{x}_{i+1} \notin L \Rightarrow x^{(i)}\bar{x}_{i+1} \notin L$.

From this discussion we see that in this minDFA the state set Q can be divided, with the exception of q_d, into two subsets: one "transient" subset with the states $\{q_0, q_1, \cdots q_{n'-1}\}$ and another "attractor" subset with the states $\{q_{n'}, \cdots q_{n-1}\}$, which form a loop in minDFA. This means that if an itinerary contains a string y, then it will repeat infinitely many times. Of course, these are combinatorial results, which are independent of analytical discussion of unimodal maps.

4.4 Minimum States DFA for Eventually Periodic Kneading Sequence

At first it is evident that for a given eventually periodic kneading sequence which is not periodic, there exist infinitely many ways to write it in the form of $\rho\lambda^\infty$. Note that if it is required that the string λ is even and minimal as above, then although the selection of λ is not unique, its length $|\lambda|$, however, is independent of such selections, and is decided entirely by the periodic part of kneading sequence. As a matter of fact, we can find the primitive period of this part first, and take it as $|\lambda|$ if the primitive string, say λ, whose repetition constitutes the periodic part of kneading sequence, is even, and double it if λ is odd.

4.4.1 Some Lemmas for Eventually Periodic Kneading Sequences

Now we need some lemmas as the preparation for the discussion in the sequel. These results appeared in Wang and Xie (1994).

Lemma 4.4.1 If a given kneading sequence $KS = \rho\lambda^\infty$ is not periodic and $k = \lfloor|\rho|/|\lambda|\rfloor + 2$, then for each $i \geq k$ the string $\rho\lambda^i$ has no proper PS with respect to KS, whose length is longer than $|\lambda^i|$.

Proof. For simplicity, we only give the proof for the case of λ being primitive. The extension for the general case is not difficult.

Assume the contrary that there exists a proper PS of $\rho\lambda^i$ with the form $\rho'\lambda^i$, where ρ' is a proper suffix of ρ. Since $|\lambda^{k-2}| \leq |\rho| < |\lambda^{k-1}|$ by the selection of k, we have string ρ is a proper prefix of $\rho'\lambda^{k-1}$. Using the primitivity of λ, we have $\rho'\lambda^i = \rho\lambda^l$ for some integer l, $0 < l < i$. But this leads to $\sigma^{|\rho|-|\rho'|}(\rho\lambda^\infty) = \rho\lambda^\infty$ and, consequently, $KS = \rho\lambda^\infty$ would be periodic, a contradiction. ∎

Lemma 4.4.2 If $KS = \rho\lambda^\infty$ is not periodic, where λ is even, then the string λ and its every cyclic shift $\sigma^i(\lambda), 0 < i < |\lambda|$, cannot be a prefix of KS.

Proof. It is clear that we can restrict our discussion to the case of $\lambda = M(\lambda)$ (see Subsection *1.2.2*).

If, assuming the contrary, we have $KS = \rho\lambda^\infty = \lambda z$, where z is an infinite suffix of KS, then, because of the maximal property of KS (see *3.2.3*), $z \leq \lambda$. On the other hand, from $\lambda^\infty \leq \rho\lambda^\infty = \lambda z$ and λ being even, we also have $\lambda \leq z$. Combining these facts, we see that the string λ must be a prefix of the infinite string z. Then KS has λ^2 as its prefix. Arguing in this way leads to $KS = \lambda^\infty$ and this contradiction completes our proof. ∎

Lemma 4.4.3 If $KS = \rho\lambda^\infty$ is not periodic, where λ is even and maximal, then λ has no even PS with respect to KS.

Proof. Without loss of generality, we can assume that $|\lambda|$ is minimal already, that is, λ is even and minimal. Assume the contrary that λ has an even PS, denoted by λ', with respect to KS. Using Lemma 4.4.2 we see that $\lambda' \neq \lambda$. Now since λ' is a prefix of KS, so we must have $\lambda \le \lambda'$. But the condition $\lambda = M(\lambda)$ means also that $\lambda' \le \lambda$. Then the string λ' must be an even proper PS with respect to λ itself. Using Theorem 1.2.10, our proof is completed. ∎

Lemma 4.4.4 If $KS = \rho\lambda^\infty$ is neither periodic nor 10^∞, where λ is even and maximal ($\lambda = M(\lambda)$), and the string λ' is the LPS of λ with respect to KS, then we have a decomposition $\lambda = \lambda'\lambda''$ such that the string $\lambda''\lambda'$ has no even PS with respect to KS.

Proof. Under our conditions the string λ contains at least twice the symbol 1, which implies that the λ', its LPS, is nonempty (see the proof of Proposition 3.4.2). Using Lemma 4.4.3, we know that λ' is odd and also a prefix of λ, so we have a decomposition $\lambda = \lambda'\lambda''$. Using this expression we have the string $\lambda''\lambda'$ and consider its PS with respect to KS. The conclusion of Lemma 4.4.2 means that we need not consider $\lambda''\lambda'$ itself. Again using Lemma 4.4.3, the string λ' (as a suffix of λ) has no even PS with respect to KS.

Therefore, if the statement of our Lemma is wrong, then $\lambda''\lambda'$ can only have an even PS of the form $p\lambda'$, where p is a non-empty proper suffix of λ''. Since λ' is odd, p is odd too.

From $\lambda = \lambda'\lambda''$ we see that p is also a PS of λ, so that we must have $|p| \le |\lambda'|$, as λ' =LPS. On the other hand, all of the three strings $p\lambda'$, p and λ' are prefixes of KS. These considerations lead to that p^2 is a prefix of KS and $KS = p^\infty$ by Lemma 1.2.4, a contradiction. ∎

4.4.2 Calculation of the Index of R_L

First of all we need some definitions for our main result (Wang and Xie 1994).

Definition 4.4.5 Let KS be an eventually periodic kneading sequence. An expression $KS = \rho\lambda^\infty$ is called an *exact form* if the strings ρ and λ satisfy conditions: (1) ρ is odd and λ is even, (2) both ρ and λ have no even PS (prefix-suffix) with respect to the given KS.

An exact form $KS = \rho\lambda^\infty$ is called *minimal* if the length of string $\rho\lambda$, $|\rho\lambda|$, reaches the minimum value among all exact forms of the given KS.

The main result of this section is as follows.

Theorem 4.4.6 If a kneading sequence KS is eventually periodic, but neither periodic nor 10^∞, then the minimal exact form of KS

$$KS = \rho\lambda^\infty$$

exists, and the index of the equivalence relation R_L, that is, the regular complexity of the language $\mathcal{L}(KS)$, is

$$N = |\rho\lambda| + 1.$$

Lemma 4.4.7 If KS is eventually periodic, but not periodic, then the minimal exact form $KS = \rho\lambda^\infty$ of KS exists.

Proof. For $KS = 10^\infty$ it suffices to select $\rho = 1$ and $\lambda = 0$. Otherwise, we need only to prove that there exists an exact form $KS = \rho\lambda^\infty$ for the given KS.

At first we can write the given KS in the form of $KS = \rho_1\lambda_1^\infty$, where λ_1 is even and maximal, $\lambda_1 = M(\lambda_1)$. Then we discuss two cases respectively.

(1) ρ_1 is odd. From Lemmas 4.4.1, 4.4.2, and 4.4.3 the string $\rho = \rho_1\lambda_1^k$, where $k = [|\rho_1|/|\lambda_1|] + 2$, has no even PS with respect to KS. Take $\lambda = \lambda_1$, the expression $KS = \rho\lambda^\infty$ is an exact form as required.

(2) ρ_1 is even. At first we use Lemma 4.4.4 to obtain the LPS λ' of λ_1 with respect to KS, and write $\lambda_1 = \lambda'\lambda''$. Writing $KS = \rho_1\lambda'(\lambda''\lambda')^\infty$, where $\rho\lambda'$ is odd, and $\lambda''\lambda'$ has no even PS with respect to KS. The remaining part is similar to (1). ∎

Lemma 4.4.8 If $KS = \rho\lambda^\infty$ is an exact form, then $\rho R_L \rho\lambda$ holds.

Proof. We have to prove that for each $\alpha \in S^*$, $\rho\alpha \in L$ if and only if $\rho\lambda\alpha \in L$. It suffices to write out the proof for the "only if" part, while the "if" part is easier.

Using Proposition 3.4.2, we see if every suffix of $\rho\lambda\alpha$, say z, satisfies $z \le KS$.

Since $\rho\alpha \in L$ implies $\rho\alpha \le KS$ and λ is even, we have $\rho\lambda\alpha \le KS = \rho\lambda^\infty$ too.

Because $\rho\alpha \in L \Rightarrow \alpha \in L$, we need only to consider those proper suffixes of $\rho\lambda\alpha$, which are of length $> |\alpha|$. Write it as $p\alpha$, where p is a non-empty proper suffix of $\rho\lambda$.

If p is not a prefix of KS, then from $\rho\lambda \in L$, we already have $p < KS$, which leads directly to $p\alpha < KS$.

If p is a prefix of KS, then by the condition (1) of Definition 4.4.5 p is odd. Hence we have $\rho\lambda = ap$, where the string a is even. Now we can argue by

$$\rho\alpha \le \rho\lambda^\infty \Rightarrow \rho\lambda\alpha \le \rho\lambda^\infty \Rightarrow p\alpha \le p\lambda^\infty \le \rho\lambda^\infty,$$

which completes our proof. ∎

Remark. From Lemmas 4.2.7 and 4.4.8 we obtain a new proof for Theorem 4.2.16 and know that for the index of R_L, the inequality $N \le |\rho\lambda| + 1$ holds for every exact form, and, consequently, for the minimal exact form too. The remaining part of the proof of Theorem 4.4.6 is to establish that the equality holds in this expression. But we still need several lemmas before we can complete it.

Lemma 4.4.9 If $KS = \rho\lambda^\infty$ is an exact form, then

$$\overline{\rho\lambda^l} \in L \text{ for each } l \ge 0,$$

and has no even PS with respect to KS.

Proof. This is a direct consequence of Lemmas 3.4.5 and 3.4.4 by applying them to $\rho\lambda^l$ and using the condition of ρ being odd and λ being even. ∎

Lemma 4.4.10 Let y be an odd prefix of KS, and has no even PS with respect to KS. If $\overline{y} = y_1x$, where x is the LPS of $\overline{y}$ with respect to KS, then y_1 has the same property with y, i.e. y is an odd prefix of KS and has no even PS with respect to KS.

Proof. Using lemmas 3.4.5 and 3.4.4, we find that $\overline{y} \in L$ and has no even PS, so x is odd, and hence y_1 is odd too.

Assume the contrary that if y_1 has an even PS, say p, then from $\overline{y} = y_1 x \in L$ and Proposition 3.4.2 we know that $px \leq KS$. But as x is the LPS of $\overline{y}$, we must have $px < KS$. Since p is an even prefix of KS, we obtain $x < \sigma^{|p|}(KS) \leq KS$, a contradiction to the fact that x is a prefix of KS too. ∎

Lemma 4.4.11 If $KS = \rho\lambda^\infty$, where ρ has no even PS with respect to KS and λ is even, then $\rho\overline{\lambda}$ and $\overline{\lambda}$ have the same LPS with respect to KS.

Proof. It is enough if we can prove that the LPS of $\rho\overline{\lambda}$ is of length $\leq |\lambda|$. If this is not true, writing it as $p\overline{\lambda}$, where p is a nonempty PS of ρ, and using Lemma 4.4.9, then $p\lambda$ is odd and, consequently, p is even, a contradiction. ∎

Lemma 4.4.12 If $KS = \rho\lambda^\infty$ is an exact form, then it is the minimal exact form if and only if the LPS of $\overline{\rho}$ and $\overline{\lambda}$ with respect to KS are not equal.

Proof. For simplicity we only write out the proof of the necessary part, which is enough for the proof of Theorem 4.4.6.

Assume the contrary that $\overline{\rho}$ and $\overline{\lambda}$ have the same LPS, denoted by x, then we can write $\overline{\rho} = \rho_1 x$ and $\overline{\lambda} = \lambda_1 x$.

By Lemma 4.4.10, ρ_1 is odd and has no even PS. By Lemma 4.4.11 x is also the LPS of $\rho\overline{\lambda}$. Writing $\rho\overline{\lambda} = \rho\lambda_1 x = \rho_1\overline{x}\lambda_1 x$ and using Lemma 4.4.10 again to $\rho\lambda$, we know that $\rho\lambda_1 (= \rho_1\overline{x}\lambda_1)$ is odd and has no even PS. This implies that $\overline{x}\lambda_1$ also has no even PS.

Rewriting $KS = \rho\lambda^\infty$ as $KS = \rho_1(\overline{x}\lambda_1)^\infty$, we have a new exact form of KS. From $|\rho_1\overline{x}\lambda_1| < |\rho\lambda|$, we have a contradiction to the minimal property of the expression $KS = \rho\lambda^\infty$. ∎

The following Lemma is the final step of our preparation for the proof of Theorem 4.4.6.

Lemma 4.4.13 Let $KS = \rho\lambda^\infty$ be the minimal exact form. If $aR_L b$ holds, where strings a and b are two distinct odd prefixes of KS, and both have no even PS with respect to KS, then

$$|\rho\lambda| \leq \max\{|a|, |b|\}.$$

Proof. Without loss of generality, we assume that $|a| < |b|$ and $b = ac$, where c is even and has no even PS. From the Lemma 4.2.10 we obtain $KS = ac^\infty$. Since this expression is an exact form of KS, and $KS = \rho\lambda^\infty$ is the minimal one, the conclusion $|\rho\lambda| \leq |ac| = |b| (= \max\{|a|, |b|\})$ is obviously true. ∎

The proof of Theorem 4.4.6. Lemma 4.4.7 ensures the existence of the minimal exact form of $KS = \rho\lambda^\infty$. Using Lemmas 4.4.8 and 4.2.7, we have the estimation

$$N \leq |\rho\lambda| + 1$$

already. In order to prove that the equality holds in it, we need only to show that any two nonempty prefixes of KS, which are of length less than $|\rho\lambda|$, do not belong to one equivalence class of R_L. This can be done in several steps below.

(1) $\rho\pi R_L \rho\lambda\pi$ is false (the effect of π after a nonempty string is to delete its last symbol.).

Assume the contrary that $\rho\pi R_L \rho\lambda\pi$ is true.

If ρ and λ have the same last symbol, then using the right-invariant property for R_L, the relation $\bar{\rho} R_L \rho\bar{\lambda}$ also holds. Taking the LPS of $\bar{\rho}$ and $\rho\bar{\lambda}$ with respect to KS and using Lemmas 4.4.11 and 4.4.12, we obtain two distinct prefixes of KS. By Lemma 4.4.9 both of them are odd and satisfy the conditions of Lemma 4.4.13. Since these prefixes are of length less than $|\rho\lambda|$, using Lemma 4.2.5 leads to a contradiction with the condition that $KS = \rho\lambda^\infty$ being in the minimal exact form.

If the last symbols of ρ and λ are not the same, then we have $\bar{\rho} R_L \rho\lambda R_L \rho$ instead. Starting from $\bar{\rho} R_L \rho$ and using Lemmas 4.4.9 and 4.2.5 we can argue as before.

(2) $pR_L \rho\lambda\pi$ is false for each prefix p of KS with the properties: $|p| < |\rho\lambda\pi|$ and $p \neq \rho\pi$.

Assume the contrary that $pR_L \rho\lambda\pi$ holds for this p. Let α be the last symbol of λ, then the relation $p\alpha R_L \rho\lambda R_L \rho$ holds too. Consider the following alternatives:

(2.1) $p\alpha$ is odd. Since both $\rho\lambda$ and $\rho\bar{\lambda} \in L$ by Lemma 4.4.9, using $pR_L \rho\lambda\pi$ and the right-invariant property of R_L, we see that both $p0$ and $p1$ belong to L. It is obvious that the odd one among them is a prefix of KS. Using Lemma 3.4.4 and $\overline{p\alpha} \in L$, we know that $p\alpha$ has no even PS with respect to KS. Starting from $p\alpha R_L \rho$ and using Lemma 4.4.13 is enough.

(2.2) $p\alpha$ is even. Then $p\bar{\alpha}$ is a prefix of KS. If x is the LPS of $p\alpha$, then, by Lemma 3.4.4, x is odd and has no even PS.

If $x \neq \rho$, then starting from $xR_L \rho$ by Lemmas 4.2.5 and 4.4.13 is enough.

If $x = \rho$, now we use the relation $p\bar{\alpha} R_L \rho\bar{\lambda}$ instead, where $p\bar{\alpha}$ is a prefix of KS. Taking the LPS of $\rho\bar{\lambda}$, denoted by y, and using Lemma 4.2.5, we obtain $p\bar{\alpha} R_L y$.

If $y \neq p\bar{\alpha}$, then using Lemmas 4.4.9 and 4.4.13 is enough.

If $y = p\bar{\alpha}$. Observe the following facts: $x = \rho$ is a suffix of $p\alpha$, $y = p\bar{\alpha}$ is the LPS of both $\rho\bar{\lambda}$ and (by Lemma 4.4.11) $\bar{\lambda}$. Combining them leads to the conclusion that λ also has ρ as its suffix. But this means that $\sigma^{|\lambda|}(\rho\lambda^\infty) = \rho\lambda^\infty$, a contradiction to the condition that KS is not periodic.

(3) Now we can prove that any two distinct prefixes of KS which are of length less than $< |\rho\lambda|$ must belong to different equivalence classes of R_L. In fact, if $aR_L b$ holds for such two prefixes a and b, then without loss of generality we can assume that $|a| < |b| \leq |\rho\lambda\pi|$. Writing $\rho\lambda\pi = bc$, then we have $acR_L \rho\lambda\pi$. Using Lemma 4.4.13 and the conclusion of (2), our proof is completed. ∎

4.4.3 Some Examples

Example 4.4.14 Calculate the index N of R_L for $KS = t_n\bar{t}_n^\infty$ where t_n is the string $h^n(1)$ generated by the homomorphism $h = \{1 \to 10, 0 \to 11\}$.

In Subsection *6.1.1* it is explained that this homomorphism h is the symbolic representation of renormalization from f^2 to f (see Figure 6.1). Thus it is easy to understand that this KS is the kneading sequence of $2^n \to 2^{n-1}$ band-merging points in the bifurcation Diagram (see Figures 3.1 and 3.2). The name of inverse bifurcation point is also used.

The basic facts about $\{t_n\}$ is

(1) $t_{n+1} = t_n \bar{t}_n$,
(2) t_n is odd for $n \geq 0$,
(3) $|t_n| = 2^n$ for $n \geq 0$,
(4) t_n is maximal and primitive.

(see Lemma 6.1.1 for proof.) These facts imply that the expression $KS = t_n \bar{t}_n^{\infty}$ is the minimal exact form as desired (by Corollary 1.2.11 and Lemma 3.4.4), and, consequently, $N = 2^{n+1} + 1$.

Example 4.4.15 Calculate the minimal exact form of $KS = 101(10110)^\infty$

Since in this expression the strings 101 and 10110 have different last symbols, the length $\rho\lambda$ is minimal among all possible forms. But it is not an exact form. What we need to do is simply to increase the length of ρ step by step and cyclic shift λ properly. Lemma 4.4.5 ensures its success.

Rewrite this KS by $1011(0110101101)^\infty$, where ρ is odd and λ is even, but λ has even PS, for instance 101 and 101101.

Increasing ρ again, we have $KS = 10110(1101011010)^\infty$, which is an exact form and the procedure itself ensures that it is the minimal one desired.

Example 4.4.16 Calculation of the minimal exact form of

$$KS = 10^3 1(1010)^m 10^2 10^3 1(1010)^\infty,$$

where m is a fixed integer, can be determined as follows by the procedure as in last Example. The resulted minimal exact form is

$$KS = 10^3 1(1010)^m 10^2 10^3 1(1010)^m 101(0101)^\infty$$

4.4.4 Construction of minDFA for $KS = \rho\lambda^\infty$

Using the same idea as in Subsection *4.3.2* we can construct the minDFA for eventually periodic kneading sequence. Here the most important point is to use the conclusion of Lemma 4.4.8 that if $KS = \rho\lambda$ is an exact form, then

$$\rho R_L \rho\lambda.$$

Example 4.4.17 Determine the minDFA accepting $\mathcal{L}(KS)$ for $KS = 101(10)^\infty$

The first step is to obtain its minimal exact form $KS = 10110(1010)^\infty$ by the method used in Example 4.4.15. This provides us the number of states of the min DFA accepting $\mathcal{L}(KS)$, that is, the regular complexity N=10.

The remaining steps are very easy, where Lemma 4.2.5 is used to calculate the transition rules in minDFA. Using the Myhill-Nerode Theorem we can identify each state with an equivalence class as follows.

$$q_0 = [\varepsilon], q_1 = [1], q_2 = [10], q_3 = [101],$$
$$q_4 = [1011], q_5 = [10110], q_6 = [101101],$$
$$q_7 = [1011010], q_8 = [10110101] \text{ and } q_9 = L'.$$

where the notation $[x]$ denotes the equivalence class of R_L containing the string x. We also call x a representative element of this class.

Write $q_i = [x_i], 0 \le i \le 8$, where x_i is the prefix of KS of the length $|x_i| = i$ as before. Let a be 0 or 1. For $0 \le i \le 7$, if $x_i a \notin L$, which can be decided by using Proposition 3.4.2, then let $\delta(q_i, a) = q_9$. Otherwise, if $x_i a \in L$, then using Lemma 4.2.5 to decide the LPS of $x_i a$, say y, then from $|y| = j$ we have $\delta(q_i, a) = q_j$. A special case is $x_i a = x_{i+1}$ and $\delta(q_i, a) = q_{i+1}$. For q_8 and $a = 1$, using Lemma 4.2.5 leads to $\delta(q_8, 1) = q_4$. But for $a = 0$, the relation $\rho\lambda R_L \rho$ provides us with $\delta(q_8, 0) = q_5$. Finally, for the dead state q_9 we have $\delta(q_9, a) = q_9$ for both 0 and 1. The transition diagram of M is shown in Figure 4.6 (where the number in each circle is the subscript i of the state q_i).

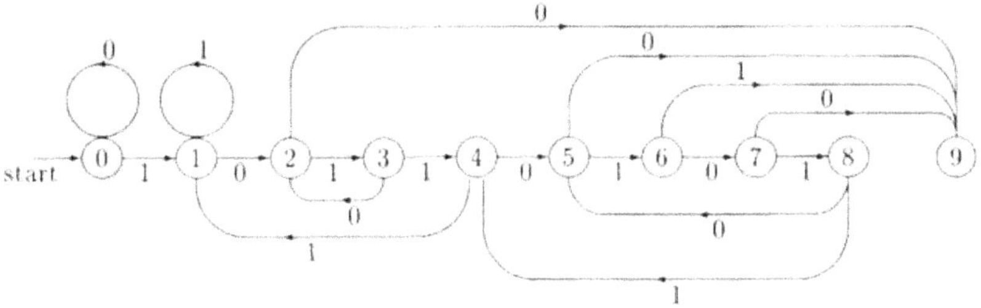

Figure 4.6. The minDFA for $KS = 101(10)^\infty$

In Figure 4.7 we present the minDFA's for $\mathcal{L}(\rho\lambda^\infty)$ where the eventually periodic kneading sequences are neither periodic nor 10^∞, and are written in the minimal exact form. Thus the regular complexity of them are $N = |\rho\lambda| + 1$. All possibilities of $N \le 7$ are included in Figure 4.7. The first example is $KS = 10(11)^\infty$, the kneading sequence of the first inverse bifurcation point, or the 2-1 band-merging point (cf. Figure 3.1). We also omit the non-accepting state and arcs involved as in Figures 4.4 and 4.5.

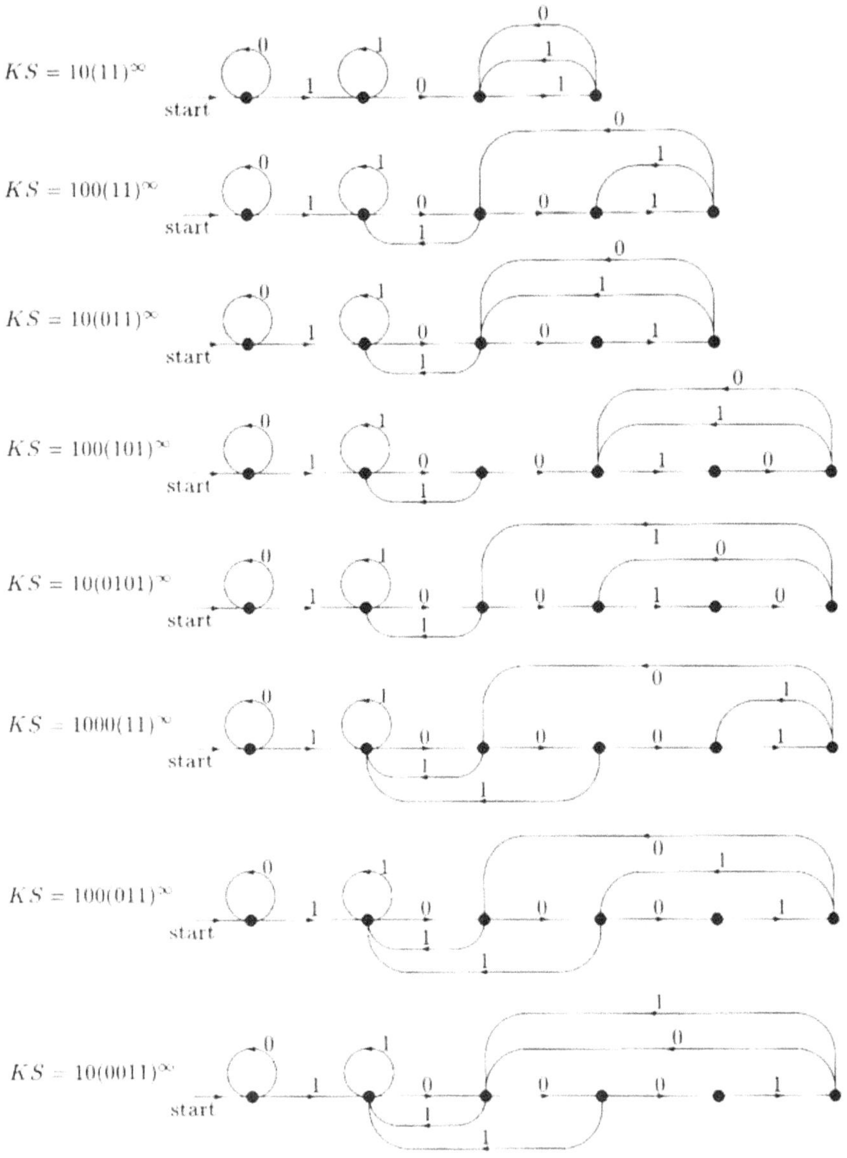

Figure 4.7: The minDFA's for $KS = \rho\lambda^\infty$ with $4 \leq |\rho\lambda| \leq 6$.

4.5 Composition Laws and Self-Similarity

It turns out that the regular languages generated from unimodal maps lead directly to the $*$-composition laws (Derrida, Gervois and Pomeau 1978) and the generalized composition laws (Zheng 1989b). It is well-known that each periodic window corresponds a certain $*$-composition law, and it also can generate infinitely many generalized composition laws. In Wang and Xie (1994) it was discussed how to generate all possible generalized composition laws.

4.5.1 The $*$-Composition Law

For the logistic maps (3.1), we can see some windows in its bifurcation diagram, the Figure 3.1. This phenomenon is due to the existence of stable periodic orbits for some parameter values. Of course, in order to be a visible window on computer's screen, it must have a non-zero widths and big enough. It is easy to prove that if for a certain value of parameter b, say b_0, the map $x \mapsto b_0 x(1-x)$ has a superstable periodic orbit, that is, the point c is periodic, then there exist two values of b, say $b_1 < b_0 < b_2$, such that for each $b \in (b_1, b_2)$, the corresponding map has a stable periodic orbit which has the same primitive period with that of the superstable periodic orbit for b_0. Moreover, at the point b_1 the tangent bifurcation happens while at the point b_2 the pitchfork bifurcation happens (see, e.g., May 1976, Devaney 1989).

It can be shown that, if the kneading sequence for b_0 is written as $(e_1 \cdots e_{n-1}c)^\infty$, then for $b \in [b_1, b_0)$, we have $KS = ((e_1 \cdots e_{n-1}c)_-)^\infty$, and for $b \in (b_0, b_2]$, $KS = ((e_1 \cdots e_{n-1}c)_+)^\infty$ (see (3.3) and (3.4)).

Let $x = (e_1 \cdots e_{n-1}c)_-$, then x is even, and $\overline{x} = e_1 \cdots e_{n-1}c)_+$. As to the problem of primitivity of x and $\overline{x}$, we have the following fact, where the notation $t|_c$ means a string obtained from a nonempty string $t \in \Sigma^*$ by changing its last symbol to c.

Lemma 4.5.1 If $x|_c$ is maximal, and x, $\overline{x}$ are defined as above, then $\overline{x}$ is maximal and primitive, and x is either maximal and primitive or $x = yy$ with an odd maximal primitive string y.

Proof. It is a simple consequence of Theorem 1.2.10. We need only to verify the conditions (P1) and (P2) for x (see Definition 1.2.7).

First discuss x. If u is a suffix of x and $u > x$ holds, then $u|_c > x|_c$ holds too and contradicts the fact that $x|_c = e_1 \cdots e_{n-1}c$ is maximal. Hence the first condition of Theorem 1.2.10, that is, each suffix of x, say u, satisfies $u \le x$, the property (P1) holds. For the condition (P2), assume the contrary that x has a even proper PS, say v, then $v|_c$ is a suffix of $x|_c$ and $v|_c > x = v \cdots$ holds, a contradiction again.

Using Theorem 1.2.10 and Lemma 1.2.9, both x and $\overline{x}$ satisfy the condition (P3). Since $\overline{x}$ is odd, our conclusion is true. ∎

For example, for the period 3 window we have $x = 101$ and $\overline{x} = 100$. But for the periodic $KS = (10010c)^\infty$ we have $x = 100100$ and $\overline{x} = 100101$.

The $\star$-composition law was introduced by Derrida, Gervois and Pomeau (1978). It can also be seen as a special class of homomorphisms of languages (see Subsection *A.4.2*).

Definition 4.5.2 A $\star$-composition law is a homomorphism h from Σ^* to itself if the string x in

$$h = \{1 \to \bar{x}, 0 \to x\}$$

satisfies conditions:

$$\begin{array}{l} (1)\ x \text{ is even,} \\ (2)\ x \text{ is maximal,} \\ (3)\ x \text{ is either primitive or } x = yy \text{ where } y \text{ is odd and primitive.} \end{array} \tag{4.1}$$

Evidently, this h preserves both order relation between and parity of strings. In particular, if $KS = e_1 \cdots e_n \cdots$ and let $h(KS)$ be $h(e_1) \cdots (e_n) \cdots$, then we can see that $h(KS)$ is also a maximal infinite sequence, and hence a kneading sequence which can be realized in a full family (Proposition 3.5.3).

4.5.2 Self-Similarity

Denote by the notation

$$[KS_1, KS_2] = \{KS \mid KS_1 \le KS \le KS_2\} \tag{4.2}$$

the set of all kneading sequences between two given kneading sequences KS_1 and KS_2, and call it henceforth an *interval of kneading sequence*, which has KS_1 and KS_2 as its left and right edges.

Thus $W_0 = [0^\infty, 10^\infty]$ is the largest interval of kneading sequences, that is, the set of all kneading sequences that a full family of unimodal maps can generate.

Of course, the image of the set W_0 under a $\star$-composition law h is a subset of W_0, $h(W_0) \subset W_0$. Furthermore, we can prove more that $h(W_0)$ is an interval of kneading sequences.

Definition 4.5.3 If x is an even, maximal and primitive string, then the set

$$W_x = [x^\infty, \bar{x}x^\infty]$$

is said to be a *window of kneading sequences generated by x*.

Proposition 4.5.4 If W_x is a window of kneading sequences generated by an even maximal primitive string x, then

$$W_x = h(W_0), \text{ where } h = (1 \to \bar{x}, 0 \to x).$$

Proof. If $KS \in W_x$ then we can write it into the form of

$$KS = y_1 y_2 \cdots y_n \cdots, \text{ where each } y_i \text{ is of length } |x|$$

We will claim that each y_i is either x or $\bar{x}$.

Proceed inductively on n. For $n = 1$, since $x \le y_1 \le \bar{x}$, the claim is true for $n = 1$. Suppose the claim is true for n and consider the case of $n+1$. Since KS is maximal, $y_{n+1} \le y_1 \le \bar{x}$ holds. It suffices to prove that $y_{n+1} \ge x$. Assume the contrary that $y_{n+1} < x$, then observe $y_1, \ldots, y_n$. If there exists a string $y_i = \bar{x}$ among them, then, without loss of generality, we can assume that $y_i y_{i+1} \cdots y_{n+1} = \bar{x}x \cdots x y_{n+1}$. Then we have

$$y_i \cdots y_{n+1} > \bar{x}x^\infty,$$

which contradicts

$$y_i \cdots y_{n+1} \le KS \le \bar{x}x^\infty$$

On the other hand, if it happens that $y_1 \cdots y_{n+1} = x^n y_{n+1}$, then

$$y_1 \cdots y_{n+1} < x^\infty \le KS,$$

a contradiction again. Therefore, our induction is finished.

Using the inverse homomorphism of h, we can write

$$KS = h(s), \text{ where } s = s_1 \cdots s_n \cdots, \text{ and } s_n = h^{-1} y_n \text{ for } n \ge 1.$$

From KS is maximal and h is order preserving, the sequence s found thus far is also a maximal sequence, that is, a kneading sequence. ∎

For instance, all kneading sequence in the period 3 window is just

$$h([0^\infty, 10^\infty]) = W_{101} = [(101)^\infty, 100(101)^\infty].$$

Thus the self-similarity of bifurcation diagram (3.1) has its root in symbolic dynamics.

Remark. In defining windows of kneading sequences we only consider those $*$-composition laws in which x is primitive, while in Definition 4.5.2 $x = yy$ is allowed. In fact, in this case the string y is odd maximal and primitive, and

$$\bar{y}^\infty \le y^\infty \le y\bar{y}y^\infty \le y\bar{y}^\infty,$$

thus we have

$$[y^\infty, y\bar{y}y^\infty] \subseteq W_{\bar{y}} = [\bar{y}^\infty, y\bar{y}^\infty]$$

Therefore, what we obtain is a self-similarity in a window, such as

$$[1^\infty, 101^\infty] \subseteq [0^\infty, 10^\infty],$$

but not a new window.

4.5.3 The Generalized Composition Laws

While the $*$-composition law can be seen as a substitution of equal length (cf Queffélec 1987), the generalized law was found by Zheng (1989b). It was pointed out by Wang and Xie (1994) that this law has close connection with the exact forms defined in Definition 4.4.5. As a result of study of this connection, we can answer the question that where all these laws come from, and provide many new mappings from the set of all KS, $[0^\infty, 10^\infty]$ (or $[1^\infty, 10^\infty]$), into itself.

Definition 4.5.5 A generalized composition law is a homomorphism h from Σ^* to itself if the strings ρ and λ in

$$h = \{1 \to \rho, 0 \to \lambda\}$$

satisfy the following five conditions:

$$\begin{aligned} &(1)\ \rho \text{ is odd and } \lambda \text{ is even,} \\ &(2)\ \rho > \lambda, \\ &(3)\ \rho|_c \text{ is maximal,} \\ &(4)\ \rho\lambda|_c \text{ is maximal,} \\ &(5)\ \rho\lambda^\infty \text{ is a kneading sequence.} \end{aligned} \tag{4.3}$$

Remark. A $*$-composition law is obtained from a generalized composition law if $\rho\pi = \lambda\pi$.

Like the $*$-composition laws, the generalized composition laws also preserve both order relation and parity of strings. Hence if $s \neq 0^\infty$ is a KS, then $h(s)$ is also a KS. But $h(0^\infty) = \lambda^\infty$ need not be a kneading sequence. The other point which should be noted is that although we can also call a generalized composition law h a mapping from $[1^\infty, 10^\infty]$ into itself, but its image $h([1^\infty, 10^\infty])$ need not be an interval, and thus it is difficult to see this mapping in a bifurcation diagram generated from a PC.

See Derrida, Gervois and Pomeau (1978), Collet and Eckmann (1980), Zheng (1989b), Hao (1989) for varied applications of the $*$-composition laws and generalized composition laws.

In the sequel of this section we investigation the connection between generalized composition laws and exact forms in Definition 4.4.5.

Lemma 4.5.6 Let a kneading sequence KS be an eventually periodic but not periodic. Its expression $KS = \rho\lambda^\infty$ is an exact form if and only if the strings ρ and λ satisfy the conditions in Definition 4.5.5.

Proof. The proof of the "if" part is simple is omitted.

The "only if" part. If $KS = \rho\lambda^\infty$ is an exact form, then the conditions (1) and (5) of Definition 4.3 hold already.

Since $KS = \rho\lambda^\infty$ is maximal, we have $\rho \geq \lambda$. If ρ has λ as its prefix, then by Lemma 4.4.2 it leads to a contradiction. Otherwise, if λ has ρ as its prefix, then KS

has ρ^2 as its prefix, and we can use Lemma 1.2.4 to contradict the condition that KS is not periodic. These observation means that we must have $\rho > \lambda$, that is, the condition (2) is true.

Now consider $\rho|_c$. If it is not maximal, then we have decompositions $\rho = \beta\alpha = \alpha'\beta'$, $0 < |\alpha| = |\alpha'| < |\rho|$, such that $\alpha|_c > \alpha'$. Since $\rho\lambda^\infty$ is a kneading sequence, $\alpha \leq \alpha'$ is true. Combining these facts we have $\alpha\pi = \alpha'\pi$. If $\alpha < \alpha'$, then α' is odd, and this contradicts to the fact that $\alpha|_c > \alpha'$. If $\alpha = \alpha'$, then since ρ has no even PS, so $\alpha = \alpha'$ is odd too, and $\alpha|_c > \alpha'$ is impossible again. So the condition (3) holds.

The proof for the condition (4) is quite the same as for (3), where the condition that $\rho\lambda$ has no even PS is used. ∎

Lemma 4.5.7 If $KS = x^\infty$, where the string $x \neq 0$ is even and primitive, then there exist two strings ρ and λ satisfying the conditions (1)–(5) in Definition 4.5 such that $KS = x^\infty = \rho\lambda^\infty$ is an exact form.

Proof. From $KS = x^\infty$ we know that x is maximal, namely, $x = M(x)$. Using Corollary 1.2.11, x has no proper even PS with respect to KS. Since $x \neq 0$, the LPS of x, denoted by x', is nonempty and odd. Writing $x = x'x''$ and $KS = x'(x''x')^\infty$, we can select $\rho = x'$ and $\lambda = x''x'$.

Of course, x' has no even PS too. Considering $\lambda = x''x'$, since $x = x'x''$ is primitive, we have $\lambda < x$ (see Lemma 1.1.11). If λ has an even PS, then this PS is of the form px', where p is a proper suffix of x''. But since p is also a PS of $x(= x'x'')$, so p is odd too. From the fact that x' is the LPS of x, we have $|p| \leq |x'|$. Since both p and x' are prefixes of KS, so p is a prefix of x'. Now px' has p^2 as its prefix and px' is KS's prefix too. Using Lemma 1.2.4 leads to $KS = x^\infty = p^\infty$, a contradiction to the primitivity of x. Therefore, $KS = \rho(\lambda)^\infty$ is an exact form of KS. The remaining part that the conditions (1)–(5) of (4.5) are satisfied is in the same way as in the previous Lemma. ∎

Lemma 4.5.8 If $KS = x^\infty$, where x is odd and primitive, then there exist no ρ and λ, which satisfy both conditions (1)–(5) of (4.5) and $x^\infty = \rho\lambda^\infty$.

Proof. Assume the contrary that we have $x^\infty = \rho\lambda^\infty$ for a pair of strings ρ and λ, which satisfy all conditions of (4.5).

From $\rho > \lambda$ we can find that both ρ and λ cannot be a power of x. So we can write $x = x'x''$, where both x' and x'' are nonempty and $\rho = x^a x'$, $\lambda = x''x^b x'$ for some integers $a, b \geq 0$.

From the condition (3) $\rho|_c$ is maximal, so the string x', which being ρ's PS, must be an odd string. From the condition that λ is even and x is odd, we see that $b > 0$. But now the string λ has xx' as its even PS, which contradicts to the maximal property of $\rho\lambda|_c$ and the definition of exact form. ∎

Summarizing these lemmas, we have the following Theorem.

Theorem 4.5.9 If $h = (1 \to \rho, 0 \to \lambda)$ is a generalized composition law, then the corresponding kneading sequence $KS = \rho\lambda^\infty$ is either (1) eventually periodic but not periodic or (2) $\rho\lambda^\infty = x^\infty$, where x is even and primitive, and every such kneading sequence can generate infinitely many generalized composition laws, $(1 \to \rho, 0 \to \lambda)$ such that $KS = \rho\lambda^\infty$

Proof. Here we need only to prove the last point. As a matter of fact, for each KS of either (1) or (2), we have ρ and λ such that $KS = \rho(\lambda)^\infty$ is in exact form. But then we can take any pair of integers $a \geq 0$ and $b > 0$, and $KS = (\rho\lambda^a)(\lambda^b)^\infty$ is also an exact form. Thus $(1 \to \rho\lambda^a, 0 \to \lambda^b)$ is also a generalized composition law. ∎

Every window in the bifurcation Diagram is generated by a tangent bifurcation from the left edge (of the window) with the $KS = x^\infty$, where x is even and primitive. The right edge of the window has the $KS = \bar{x}x^\infty$, which corresponds to the phenomena of crisis. We also call it the band-merging point of $|x| \to 1$.

From the discussion above, from $KS = x^\infty$ we can have other kind of generalized composition laws, which also provide mappings from $[1^\infty, 10^\infty]$ into itself. The new feature of this mapping is easy to see that $h(10^\infty) = x^\infty$, it means that the rightmost member of the set $h[1^\infty, 10^\infty]$ is just the left edge of the window, which corresponds to $KS = x^\infty$. This point was used in Zheng (1989b) and Hao (1989) to study the intermittency through symbolic dynamics.

CHAPTER 5

A General Discussion of Kneading Sequences

From the conclusion of Theorem 4.2.1 we see that in order to obtain non-regular languages generated by unimodal maps we have to consider those kneading sequences that are neither periodic nor eventually periodic. Thus it is necessary to have a general discussion about all kneading sequences.

Section 5.1 is devoted to the theory of kneading maps of Hofbauer (1980). Here we begin from a discussion of maximal primitive prefixes of a given kneading sequence. Proposition 5.1.1 asserts that there exist two classes of those prefixes for each non-periodic kneading sequence. The first class consists of all odd maximal primitive prefixes for a given kneading sequence. The discussion of this class of prefixes leads naturally to the theory of kneading maps, while the second class of even maximal primitive prefixes will play a basic role in Chapter 7 that is devoted to a discussion about distinct excluded blocks for unimodal maps.

It is pointed out in Section 5.2 that the knowledge of kneading maps gives explicitly the structure of infinite automata accepting $L = \mathcal{L}(KS)$ for the difficult case of non-regular languages. Two examples are discussed by this idea, that is, intermittency and crisis in unimodal maps.

In Section 5.3 it is proved that each kneading sequence can be regarded as a limit of finite strings. This is a generalization of Lemma C.1.3 of Appendix C in the case of unimodal maps. The density problem of periodic kneading sequences in the set of all kneading sequences is solved here in the sense of product topology of symbolic sequences.

Section 5.4 is used to discuss the existence of uncountably many kneading sequences, and thus of the uncomputable complexity in unimodal maps. The necessary and sufficient condition of the existence of uncountably many kneading sequences between two given kneading sequences is obtained.

5.1 Maximal Primitive Prefixes of Kneading Sequences

Although each kneading sequence is a maximal sequence, but not all its prefix is a maximal string. For instance, 10110101 is a prefix of $KS = (10110)^\infty$, but

$$\sigma^5(10110101) = 10110110 > 10110101.$$

It turns out that the maximal primitive prefixes of a given kneading sequence KS play an indispensable role in analyzing the complexity of language $\mathcal{L}(KS)$.

5.1.1 *Existence of Special Prefixes of Kneading Sequences*

First we need to prove the existence of maximal primitive prefixes for a given kneading sequence. We begin our discussion from the non-periodic case. The discussion of periodic kneading sequences is easy and in the next subsection.

Note that in the sequel the tools developed in Subsection *1.2.3* and Section 3.4 are very useful, and if x is a prefix of a kneading sequence KS, then the notions of the PS of x to itself and the PS of x with respect to KS is quite the same.

Proposition 5.1.1 If a kneading sequence is not periodic, then it has infinitely many prefixes which are maximal and primitive.

Proof. If $KS = 10^\infty$, then simply take 10^i for all $i \geq 0$. Otherwise, this KS begins from 10^n1 for some $n > 0$. It is obvious that we can take 1 as the first prefix, and 10 as the second one as desired and so no. Now if x is the n-th prefix in the order of size which is maximal and primitive, we need to prove that there exists a longer prefix having the same property.

As a matter of fact, from Theorem 1.2.10 we know that this maximal and primitive prefix x has the properties (P1) and (P2), that is, each suffix of x, say s, satisfies $s \leq x$, and x has no even proper PS to itself (see Definition 1.2.7). Now if the next symbol in KS after the prefix x is 0, then it is easy to verify that the string $x0$ still has these properties. Otherwise, since KS is not periodic, the sequence $\sigma^{|x|}(KS)$, namely, the (infinite) suffix of KS after x, cannot coincide with KS completely. Comparing KS with $\sigma^{|x|}(KS)$, we can find a string y such that both xy and $\overline{y}$ are prefixes of KS. From $y < \overline{y}$ the string y is even. (thus $y = 0$ is the simplest case.)

We claim that xy, the new prefix of KS, also has the properties (P1) and (P2). Since xy is a prefix of KS, the first one is trivial. Assume the contrary that there exists an even PS of xy, denoted by u. If $|u| > |y|$, then it can be written as $u'y$ in which u' becomes an even PS of x, a contradiction. Otherwise, y has u as its even PS with respect to KS. But since $\overline{y}$ is a prefix of KS, we have $|u| < |y|$. Thus u is a prefix of $\overline{y}$ and KS. Since $\overline{y}$ has $\overline{u}$ as its odd suffix, the inequality $\overline{u} > KS = u\cdots$ contradicts the maximal property of KS. Using Theorem 1.2.10 we see that xy has the property (P3), thus xy is maximal.

Finally, if it happens that $xy = zz$ where z is odd maximal and primitive, then the kneading sequence K has zz as its prefix. By Lemma 1.2.4 we would have $KS = z^\infty$, a contradiction. Therefore, because KS is not periodic, the prefix xy found thus far is primitive. Hence our proof is completed by induction. ∎

Remark. Since the string y is even, the two prefixes x and xy have the same parity. Thus for $KS = 10^n1\cdots$ $(n > 0)$ we can take either 1 or 10^n1 as the first prefix, and obtain the following consequence. Here the unique exception is $KS = 10^\infty$, which has no even prefix at all.

Corollary 5.1.2 If a kneading sequence KS is neither periodic nor 10^∞, then there exist two sequences of prefixes of KS, $\{s_n\}$ and $\{s'_n\}$, where each s_n is odd maximal and primitive, and each s'_n is even maximal and primitive.

5.1.2 *Odd Maximal Primitive Prefixes of Kneading Sequences*

Two special classes of prefixes have been found in Corollary 5.1.2 for each non-periodic kneading sequence. In this subsection we discuss the first class, which consists of all odd maximal primitive prefixes of a given kneading sequence which may be periodic or not.

Proposition 5.1.3 If y is a prefix of a given kneading sequence, then the following statements are equivalent:

(1) the dual string $\bar{y} \in L = \mathcal{L}(KS)$,
(2) y has no even PS with respect to KS,
(3) y is odd maximal and primitive.

Proof.
(1) $\Rightarrow$ (2). This is a consequence of Lemma 3.4.4 applying to the string $\bar{y}$.
(2) $\Rightarrow$ (3). Since y is a prefix of KS, each its suffix, say u, satisfies $u \le y$. The condition (2) implies that the string y itself is odd. Using Corollary 1.2.12 leads to (3) directly.
(3) $\Rightarrow$ (1). By Corollary 1.2.11 each proper PS of y is odd, then it suffices to apply Lemma 3.4.5. ∎

Remark. In Definition 4.2.8 of Chapter 4 a prefix z of KS is called of the first kind if both $z0, z1 \in L = \mathcal{L}(KS)$. By the condition (1) of Lemma above we see that for a prefix x of KS,

$$x \text{ is odd maximal primitive} \iff \text{the prefix } x\bar{x} \text{ is of the first kind.} \qquad (5.1)$$

Proposition 5.1.4 Let y be an odd maximal primitive prefix of a given KS. For a longer prefix yz of KS, the following statements are equivalent:

(1) the string yz is the next odd maximal primitive prefix of KS, namely, $|z| > 0$ is minimal for yz being so.
(2) $\bar{z}$ is a prefix of KS,
(3) $\bar{z}$ is an odd maximal primitive prefix of KS.

Proof. (1) $\Rightarrow$ (2). Since both y and yz are odd, the string z is even. By applying Corollary 1.2.12 to yz, z cannot be a prefix of KS. Let u be the longest common prefix of z and KS ($u = \varepsilon$ is allowed), then we have

$$KS = ua\cdots, \quad z = u\bar{a}\cdots,$$

where $a \in \Sigma$, and $u\bar{a}$ is even (by $z \le KS$).

By the same argument used in the proof of Proposition 5.1.1 we know that $yu\bar{a}$ is an odd maximal primitive prefix of KS.

Since z is the shortest nonempty string which makes yz odd maximal primitive, we have $z = u\bar{a}$ exactly. Thus $\bar{z} = ua$ is a prefix of KS.

(2) ⇒ (3). It suffices by applying Lemma 3.4.4 to z first, and then Proposition 5.1.3 to $\bar{z}$.

(3) ⇒ (1). Since $\bar{z}$ is an odd maximal primitive prefix of KS, by applying Lemma 3.4.4 to $\bar{z}$, z has no even PS with respect to KS. This implies that the prefix yz also has no even PS and, again by Proposition 5.1.3, is odd maximal primitive.

Finally, we show that each prefix of KS whose length between $|y|$ and $|yz|$ cannot be odd maximal primitive. In fact, we can write this prefix by yu such that

$$|y| < |yu| < |yz|.$$

If yu is odd, then since y is also odd, u is even. From the facts that u is a proper prefix of z, and $\bar{z}$ is a prefix of KS, we find that u is an even prefix of KS. Thus yu has u as its even PS with respect to KS. By Corollary 1.2.11 yu cannot be maximal and primitive. ∎

Proposition 5.1.5 If both strings y, yz are odd maximal primitive prefixes of KS, and there exists no such prefix whose length between those of y, yz, then $|z| \le |y|$.

Proof. If $|z| > |y|$, then from Proposition 5.1.4 the string $z\pi$ is a prefix of KS. From $|z\pi| \ge |y|$ and both of them are prefixes of KS, the string $z\pi$ has y as its prefix. Thus $KS = yz\cdots$ would have yy as its prefix. Since y is odd, by Lemma 1.2.4 we have $KS = y^\infty$. But then $\sigma^{y}(KS) = KS$ and there exists no z such that both yz and $\bar{z}$ are prefixes of KS. ∎

Proposition 5.1.6 If z and yz are the same as above, then $\overline{yz}$ has $\bar{z}$ as its LPS with respect to KS.

Proof. From Proposition 5.1.4 $\bar{z}$ is a PS of $\overline{yz} = y\bar{z}$ with respect to KS. Assume the contrary that the string $y\bar{z}$ has a longer PS, then we can write it in the form of $u\bar{z}$. Thus u is a PS of y with respect to KS. Using Lemma 3.4.4 we find that $y\bar{z}$ has no even PS, and hence $u\bar{z}$ is odd. Since z is even, thus u is also even. But by Corollary 1.2.11 y cannot have any even PS, a contradiction. ∎

5.1.3 Kneading Maps

Now we introduce the notion of kneading maps which is due to Hofbauer (1980).

Let $KS = e_1 \cdots e_n \cdots$ over Σ be a given kneading sequence.

First we consider the existence of infinitely many odd maximal primitive prefixes of kneading sequence in general case. By Proposition 5.1.1 we already know that for each non-periodic kneading sequence there exist infinitely many odd maximal primitive prefixes, but its converse is not true. A periodic kneading sequence can also have infinitely many such prefixes. For instance, if $KS = (101)^\infty$ then it has infinitely many odd maximal primitive prefixes:

$$1, 10, 1011, 10110, \ldots, (101)^n 1, (101)^n 10, \ldots$$

Lemma 5.1.7 There exist only finite odd maximal primitive prefixes of a given kneading sequence KS if and only if this KS is either 0^∞ or x^∞ with an odd x.

Proof. The "only if" part. If the KS begins from 1, then this 1 is the shortest odd maximal primitive prefix of KS. Thus $KS = 0^\infty$ is the only case that there exists no such prefix at all. Otherwise, we denote the longest odd maximal primitive prefix of KS by y, and write

$$KS = ys,$$

where s is an infinite suffix of KS. Starting from the first symbol of s, by the condition (2) in Proposition 5.1.4, we see that if $KS \neq s = \sigma^{|y|}(KS)$ then we would find a longer prefix than y that has the desired property and contradicts our hypothesis that y is the longest one. Thus it follows that $KS = y^\infty$ and y is odd.

The "if" part. Without loss of generality, we can suppose that x is already an odd maximal primitive string. By applying (2) of Proposition 5.1.4 it is impossible to find longer prefix with desired property. ∎

Now assume that for a given KS there are infinitely many odd maximal primitive prefixes listed by the order of increasing size:

$$y_1, y_2, \ldots, y_n, \ldots.$$

Here we always have $y_1 = 1$ and $y_2 = 10$.

Take the notation as in de Melo and van Strien (1993).

$$y_1 = B_0, \text{ and } y_{n+1} = y_n B_n \text{ for } n > 0.$$

Then we have $KS = B_0 B_1 \cdots B_n \cdots$.

We call the sequence of strings $\{B_n\}$ *eventually periodic* if there exist two integers N and p such that $B_{n+p} = B_n$ hold for each $n > N$.

Lemma 5.1.8 The sequence of strings $\{B_n\}$ is infinite and eventually periodic if and only if the kneading sequence belongs to one of the following class:

(1) $KS = x^\infty$, where x is even and primitive.
(2) KS is eventually periodic but not periodic.

Proof. The "only if" part is trivial. For the "if" part, using Lemmas 4.4.7 and 4.5.7 of Chapter 4 we know that for KS belonging to both classes it can be written into an exact form

$$KS = \rho\lambda^\infty,$$

where the strings ρ and λ satisfy the conditions of Definition 4.4.5 of Chapter 4. Since each $\rho\lambda^i$ has no even PS with respect to KS for $i \geq 0$, thus we obtain infinitely many odd maximal primitive prefixes. Although these may not include those prefixes completely, by using Proposition 5.1.4 we can find the others (if any) and see that the string $\{B_n\}$ is eventually periodic. ∎

In the sequel we focus our attention on those kneading sequences which are neither periodic nor eventually periodic, hence there exist infinitely many odd maximal primitive prefixes and the corresponding sequence $\{B_n\}$ is not eventually periodic.

Therefore, we can write

$$KS = e_1e_2\cdots e_n\cdots = B_0B_1\cdots B_n\cdots$$

Denote $B_n = e_{s_{n-1}+1}\cdots e_{s_n}$, and hence the $(n+1)$-th odd maximal primitive prefix is $y_{n+1} = e_1\cdots e_{s_n}$. Here we always have $s_0 = 1, s_1 = 2$.

From Proposition 5.1.4 the string $\overline{B}_n$ is also an odd maximal primitive prefix, thus there exists an integer denoted by $Q(n)$ such that

$$\overline{B}_n = e_1\cdots e_{s_{Q(n)}}.$$

Hence we obtain a map Q from N, the set of natural integers, to $\{0\}\cup$N, and call it the *kneading map* (Hofbauer 1980).

From the conclusion of Lemma 5.1.5 we have a basic inequality for kneading map:

$$Q(n) < n \text{ for each } n > 0.$$

The kneading map Q provides a way to obtain the kneading sequence as follows. Define a sequence $\{y_n\}$ and its limit by

$$\begin{aligned} &y_1 = 1, \quad y_n = y_{n-1}\overline{y}_{Q(n)} \quad \text{for all } n > 0, \\ &s = \lim_{n\to\infty} y_n. \end{aligned} \tag{5.2}$$

If Q is the kneading map for a given KS, then the string y_n defined thus far is the n-th odd maximal primitive prefix of KS, and $KS = s$, the limit of $\{y_n\}$.

On the other hand, if a map Q : N$\mapsto\{0\}\cup$N is given, then what is the condition to guarantee that Q is a kneading map, namely, if the limit s in (5.2) is a kneading sequence? From Hofbauer (1980), de Melo and van Strien (1993) the answer is obtained as follows.

Proposition 5.1.9 A map Q from N to $\{0\}\cup$N is a kneading map if and only is

$$Q(Q(Q(k))+1)Q(Q(Q(k))+2)\cdots \preceq Q(k+1)Q(k+2)\cdots$$

holds for each k with $Q(k) > 0$.

Here the order relation $\preceq$ is the lexicographical order between integer strings.

Two examples about $Q(\cdot)$ is the Feigenbaum attractor, which is the topic of next Chapter, and the Fibonacci map (Lyubich and Milnor 1993, de Melo and van Strien 1993).

For the Feigenbaum attractor,

$$Q(n) = n-1;$$

while for the Fibonacci map,

$$Q(n) = n-2.$$

5.2 Infinite Automata of Unimodal Maps

From the discussion of Chapter 4 we know that if a kneading sequence of unimodal map is neither periodic nor eventually periodic, then the language $L = \mathcal{L}(KS)$ is non-regular, and hence there exists no finite automaton accepting L. Using the tool of kneading map, we can construct infinite automata for non-regular languages generated by unimodal maps.

5.2.1 Infinite Automata

By the discussion of natural equivalence relation R_L and the Myhill-Nerode Theorem (in Appendix A), we find that even for a non-regular language L there still exists a deterministic automaton accepting it. This automaton is similar to a finite automaton in almost every aspect. The only difference between them is the automaton accepting non-regular language has infinitely many states. In this subsection we will describe this kind of automata, which are referred to as *infinite automata*, by the theory of kneading maps developed above.

Let KS be a given kneading sequence which is not eventually periodic, and $L = \mathcal{L}(KS)$ the language generated by KS. By Theorem 4.2.1 the language L is non-regular, and the index of R_L is infinite. Using the idea in the proof of the Myhill-Nerode Theorem we can construct an infinite deterministic automaton to accept L as follows.

First of all, the fact of Lemma 2.2.11 still holds, namely, L', the complement language of L seen as an equivalence class, is the unique non-accepting state in the automaton.

Furthermore, by Lemma 4.2.6 each accepting state must contain a prefix of KS. From Lemma 4.2.7 any two distinct prefixes of KS are not of the same equivalence class. Therefore, we can simply take the equivalence classes of all prefixes of KS as the accepting states of the automaton, and we can use the same notation as in Appendix A:

$$M = (Q, \Sigma, \delta, q_0, F),$$

where $Q = F \cup \{L'\}$, and $F = \{[x] \mid x \text{ is a prefix of } KS\}$. The discussion of Example 4.2.2–4.2.4 is still effective here, and we have $q_0 = [\varepsilon]$. The only difficulty remaining is the determination of the transition function δ.

Let $e^{(0)} = \varepsilon$ and $e^{(i)} = e_1 \cdots e_i$ for $i > 0$ be the prefixes of $KS = e_1 \cdots e_n \cdots$. From the proof of the Myhill-Nerode Theorem we always have

$$\delta([e^{(i)}], e_{i+1}) = [e^{(i+1)}] \text{ for all } i \geq 0$$

for each KS. Hence, the problem of determination of function δ amounts to know all first kind of prefixes of KS, which is introduced in Definition 4.2.8.

By Remark (5.1) if for some $i \geq 0$ $e^{(i)}$ is a prefix of the first kind, then the next prefix $e^{(i+1)}$ is odd maximal primitive prefix, and vice versa. Thus if we can find all odd maximal primitive prefixes for a given kneading sequence by Proposition 5.1.4, then by Lemmas 4.2.5 and 5.1.6 the transition is determined by kneading map completely.

Therefore, the knowledge of kneading map for a given kneading sequence is just sufficient to construct an infinite automaton for the language $L = \mathcal{L}(KS)$. It is well-known that there are uncountably many distinct kneading sequences, and only countably many among them are accepted by Turing machine (see Section 5.4). Hence most of them are uncomputably complex. But as long as we can really compute a kneading sequence or a sufficiently long prefix of it, then we can use the above procedure to know its symbolic behavior. In the next subsection we will discuss some examples to explain this idea.

It is clear that if the language $L = \mathcal{L}(KS)$ is regular, the infinite automaton constructed above is still correct in that it accepts L exactly. But in this case it is not a minimal description of L, since there are only finite distinct equivalence classes which contain all prefixes of kneading sequence, and the states of the infinite machine will merge into finite states to make the automaton a finite state machine. In other words, Lemmas 2.2.11, 4.2.6 and 5.1.6 and the knowledge of kneading map are enough to construct infinite automata for a given kneading sequence, but for the cases treated in Chapter 4 we can use Theorems 4.3.1 and 4.4.6 to obtain minDFA accepting $\mathcal{L}(KS)$.

5.2.2 *Some Examples*

Note: in the following Examples we only construct the initial part of an automaton, which may be infinite or not. If the kneading sequence considered is periodic or eventually periodic, then the construction is indeed still correct, but the automaton described is not the minDFA for the regular language $\mathcal{L}(KS)$ as explained in the previous Subsection.

Example 5.2.1 *Symbolic behavior of intermittency*

Consider only the intermittency near the left edge of period 3 window (in bifurcation diagram 3.1) whose kneading sequence is $(101)^\infty$. In this case we know the kneading sequence is of the form

$$(101)^N 0 \cdots \quad \text{or } (101)^N 11 \cdots,$$

where N is a large integer. Using the theory of kneading maps it is easy to construct the automaton accepting the language $\mathcal{L}(KS)$. Of course, we can only determine the initial part corresponding the prefix $(101)^N$ as shown in Figure 5.1.

It is clear that all odd maximal primitive prefixes of length less than $(101)^N$ are

$$1, 10, 1011, 10110, 1011011, \ldots, (101)^{N-1}1, (101)^{N-1}10.$$

Denote the last one of them by $y = (101)^{N-1}10$. The dashbox on the right-hand side of Figure 5.1 represents the part of automaton which corresponds the suffix of KS

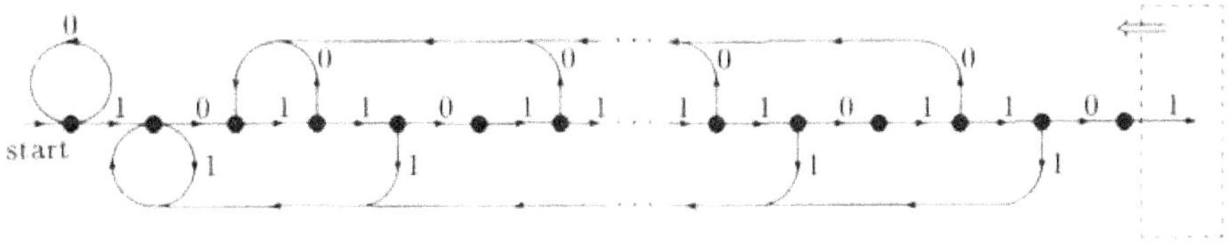

Figure 5.1 An automaton for intermittency with $KS = (101)^N \cdots$.

after y. If the next odd maximal primitive prefix after y is denoted by yz, then from Lemma 5.1.5 we see that $|z| \leq |y|$. By Lemma 5.1.6 the LPS of $y\bar{z}$ is $\bar{z}$, which is also a odd maximal primitive prefix of KS. Hence there must exists some path from somewhere of this suffix to some states corresponding to the odd maximal primitive prefixes listed above. This possibility is shown in Figure 5.1 by an arrow "$\Longleftarrow$" Thus the string $(101)^i$ $(i \leq N)$ will appear again and again in some itineraries.

Discuss possibilities about $|z|$. Since the last symbol of y is 0, and KS is a maximal sequence, the shortest z is 11. On the other hand, it $|z| = |y|$, then we have $KS = (101)^{N-1}10(101)^{N-1}11\cdots$.

An exception is the case of

$$KS = y^2 = ((101)^{N-1}10)^2,$$

for which there exists no odd maximal primitive prefix after y (see Lemma 5.1.7), and the automaton accepting $\mathcal{L}(KS)$ is finite as explained in Chapter 4 (see the analysis after Figure 4.5).

Example 5.2.2 Symbolic behavior near the point of interior crisis.

We discuss the crisis at the right edge of period 3 window to elucidate our analysis. For the point of crisis itself the kneading sequence is $100(101)^\infty$ If we consider each kneading sequence which is near crisis, then there are three possibilities:

(1) $KS_1 = 100(101)^N0\cdots$,
(2) $KS_2 = 100(101)^N11\cdots$,
(3) $KS_3 = 100(101)^N100\cdots$.

The prefixes being odd maximal primitive for $KS = 100(101)^N \cdots$ are

$$1, 10, 100, 100101, 100(101)^2, \ldots, 100(101)^N$$

Hence it is easy to construct the initial part of automaton as shown in Figure 5.2. Here there exists differences between $\mathcal{L}(KS_1)$, $\mathcal{L}(KS_2)$, and $\mathcal{L}(KS_3)$. It is easily to know

$$KS_3 < 100(101)^\infty < KS_2 < KS_1.$$

Therefore, using Lemma 5.1.6 we can find that if a word $x \in \mathcal{L}(KS_3)$ and has 100 as its subword, then its suffix after 100 must belong to $(100+101)^*$ As a matter of fact, if a prefix y being odd maximal primitive is longer than 2 and belongs to $(100+101)^*$,

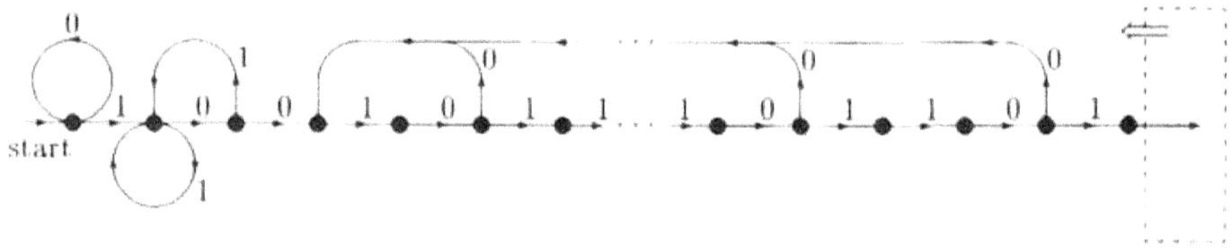

Figure 5.2 An automaton for kneading sequence near the crisis with $KS = 100(101)^\infty$

then the suffix z of its next prefix yz, which is the next odd maximal primitive prefix, must have the same property. This phenomenon reflects the existence of an attractor which is contained in three separate subinterval. But for the other two languages, $\mathcal{L}(KS_1)$ and $\mathcal{L}(KS_2)$ there is no such phenomenon happens. In fact the attractor collapses, and bandmerging from 3 to 1 happens. This change is due to the so-called interior crisis.

5.3 Density of Periodic Kneading Sequences

In this section we will show that the set of all periodic kneading sequences is density in the set $[0^\infty, 10^\infty]$, the set of all kneading sequences (see (4.2) for this notation). Here the topology is introduced as in Section 2.3 for all one-sided infinite sequences over $\Sigma = \{0, 1\}$.

5.3.1 Regard Kneading Sequence as a Limit

In discussion of kneading sequences we often treat them as some kind of limit. As a matter of fact, if a sequence of strings, say $\{s_n\}$, satisfies the condition that each s_n is a proper prefix of the next member s_{n+1}, then there exists a unique infinite sequence over Σ whose prefix of length $|s_n|$ equals to s_n. It is reasonable to define this infinite sequence as the limit of $\{s_n\}$, and denote it as $\lim_{n\to\infty} s_n$. An example is provided by (5.2) through the kneading map.

Remark. A special situation is the infinite sequence generated by a D0L system in Appendix C. See Lemma C.1.3 there for comparison with the following fact.

Proposition 5.3.1 An infinite sequence s over Σ is a kneading sequence if and only if there exists a sequence of strings $\{s_n\}$ which satisfies the following conditions:

(1) each s_n is a proper prefix of the next s_{n+1},
(2) each s_n is maximal,
(3) $s = \lim_{n\to\infty} s_n$.

Proof. The "only if" part. If $s = KS$ is periodic, then write it as x^∞ in which the string x is maximal. Take $s_n = x^n$ is sufficient for our purpose. Otherwise, it suffices to use Proposition 5.1.1 to obtain the sequence of prefixes $\{s_n\}$ in the order of size, each of them is maximal (and primitive).

The "if" part. Let s be an infinite sequence which satisfies the conditions (1)–(3). Assume the contrary that s is not maximal, that is, there is a shift of s such that

$$\sigma^i(s) > s \text{ for some } i > 0.$$

Consider the longest common prefix of s and $\sigma^i(s)$, say u, then we have

$$s = ua\cdots, \quad \sigma^i(s) = u\bar{a}\cdots,$$

where $a \in \Sigma$. Denote v the prefix of s and of length i, then we also have $s = vu\bar{a}\cdots$. Since the condition (1) implies that $\lim_{n\to\infty}|s_n| = \infty$, thus if n is large enough, we have by condition (3) that

$$s_n = vu\bar{a}\cdots$$

But then $\sigma^i(s_n) = u\bar{a}\cdots > s_n = ua\cdots$, a contradiction with condition (2). ∎

5.3.2 *Density of Periodic Kneading Sequences*

Using the result of Proposition 5.1.1 we can establish the density of periodic kneading sequences in $[0^\infty, 10^\infty]$. As a fact, we obtain more result than the conclusion of density as follows. (the notation $x\pi$ is defined in Subsection *1.1.5*.)

Proposition 5.3.2 If a kneading sequence KS is neither periodic nor 10^∞, then there exist two periodic kneading sequences $\{s_n^\infty\}_{n>0}$ and $\{t_n^\infty\}_{n>0}$ having the following properties for all $n > 0$:

(1) each s_n and t_n is odd maximal primitive string,
(2) each s_{n+1} ($t_{n+1}\pi$) has s_n ($t_n\pi$) as its proper prefix,
(3) each s_n is a prefix of KS, and each $t_n\pi$ is a prefix of KS,

and

$$s_1^\infty < s_2^\infty < \cdots < s_n^\infty < \cdots < KS < \cdots < t_n^\infty < \cdots < t_2^\infty < t_1^\infty$$

Proof. Construct $\{s_n\}_{n>0}$ and $\{t_n\}_{n>0}$ respectively.

Using Corollary 5.1.2 and denoting by $\{s_n\}$ all odd maximal primitive prefixes of KS in the order of size, then we obtain a sequence $\{s_n\}$ satisfying the conditions (1)–(3).

Write $s_{n+1}^\infty = s_n y\cdots$, where $|y| = |s_n|$. Since s_{n+1}^∞ is maximal, $y \le s_n$. If $y = s_n$, then by Proposition 1.2.4 we would have $s_{n+1}^\infty = s_n^\infty$. Because both s_{n+1} and s_n are primitive, this is impossible. Thus we have $y < s_n$, which leads to

$$s_n^\infty < s_{n+1}^\infty.$$

Similarly we have

$$s_n^\infty < KS$$

for each $n > 0$ since KS is non-periodic.

Again using Corollary 5.1.2 we obtain another sequence $\{s'_n\}$, the set of all even maximal primitive prefixes of KS in the order of size. Let $t_n = \overline{s'_n}$ for each $n > 0$. Using Proposition 1.2.13 each t_n is odd maximal primitive. Hence $\{t_n\}_{n>0}$ satisfies the conditions (1)–(3). From $t_n > s'_n$ it follows that $t_n^\infty > KS$. Similarly, $t_n > t_{n+1}$ holds for each $n > 0$. ∎

Remark. Since both $|s_n|$ and $|t_n|$ are monotone increasing to infinity, thus in each neighborhood of the kneading sequence KS there exist periodic kneading sequences.

For $KS = 10^\infty$ or x^∞ with an even maximal primitive x, using the same method and Lemma 5.1.8, we can obtain a sequence $\{s_n\}$ which satisfies all conditions in Proposition 5.3.2, but there exists no $\{t_n\}$ as stated there.

5.4 Existence of Uncomputable Complexity

5.4.1 Uncountable Infinity of Kneading Sequences

Using the Church-Turing thesis, a formal language L is said to be of uncomputable complexity, if there exists no Turing machine accepting L (Hopcroft and Ullman 1979, Friedman 1991). Of course, we cannot describe such a language by finite rules. Here we have to rely on abstract reasoning. Since there are only countably many Turing machines, it is easy to understand the existence of uncomputable complexity in any uncountable set of languages. In this sense it is very easy to answer the question of existence of uncomputable complexity in unimodal maps if we can find uncountable kneading sequences.

Since each maximal infinite sequence can be a kneading sequence for some unimodal maps by Proposition 3.5.3, it suffices to find uncountably many different maximal infinite sequences over $\{0, 1\}$.

There are many ways to do it. One of them is as follows.

Let $s_1 = 100$, $s_{n+1} = s_n c_n$ where c_n is either 11 or 101. It is easy to prove that each s_n is maximal (and primitive), hence the limit $s = \lim n \to \infty s_n$ is maximal. Since the selection of c_n is arbitrary for each n, we obtain uncountable prefixes already.

We can strengthen this example much more by showing that, if KS and KS' are two distinct kneading sequence, say $KS < KS'$, then there exist uncountably many kneading sequences between them if KS and KS' do not satisfy the conditions in Propositions below.

Here some notations are in need: a sequence of strings $\{t_n\}_{n\geq 0}$ defined by $t_0 = 1$, $t_n = t_{n-1}\overline{t}_{n-1}$ for $n > 0$, and $t_\infty = \lim_{n\to\infty} t_n$, a non-periodic kneading sequence (see Subsection *6.1.1* for detail).

Proposition 5.4.1 There exist only finite kneading sequences between KS' and KS'' if and only if there exists a $*$-composition law h such that

$$KS' = h(0)^\infty, KS'' = h(t_n)^\infty \text{ for some integer } n \geq 0.$$

Proposition 5.4.2 There exist exactly countably many kneading sequences between KS' and KS'' if and only if there exists a $*$-composition law h such that

$$KS' = h(0)^\infty, KS'' = h(t_\infty).$$

5.4.2 Some Lemmas

In order to prove Propositions in the previous Subsection some Lemmas are needed. First we establish a basic fact, which concerns how to insert a kneading sequence between two given ones. Similar idea was used in Zheng (1989a) to improve the result in Metropolis, Stein and Stein (1973).

Lemma 5.4.3 If KS_1 and KS_2 are two kneading sequences, $KS_1 < KS_2$, and $x\overline{x}$ is the longest common prefix of them, then $\overline{x}$ is odd maximal primitive, and

$$KS_1 \leq x^\infty < \overline{x}^\infty \leq KS_2.$$

Proof. By the condition about $x\overline{x}$ the string x ($\overline{x}$) is a prefix of KS_1 (KS_2), and we have

$$KS_1 = x \cdots < KS_2 = \overline{x} \cdots$$

Since both x and $\overline{x}$ have the property (P1) in Definition 1.2.7, using Lemmas 1.2.8 and 1.2.9 to them leads to the conclusion that both strings have the property (P3). Since $\overline{x}$ is odd, it is maximal primitive.

Write $KS_2 = \overline{x}y\cdots$ with $|y| = |x|$. Since KS_2 is maximal, it follows that $y \leq \overline{x}$. If $y = \overline{x}$, then $KS_2 = \overline{x}^\infty$ by Proposition 1.2.4. Otherwise, $y < \overline{x}$ and hence $\overline{x}^\infty < KS_2$.

On the other hand, if $KS_1 > x^\infty$, then we can write $KS_1 = x^n z \cdots$ for some integer $n > 0$, and a string z, such that $|z| = |x|$ and $z > x$. But this contradicts the fact that KS_1 is maximal. Thus $KS_1 \leq x^\infty$ is true. ∎

Lemma 5.4.4 If KS_1 and KS_2 are two kneading sequences, KS_1 is non-periodic, and $KS_1 < KS_2$, then there exist uncountably infinite kneading sequences between KS_1 and KS_2.

Proof. Using Lemma 5.4.3 we have

$$KS_1 < x^\infty < \overline{x}^\infty \leq KS_2,$$

where $KS_1 \neq x^\infty$ by the condition about KS_1.

We claim that the string x is even maximal primitive. As a fact, we already know (from the proof of Lemma 5.4.3) that x satisfies the condition (P3) of Definition 1.2.7. If $x = yy$ with y being odd maximal primitive, then since KS_1 has $x = yy$ as its prefix, by using Proposition 1.2.4, KS_1 would be y^∞, a contradiction.

Using the same argument to KS_1 and x^∞, we obtain

$$KS_1 < z^\infty < x^\infty,$$

where z is also an even maximal primitive prefix of KS_1, and $\bar{z}$ is an odd maximal primitive prefix of x^∞. Of course, $|z| > |x|$ holds. Writing $\bar{z} = x^n y$ for some integer $n > 0$ such that $|y| \le |x|$, then we have $0 < |y| < |x|$.

Writing $x^\infty = \bar{z}x' \cdots$ where $|x'| = |x|$, and using the fact that x is maximal primitive, and $0 < |y| < |x|$, it follows that $x' < x$ (see Subsection *1.2.3*). Therefore, we have

$$\bar{z}z^\infty < x^\infty$$

Thus we have found a whole window of kneading sequence $[z^\infty, \bar{z}z^\infty]$ between KS_1 and KS_2 (see Proposition 4.5.4), and our proof is completed. ∎

5.4.3 Proof of Propositions 5.4.1 and 5.4.2

Since the proof of "if" part of them is straightforward, we just give the proof of "only if" part as follows.

Proof of Proposition 5.4.1. Write out all finite kneading sequences between KS' and KS'' as follows:

$$KS' = KS_0 < KS_1 < KS_2 < \cdots < KS_{n-1} < KS'' = KS_n.$$

Using Lemma 5.4.3 to $KS_0 < KS_1$, and since there is no kneading sequence between them, we obtain

$$KS = x_1^\infty < \bar{x}_1^\infty = KS_1,$$

where $\bar{x}_1$ is odd maximal primitive.

Using Lemma 5.4.3 to $KS_1 < KS_2$ again, we obtain

$$KS_1 = x_2^\infty < \bar{x}_2^\infty = KS_2,$$

where $\bar{x}_2$ is odd maximal primitive. Therefore, $x_2 = \bar{x}_1\bar{x}_1$ and $\bar{x}_2 = \bar{x}_1 x_1$.

Proceed inductively on all KS_i, we obtain

$$KS_i = \bar{x}_i^\infty, \text{ for } i = 1, \ldots, n,$$

where each $\bar{x}_i$ is odd maximal primitive, and

$$\bar{x}_{i+1} = \bar{x}_i x_i$$

hold for $i = 1, \ldots, n-1$.

Using the fact that x_1 satisfies the property (P3) we can define a $*$-composition law $h = (1 \to \bar{x}_1, 0 \to x)$ and complete the proof. ∎

Proof of Proposition 5.4.2. Similarly, we only give the proof of "only if" part. By Lemma 5.4.4 KS' have to be periodic. Write $KS' = x^\infty$ then we can define h as above. If $KS'' < h(t_\infty)$, then by Proposition 5.4.1 there exist only finite kneading sequences between them. On the other hand, if $KS'' > h(t_\infty)$, then by Lemma 5.4.4 there already exist uncountably infinite kneading sequences between $h(t_\infty)$ and KS''. Therefore, we can only have but $KS' = h(t_\infty)$. ∎

CHAPTER 6

NON-REGULAR LANGUAGES OF UNIMODAL MAPS

In this chapter we discuss non-regular languages of unimodal maps. This is a field in which there are more open problems than established results.

Using the notion of Chomsky hierarchy (see Appendix B), here it is theoretically possible to encounter context-free languages (CFL), context-sensitive languages (CSL), recursive languages (RL), and recursively enumerable languages (REL). From Section 5.4 there also exists uncomputable complexity in unimodal maps, namely, formal languages which cannot be accepted by Turing machines. But until now there is no systematic study about the grammatical complexity of unimodal maps beyond the class $\mathcal{L}$(REL). We even do not know if there exists any language which is generated by unimodal maps and is strictly CFL, RL or REL.

Nevertheless, there are already some results in this field. The material of this Chapter is largely obtained by Hao (1991), Chen, Lu and Xie (1993b), Xie (1995b).

In Sections 6.1 and 6.2 the language of Feigenbaum attractor is defined and analyzed. It is found that this language is a CSL, but not a CFL. As a fact, we have proved that it may be generated by a developmental system, that is, an ET0L system, and hence occupies a rather precise position between CFL and CSL.

The Section 6.3 is devoted to the approach of Fibonacci sequences proposed by Hao (1991). It is proved that except in a trivial case this approach is always successful in finding non-regular complexity for unimodal maps.

A more powerful route is developed in Section 6.4, where we rely on the notion of homomorphisms on free submonoids presented in Subsection *1.1.2*. It is shown that all kneading sequences that can be obtained by the approach of Fibonacci sequences can also be generated by two special kind of homomorphisms on some free submonoids. Moreover, this new method also extends the result which can be obtained from either $*$-composition laws or generalized composition laws discussed in Section 4.5.

6.1 The Language of Feigenbaum Attractor

The Feigenbaum attractor is an important case in one-dimensional dynamical systems. There are many studies devoted to it (see, e.g., Feigenbaum 1978, Vul, Sinai and Khanin 1984). In this section we discuss the complexity of the language generated by Feigenbaum attractor, using tools from symbolic dynamics and formal languages (Chen, Lu and Xie 1993a).

6.1.1 *Renormalization and Kneading Sequence t_∞ of Feigenbaum Attractor*

Let the alphabet set Σ be $\{0, 1\}$ as before. First we define a homomorphism h on Σ^* as follows:

$$\begin{aligned} &h(\varepsilon) = \varepsilon, \\ &h(1) = 10, \quad h(0) = 11, \\ &h(xy) = h(x)h(y) \text{ for } x \text{ and } y \text{ over } \Sigma. \end{aligned} \tag{6.1}$$

Then we use this homomorphism h to generate a sequence of finite strings in the following way. Let $t_0 = 1$ and define the others inductively:

$$t_n = h(t_{n-1}) \text{ for all } n \geq 1.$$

Thus we have $t_n = h^n(1)$ for $n \geq 0$. Here we use the convention that $h^0(w) = w$ for each w.

The first few members of $\{t_n\}$ are

$$t_0 = 1,\ t_1 = 10,\ t_2 = 1011,\ t_3 = 10111010, \cdots$$

Lemma 6.1.1 The sequence $\{t_n\}_{n\geq 0}$ has the following properties:

(1) $t_{n+1} = t_n\bar{t}_n$,
(2) t_n is odd for $n \geq 0$,
(3) $|t_n| = 2^n$ for $n \geq 0$,
(4) $\bar{t}_n = h(\bar{t}_{n-1}) =$ and $\bar{t}_n = h^n(0)$ for $n \geq 0$,
(5) t_n is maximal and primitive.

Proof. The proof of (1)–(4) is very simple, so we only prove (5) by induction. it is trivial for $n = 0$ and $n = 1$. Suppose the claim is true for n and consider the case of $n + 1$. Since each t_n is odd, it suffices here to apply Corollary 1.2.12 (see Definition 1.2.7 there) as follows.

By the inductive hypothesis, t_n satisfies the conditions (P1) and (P2), that is, it has no even PS to itself, and each suffix of t_n, say u, satisfies $u \leq t_n$.

First we prove that there exists no even PS of t_{n+1} (to itself). Assume the contrary that x is an even PS of t_{n+1}. If $|x| > |t_n|$, then we can write $x = x'\bar{t}_n$, and x' becomes an even PS of t_n. By Corollary 1.2.12, this is impossible. If $|x| \leq |t_n|$, then x is an even suffix of $\bar{t}_n$, and t_n will have $\bar{x}$ as its odd suffix. But then $\bar{x} > x$ contradicts the inductive hypotheses of t_n being maximal.

Next we consider any suffix of t_{n+1} and denote it by x. Since t_n is maximal and primitive, if x is a proper suffix of t_{n+1} and $|x| \geq |t_n|$, then x's prefix of length $|t_n|$ must be either some $\sigma^i(t_n)$ with $0 < i < |t_n|$ or $\bar{t}_n$, and $x < t_n$ holds. If $|x| < |t_n|$, then $\bar{x}$ is a suffix of t_n and $x \leq t_n$ holds. If $x > t_n$ happens, then $\bar{x}$ would be an even PS of t_n, and contradicts Corollary 1.2.12 again.

Therefore, there exists no even PS of t_{n+1} and each suffix of t_{n+1}, say x, satisfies $x \leq t_{n+1}$. Using Corollary 1.2.12 again completes our proof. ∎

Since the sequence $\{t_n\}$ satisfies the condition that each t_n is a proper prefix of t_{n+1}, the limit $\lim_{n\to\infty} t_n$ exists and is denoted by t_∞ in the rest of the book (cf. Lemma C.1.3 of Appendix C). More precisely, t_∞ is uniquely determined by the condition that whose prefix of length 2^n equals t_n for all $n \geq 0$. By Lemma 5.3.1 we obtain the following conclusion.

Lemma 6.1.2 t_∞ is maximal, and hence a kneading sequence.

It is easy to show that this kneading sequence t_∞ corresponds the limit of period-doubling sequence of unimodal maps (Feigenbaum 1978, Collet and Eckmann 1980). Here we explain the concept of renormalization for unimodal maps by Figure 6.1.

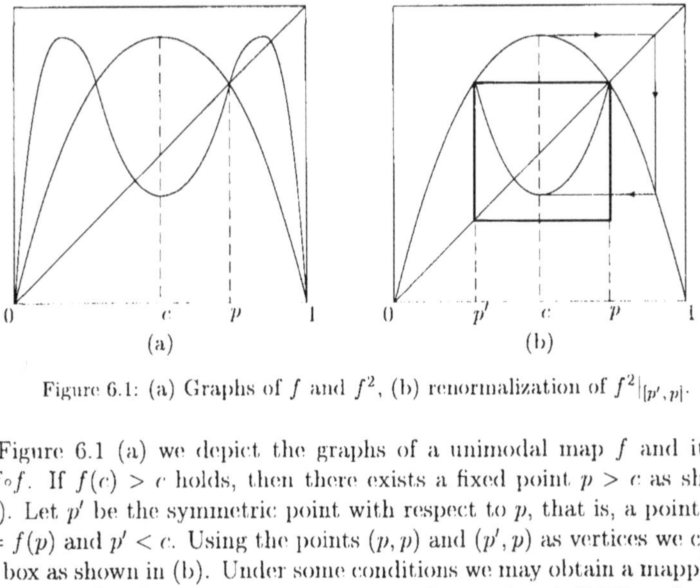

Figure 6.1: (a) Graphs of f and f^2, (b) renormalization of $f^2|_{[p',p]}$.

In Figure 6.1 (a) we depict the graphs of a unimodal map f and its iteration $f^2 = f \circ f$. If $f(c) > c$ holds, then there exists a fixed point $p > c$ as shown in (a) and (b). Let p' be the symmetric point with respect to p, that is, a point defined by $f(p') = f(p)$ and $p' < c$. Using the points (p, p) and (p', p) as vertices we can depict a square box as shown in (b). Under some conditions we may obtain a mapping f^2 from the subinterval $[p', p]$ to itself. It requires to take $2 < b \leq 3.67857\cdots$ for the family of logistic map (3.1), and $1 < s \leq \sqrt{2}$ for the family of tentmap (3.2). Moreover, we can see from Figure 6.1 (b) that this mapping f^2 on $[p', p]$ is also a unimodal map but in reverse orientation. If we introduce a renormalization operator R as in Devaney (1989), then a new unimodal map satisfying all conditions in Chapter 3 is obtained.

First let L_f be the linear map which takes p to 0 and p' to 1, that is

$$L_f(x) = \frac{1}{p' - p}(x - p).$$

It expands the subinterval $[p, p']$ onto $[0, 1]$ with a change of orientation. Of course we have the inverse of L_f by

$$L_f^{-1}(x) = (p' - p)x + p.$$

Now define the renormalization of f by

$$(Rf)(x) = L\circ f^2\circ L_f^{-1}(x).$$

Consider the relationship between $KS(f)$ and $KS(Rf)$. From Figure 6.1 (b) we have $f([p', p]) = [p, f(c)]$ and hence $f^{2k-1}(c) > c$ for each $k > 0$. For $f^{2k}(c)$ $(k > 0)$, if $f^{2k}(c) > c$, then it is in the subinterval $[c, p]$, and thus $(Rf)^k(c) < c$ holds, and vice versa. Thus we have

$$KS(f) = 1e_2 1e_4 \cdots 1e_{2k} \cdots \text{ and } KS(Rf) = \bar{e}_2\bar{e}_4 \cdots \bar{e}_{2k} \cdots$$

Using the homomorphism h defined in (6.1), this connection can be expressed by

$$KS(f) = h(KS(Rf)).$$

Therefore, we can say that the homomorphism $h = (1 \to 10, 0 \to 11)$ is the symbolic representation of renormalization from f^2 to f. A similar discussion can be done for each periodic window by the $*$-composition law defined in (4.3).

If f is infinitely renormalizable by this homomorphism h, then its kneading sequence $KS(f)$ is invariant under h, namely,

$$KS(f) = h(KS(f)).$$

This equality completely determines the kneading sequence, that is, $KS(f) = t_\infty$. In the language of symbolic dynamics, we say that t_∞ is a fixed point of h:

$$t_\infty = h(t_\infty).$$

The limit relations

$$t_\infty = h^\infty(1) = \lim_{n\to\infty} h^n(1) = \lim_{n\to\infty} t_n$$

has a simple dynamical meaning that the Feigenbaum attractor is the limit of the period-doubling sequence. On the other hand, we also have

$$t_\infty = h^\infty(10^\infty) = \lim_{n\to\infty} h^n(10^\infty),$$

which implies that the Feigenbaum attractor is also the limit of the $2^n \to 2^{n-1}$ band-merging sequence.

From the following order relations easily verified

$$(t_0)^\infty < (t_1)^\infty < \cdots < (t_n)^\infty < (t_{n+1})^\infty < \cdots < t_\infty$$

and

$$t_\infty < \cdots < h^{n+1}(10^\infty) < h^n(10^\infty) < \cdots < h(10^\infty) < 10^\infty,$$

the convergence to t_∞ of two limit processes are from different directions.

All these phenomena can be clearly seen from the Figure 6.2 (a) for the family of logistic map. The value of parameter $b = 3.5699456\cdots$ is the place where the Feigenbaum attractor happens in the family of logistic map, and hence is called the *Feigenbaum point*, which is denoted in Figure 6.2 by b_∞.

For the family of tentmap, which is not a full family (see Subsection *3.5.2*), there is neither the sequence of period-doubling nor the Feigenbaum attractor, but the sequence of $2^n \to 2^{n-1}$ band-merging still exists and convergence to the map corresponding the point of $s = 1$. It is easy to calculate the values of parameter s which correspond to band-merging. The $2^n \to 2^{n-1}$ band-merging happens at $s = 2^{1/2^n}$. For the map at $s = 1$ there is a continuum of fixed points filled the subinterval $[0, 0.5]$, and a jump happens when the parameter s tends 1 from left as shown in Figure 6.2 (b). The band-merging points from $2 \to 1$ and $4 \to 2$ are clearly visible as well.

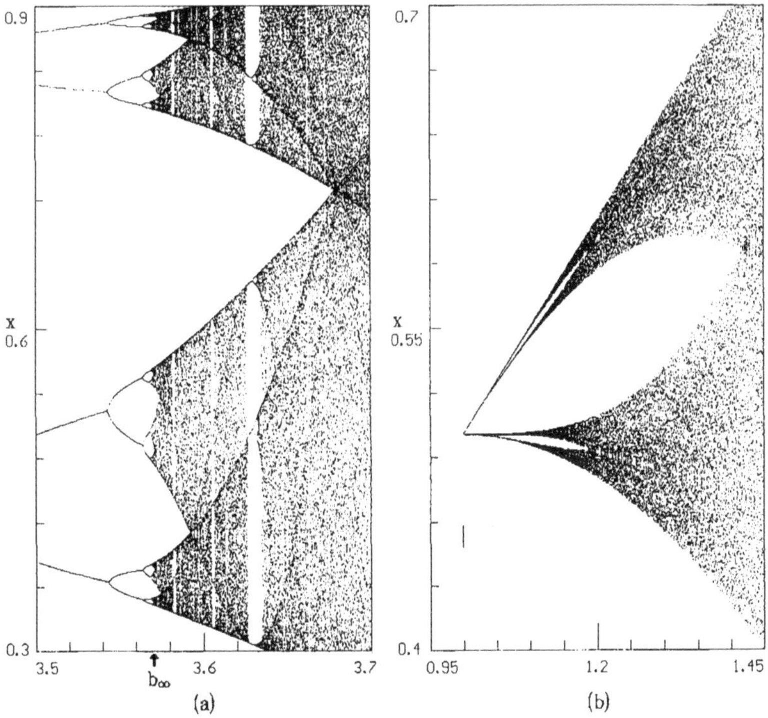

Figure 6.2: (a) Logistic map for $3.5 \le b \le 3.7$, (b) tent map for $0.95 \le s \le 1.45$.

6.1.2 Some Properties of t_∞

First we apply the theory of kneading map (of Subsection 5.1.3) to Feigenbaum attractor. It is a simple consequence of definition of t_∞ and properties of $\{t_n\}$ to obtain

$$t_\infty = B_0 B_1 \cdots B_n \cdots \text{ where } B_0 = 1,\ B_n = \bar{t}_{n-1} \text{ for } n > 0,$$

and the following result (cf. (5.2))

Lemma 6.1.3 The kneading map for $KS = t_\infty$ is given by

$$Q(n) = n - 1 \text{ for } n > 0.$$

We can establish a formula for each e_n in $t_\infty = e_1 \cdots e_n \cdots$

Lemma 6.1.4 Denote $t_\infty = e_1 \cdots e_n \cdots$, then for each $n \geq 0$

$$e_{2^k(2n+1)} = \begin{cases} 1, & \text{when } k \text{ is even,} \\ 0, & \text{when } k \text{ is odd.} \end{cases}$$

Proof. Proceed inductively on k. For $k = 0$, we need to prove $e_{2n+1} = 1$ for $n \geq 0$. From $t_\infty = h(t_\infty)$ we have $e_{2n+1}e_{2n+2} = h(e_{n+1})$ for each $n \geq 0$ and it follows by the definition of h. Assume the claim is true for $k - 1$ already and consider the case of k. Similarly we have

$$e_{2^k(2n+1)-1}e_{2^k(2n+1)} = h(e_{2^{k-1}(2n+1)}).$$

Since k and $k - 1$ always have different parity, by the definition of h and the inductive hypothesis the claim for k is true. ∎

Considering the statement of Theorem 4.2.1, the next Lemma is basic for our discussion of language generated by Feigenbaum attractor and several proofs will be given.

Lemma 6.1.5 The kneading sequence t_∞ is not eventually periodic.

Proof 1. From Lemma 6.1.3 the sequence of strings $\{B_n\}$ is not eventually periodic, thus by Lemma 5.1.8 the kneading sequence t_∞ is not eventually periodic. ∎

Proof 2. Use the formula in Lemma 6.1.4 we can directly show that t_∞ is not eventually periodic. Assume the contrary that

$$t_\infty = uv^\infty \text{ for some } u \text{ and } v.$$

Let $|v| = p$, and write it as $p = 2^m(2n+1)$ for some m and n. Consider $e_{2^k(2n+1)}$ and $e_{2^{k+1}(2n+1)}$ in t_∞ for $k > m$. Since the difference of their index is a multiple of p, these two symbols should be the same for sufficiently large k. But by the formula in Lemma 6.1.4, as k and $k+1$ are of different parity, this is impossible. ∎

Proof 3. Compare any two distinct shifts of t_∞, say $\sigma^i(t_\infty)$ and $\sigma^j(t_\infty)$ with $i < j$. It suffices to prove that any two such shifts are different to obtain our conclusion.

If we take n sufficiently large such that $i < j < |t_n| = 2^n$, then from the fact that t_∞ having t_n as its prefix for each n, we can write

$$t_\infty = t_{n+2} \cdots = t_n \bar{t}_n t_n \cdots$$

Thus both $\sigma^i(t_\infty)$ and $\sigma^j(t_\infty)$ will have some cyclic shifts, $\sigma^i(t_{n+1})$ and $\sigma^j(t_{n+1})$, as their prefix of length 2^{n+1}. Since t_{n+1} is primitive, using Lemma 1.1.11 leads to

$$\sigma^i(t_{n+1}) \neq \sigma^j(t_{n+1}),$$

and hence

$$\sigma^i(t_\infty) \neq \sigma^j(t_\infty)$$

for each $i \neq j$. ∎

The next property has clear dynamical meaning in Figure 3.1 that on the left-hand side of Feigenbaum point there exists no periodic motions other than those given by t_n^∞ $(n \geq 0)$ and 0^∞.

Lemma 6.1.6 Let x^∞ be a periodic kneading sequence containing no c. If $x^\infty < t_\infty$, then

$$\text{either } x^\infty = 0^\infty \text{ or } x^\infty = t_k^\infty \text{ for some } k \geq 0.$$

Proof. If $x^\infty \neq 0^\infty$, then x^∞ begins from the symbol 1. If it has 11 as its prefix, then by Lemma 1.2.4 $x^\infty = 1^\infty = t_0^\infty$. Otherwise x^∞ has $10 = t_1$ as its prefix. By assumption x^∞ cannot have each t_k as its prefix, hence we can find the largest integer k for which t_k is a prefix of x^∞. Write $x^\infty = t_k u v$ where $|u| = |t_k| = 2^k$ and v is an infinite suffix. Compare this form of x^∞ with $t_\infty = t_k \bar{t}_k \cdots$. Since $x^\infty < t_\infty$ and t_k is odd, we obtain $u \geq \bar{t}_k$. But since x^∞ is maximal, we must have $u \leq t_k$.

Combining these inequalities into $\bar{t}_k \leq u \leq t_k$, and using the condition on k, we see that $u \neq \bar{t}_k$. Thus we can only have $u = t_k$. Finally, it suffices to apply Lemma 1.2.4 once more to complete our proof. ∎

The following fact is essential for the study of grammatical complexity of $\mathcal{L}(t_\infty)$ later in the next section.

Lemma 6.1.7 t_∞ cannot have t_k^4 as its substring for each $k \geq 0$.

Proof. Assume the contrary that

$$t_\infty = u t_k^4 v$$

for some $k \geq 0$. From $t_\infty = h(t_\infty)$ we have $t_\infty = h^k(t_\infty) = t_k \bar{t}_k \cdots$. Hence t_∞ is a concatenation of strings t_k and $\bar{t}_k$. Since t_k is primitive, we find that the prefix u and the infinite suffix v in the expression $t_\infty = u t_k^4 v$ are also concatenations of t_k and $\bar{t}_k$. Therefore, from $t_\infty = h^k(t_\infty) = u t_k^4 v$ we find that t_∞ would have 1^4 as its substring.

Suppose $t_\infty = e_1 \cdots e_n \cdots = u' 0 1^4 v'$. Since we know $e_{2k-1} = 1$ for each $k > 0$, the symbol 0 before the block 1^4 must be an e_{2i} for some integer $i > 0$. Again by

$$t_\infty = h(t_\infty) \text{ and } e_{2i+1} e_{2i+2} e_{2i+3} e_{2i+4} = 1111,$$

we find $e_{i+1} = e_{i+2} = 0$, which contradicts the fact that $e_{2k-1} = 1$ for each $k > 0$, since one integer among $i + 1$ and $i + 2$ must be odd. ∎

6.1.3 *The Non-Regularity of* $\mathcal{L}(t_\infty)$

Taking $KS = t_\infty$ and using Definition 3.3.2, we obtain a dynamical language $L = \mathcal{L}(t_\infty)$ and call it the *language of Feigenbaum attractor*.

Using the knowledge obtained thus far it is easy to prove that $\mathcal{L}(t_\infty)$ is non-regular in several ways. We will give proofs for it from different aspects.

Theorem 6.1.8 The language $L = \mathcal{L}(t_\infty)$ of Feigenbaum attractor is not regular.

Proof 1. By Theorem 4.2.1 it suffices to recall Lemma 6.1.5 that t_∞ is not eventually periodic. ∎

Proof 2. Calculate the index of equivalence relation R_L directly (see Section A.3 of Appendix A). At first we show that $t_n R_L t_{n+1}$ is false. Let $z = t_n^4$. Since $t_n = M(t_n)$, $t_n z = t_n^5 \in L$. But from the inequality

$$t_{n+1}z = t_{n+1}t_n^4 > t_{n+1}t_n t_n t_n \bar{t}_n$$

$KS = t_{n+2}\bar{t}_{n+2}\cdots = t_{n+1}t_n t_n t_n \bar{t}_n \cdots$, we have $t_{n+1}z \notin L$.

Similarly, for any t_n and t_{n+l} ($l \geq 1$), we can use $z = t_{n+l-1}^4$ and show that $t_{n+l}z \notin L$ but $t_n z \in L$, so that $t_n R_L t_{n+l}$ is false for each $l \geq 1$.

Therefore, the equivalence relation R_L is not of finite index and, by the Myhill-Nerode Theorem A.3.2, our proof is completed. ∎

The third proof is obtained as a consequence of the calculation of L'', the set of DEB, for $\mathcal{L}(t_\infty)$ later in Chapter 7. Since the lengths of strings of L'' are 3×2^n and 5×2^n, it suffices to use the result of semilinear structure of regular $\mathcal{L}(KS)$ by Theorem 7.2.5 of Chapter 7.

By the kneading map obtained in Lemma 6.1.3 and the discussion about infinite automata in Section 5.2, it is straightforward to construct an infinite automaton accepting the non-regular language $L = \mathcal{L}(t_\infty)$ as shown in Figure 6.3, which appeared first in Grassberger (1988a p.676, Fig.4).

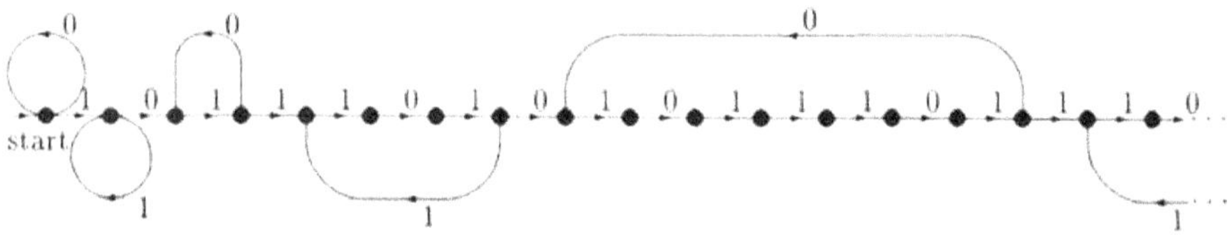

Figure 6.3: An infinite automaton for Feigenbaum attractor.

6.1.4 *The Thue-Morse Sequence*

Here we will present an observation made by, e.g., Jonker and Rand (1980), Procaccia, Thomae and Tresser (1988), Milnor and Thurston (1988) that there exists a close relationship between the kneading sequence t_∞ and the famous Thue-Morse sequence. Thus some properties of the kneading sequence t_∞ can be deduced from those of the Thue-Morse sequence and vice versa. For example, the results of Lemmas 6.1.5 and 6.1.7 above are consequences of the strong cube-freeness of the Thue-Morse sequence (Thue 1906, Morse and Hedlund 1938, Hedlund 1967 Salomaa 1981, Lothaire 1983).

The Thue-Morse sequence over $\Sigma = \{0, 1\}$ is defined similarly as for t_∞. A homomorphism g on Σ^* is defined as follows.

$$\begin{aligned} &g(\varepsilon) = \varepsilon, \\ &g(1) = 10, \quad g(0) = 01, \\ &g(xy) = g(x)g(y) \text{ for } x \text{ and } y \text{ over } \Sigma. \end{aligned} \tag{6.2}$$

Starting from $s_0 = 0$, we obtain inductively $\{s_n\}$ by $s_n = g(s_{n-1})$ for $n \geq 1$. The first few words of $\{s_n\}$ are

$$s_0 = 0, s_1 = 01, s_2 = 0110, s_3 = 01101001, \cdots$$

The *Thue-Morse sequence*, denoted here by s_∞, is the limit of $\{s_n\}$ in the same way as t_∞ or Lemma C.1.3:

$$s_\infty = 0110100110010110100101100110100 1\cdots$$

We also need the following concept. A sequence or string is said to be *strongly cube-free* if it contains no substring of the form x^2a, where x is a nonempty string and a is the first symbol of x. A basic fact about s_∞ is the following theorem.

Theorem 6.1.9 (Thue-Morse) s_∞ is strongly cube-free.

The Thue-Morse sequence has applications in a variety of quite different situations (see, e.g., Salomaa 1981). For example, in Morse and Hedlund (1938) it was used to establish the existence of non-periodic recurrent motions in dynamics. The relation of s_∞ with t_∞ can be viewed as another application of the Thue-Morse sequence in dynamical systems.

Writing $t_\infty = e_1 \cdots e_n \cdots$ as before and $s_\infty = b_0 b_1 \cdots b_n \cdots$, then it can be shown that the following relation holds:

$$\begin{aligned} &b_0 = 0, \\ &b_{i+1} = b_i \text{ if } e_{i+1} = 0, \\ &b_{i+1} = 1 - b_i \text{ if } e_{i+1} = 1 \text{ for } i \geq 0. \end{aligned}$$

It means that t_∞ can be transformed into s_∞ and vice versa.

It is easy to verify that if a substring of t_∞, say $e_{n+1} \cdots e_{n+p}$, is even, then we have $b_n = b_{n+p}$ in s_∞. Thus we obtain the following Corollary from the property of strong cube-freeness of s_∞.

Corollary 6.1.10 t_∞ contains no substring of the form x^2 for any even string x, namely, t_∞ is square-free of even string.

This is a strong result. As a matter of fact, it contains the statements of Lemmas 6.1.5 and 6.1.7 as consequences.

6.2 Grammatical Level of Feigenbaum Attractor

From the proofs given for Theorem 6.1.8 it is relatively easy to establish that the language of Feigenbaum attractor is non-regular. But in order to know more about the grammatical complexity of $L = \mathcal{L}(t_\infty)$ in Chomsky hierarchy, we have to analyze the structure of L in detail. Since the language L is formed from admissible sequences by Definition 3.3.1, we will begin our study from a simple observation about them. Here the infinite automaton in Figure 6.3 is helpful to guide us in study.

6.2.1 Structure of Language $L = \mathcal{L}(t_\infty)$

Lemma 6.2.1 If an admissible sequence $s = t_k v$ for some k is given, then the suffix v begins from either t_k or $\bar{t}_k$.

Proof. Proceed inductively on k. It's trivial for $k = 0$. Assume that the claim is true for $i \leq k$, and consider the case of $i = k + 1$. Writing

$$s = t_{k+1}v = t_k\bar{t}_k v = t_{k-1}\bar{t}_{k-1}t_{k-1}t_{k-1}v,$$

and using the inductive hypothesis to its substring $t_{k-1}v$, then v may have either t_{k-1} or $\bar{t}_{k-1}$ as its prefix. But from the ordering relation

$$t_{k+1}\bar{t}_{k-1}\cdots > t_{k+1}t_{k-1}\cdots = t_\infty,$$

it is impossible for v to have $\bar{t}_{k-1}$ as its prefix. Otherwise, s would not be admissible.

Now let $s = t_{k+1}t_{k-1}v'$. Using the inductive hypothesis to $t_{k-1}v'$, then v' may have either t_{k-1} or $\bar{t}_{k-1}$ as its prefix. By arguing as before, from the relation

$$t_{k+1}t_{k-1}t_{k-1}\cdots > t_{k+1}t_{k-1}\bar{t}_{k-1}\cdots = t_\infty,$$

v' can only have $\bar{t}_{k-1}$ as its prefix.

Finally, we have

$$s = t_{k+1}t_{k-1}\bar{t}_{k-1}v'' = t_{k+1}t_k v''$$

Again using our inductive hypothesis to $t_k v''$, we see that v'' has either t_k or $\bar{t}_k$ as its prefix. By $t_k\bar{t}_k = t_{k+1}$ and $t_k t_k = \bar{t}_{k+1}$ both cases lead to our goal. ∎

In order to describe all admissible sequences of Feigenbaum attractor, we introduce a notion of normal strings, which is a natural consequence of Figure 6.3.

Definition 6.2.2 A string u is called *normal* if it is of the form:

$$u = t_{-1}^{n_{-1}}t_0^{n_0}t_1^{n_1}\cdots t_l^{n_l},$$

where t_{-1} represents the symbol 0, n_i is nonnegative for $i = -1, 0, \cdots, l$. If $n_{-1} = n_0 = \cdots = n_l = 0$ then $u = \varepsilon$. Otherwise, we require that $n_l \geq 1$. Moreover, we allow either n_l or l be ∞, thus a normal string may be an infinite sequence.

The following result clarify the structure of all admissible sequences.

Lemma 6.2.3 There are only three kinds of admissible sequences:

(1) s is a normal string with $l < \infty$ but $n_l = \infty$,

(2) $s = ut_\infty$ for some normal string u,

(3) s is a normal string with $l = \infty$.

Proof. Begin from the first symbol of s. If $s = 0^\infty$, it's a sequence of the first kind. Otherwise, s can be written as $0^{n-1}s'$, and s' is also admissible, which starts from the symbol 1. Applying Lemma 6.2.1 to this symbol 1, there are two possibilities. If the next symbol is 1, then continue as before. Otherwise, the next symbol is 0, then we write $s' = 10s'' = t_1 s''$, and apply Lemma 6.2.1 again. Inductively, if after finite steps, we obtain $s = t^{n-1}_{-1} t_0^{n_0} \cdots t_l^{n_l} s'$ with $n_l \geq 1$, then apply the Lemma 6.2.1 to $t_l s'$. If s' has t_l as its prefix, then apply this Lemma to t_l again; but if s' has $\bar{t}_l$ as its prefix, then write first that $t_l s' = t_l \bar{t}_l s'' = t_{l+1} s''$, and apply Lemma 6.2.1 again.

Proceeding on in this way, there are three possibilities:

(1) Starting from a certain t_l, we always meet the first case in Lemma 6.2.1, whenever we use it. It means that we obtain a new t_l with the same l every time. So we have $n_l = \infty$, and s is an admissible sequence of the first kind.

(2) Starting from a certain t_l, we always meet the second case in Lemma 6.2.1. From the relation $t_{l+1} = t_l \bar{t}_l$ and the definition of t_∞ we obtain that $s = ut_\infty$.

(3) The remaining situation is that every n_l is a finite number, but with an infinite l. Thus we obtain an admissible sequence of the third kind. ∎

It is clear that the statement of Lemma 6.2.3 has dynamical meaning as well. Using the language of dynamical systems, there are only two kinds of orbit for the Feigenbaum attractor: (1) the orbits end up eventually in a periodic orbit with a power of 2 as period, (2) the distance between the orbits and the attractor tends to zero in either finite steps or not. We see that these conclusions can be derived from the knowledge of t_∞ only, without depending on any smooth conditions (beyond continuity) on unimodal maps considered.

As a consequence of Lemma 6.2.3, we can clarify the structure of the language $L = \mathcal{L}(t_\infty)$ completely.

Lemma 6.2.4 A string w belongs to the language $L = \mathcal{L}(t_\infty)$ if and only if w is a prefix of some normal string, that is, w is of the form

$$w = uP(t_k),$$

where u is normal, and $P(t_k)$ is a proper prefix of t_k for some $k \geq 1$.

Proof. If $w \in L$, then there exists an admissible s such that w is a prefix of s. Applying Lemma 6.2.3 and seeing that for all kinds of admissible sequences their prefixes are of the same form, we have w in the form desired.

On the other hand, if a string $w = uP(t_k)$ is given, then we can elongate it into an infinite sequence admissible, thus obtain the conclusion $w \in L$. ∎

6.2.2 *L is not a Context-Free Language*

Using Lemma 6.1.6 it is easy to obtain the following fact.

Lemma 6.2.5 If a nonempty string x has the property that $x^i \in L$ for all $i \geq 1$, then there exist integers $k \geq -1$ and $n_k \geq 1$ such that $M(x) = t_k^{n_k}$

Proof. From the condition it follows that

$$M(x^i) \in L \text{ for each } i \geq 1,$$

and, without loss of generality, we can assume $x = M(x)$ in the sequel of this proof.

Since every $x^i \in L$, we have $x^i \leq t_\infty$, and hence $x^\infty \leq t_\infty$. Since t_∞ is not periodic, by Lemma 6.1.6 there exists an integer $k \geq -1$ such that

$$x^\infty = t_k^\infty$$

Since t_k is primitive, hence the length of x, $|x|$, must be a multiple of $|t_k|$, and $x = t_k^{n_k}$ is obtained. ∎

Lemma 6.2.6 If $z \in L$ can be written as

$$z = t_k v t_p^n,$$

where $p < k$, then $n \leq 3$.

Proof. Suppose the contrary that there exists a word $z = t_k v t_p^4 \in L$ with $p < k$. Since t_k has $t_{k-1}, t_{k-2}, \ldots$, as its prefix, we can assume that the relation $k = p + 1$ holds.

From Definition of language L there is an admissible sequence $s = t_k v t_{k-1}^4 w$. As in the proof of Lemma 6.2.3, we can write each sufficiently long prefix of this s into the form of normal strings, so that t_{k-1}^4 becomes a subword of a normal string

$$u = t_m^{n_m} t_{m+1}^{n_{m+1}} \cdots t_l^{n_l},$$

where $m \geq k$.

Here there are two possibilities.

(1) t_{k-1}^4 is a substring of t_n^i for $1 \leq i \leq 3$ with inequalities $n \geq m \geq k$. Since $t_\infty = h^n(t_\infty) = t_n\bar{t}_n t_n t_n t_n \bar{t}_n t_n \bar{t}_n \cdots$, it turns out that t_{k-1}^4 also is a substring of t_∞, but this contradicts Lemma 6.1.7.

(2) t_{k-1}^4 is a substring of either $t_n t_{n+1}$ or $t_n^2 t_{n+1}$ with $n \geq m \geq k$. Because of $t_\infty = t_n \bar{t}_n t_n^2 t_{n+1} \cdots$, we obtain again a contradiction with Lemma 6.1.7. ∎

Now we can prove that L is not a context-free language in the Chomsky hierarchy. It turns out that here the pumping lemma B.2.7 for CFL is not strong enough. Thus a stronger tool is needed. We will use the Ogden Lemma to reach our purpose. In Example B.2.11 of Appendix B we explain its content and how to use it, and in Subsection *3.5.2* we have used this tool to prove Theorem 3.5.4.

Theorem 6.2.7 $L = \mathcal{L}(t_\infty)$ is not a context-free language.

Proof. Assume the contrary that L is a CFL. By Ogden Lemma there exists a positive integer n satisfying all conditions in the Lemma (see Lemma B.2.10). Choose k such that $2^{k-1} > n$, and take a word

$$z = t_k^4 \in L.$$

Now designate the last 2^{k-1} positions in $z = t_k^3 t_{k-1} \bar{t}_{k-1}$ as distinguished:

$$z = t_k t_k t_k t_{k-1} \overbrace{\bar{t}_{k-1}}^{2^{k-1}}.$$

By Ogden's lemma there exists a decomposition

$$z = uvwxy,$$

which satisfies all requirements in Lemma. There are two alternatives.

(1) $x \neq \varepsilon$. Observing the requirement that w contains at least one distinguished position, we have that $|x| < 2^{k-1}$. On the other hand, from $uv^i wx^i y \in L$ for $i \geq 0$ we deduce also that $x^i \in L$ for $i \geq 0$. Now applying the Lemma 6.2.5 leads to that $M(x) = t_p^{n_p}$ with $p < k - 1$ and $n_p \geq 1$. But at least one string among u, v, w must contain a substring t_k in $z(= t_k t_k t_k t_{k-1} \bar{t}_{k-1})$. Comparing these facts with Lemma 6.2.6 we obtain a contradiction and complete the proof for case (1).

(2) $x = \varepsilon$. Again from the Ogden Lemma we see that both u and v must contain distinguished positions now. It follows that $|v| < 2^{k-1}$. On the other hand, u must contain t_k as its substring. Arguing as before, and using the requirement that $uv^i wy \in L$ for $i \geq 0$ we can finish the proof now. ∎

6.2.3 *L is a Context-Sensitive Language*

We now turn to show that $L = \mathcal{L}(t_\infty)$ is a context-sensitive language. But we would like to do more than merely providing a proof of it. In the following we will find a more precise "position" for language L between $\mathcal{L}$(CFL), the family of context-free languages and $\mathcal{L}$(CSL), the family of context-sensitive languages.

In order to do that we introduce some parallel rewriting systems, which means that in each step of process, all symbols of the word considered must be rewritten. Here we need the L systems, namely, the developmental systems, which are the most-widely studied parallel rewriting systems. For readers convenience an Appendix C is included to explain some L systems used below.

Remark. As a fact we have already met some deterministic 0L (D0L) systems in 6.1 and 6.2. Let $G = (\Sigma, h, t_0)$, where Σ is $\{0, 1\}$, $h = (0 \to 11, 1 \to 10)$, $t_0 = 1$. The infinite sequence t_∞ is generated by such a D0L system as the "limit" of the language $L(G) = \{t_n\}_{n \geq 0}$. The Thue-Morse sequence s_∞ is another example of D0L systems as well. But here we need more sophisticated L systems than D0L system. In Section C.1 the D0L, 0L, T0L and ET0L systems and the corresponding languages are defined. Here the number "0" means that they are context-free, that is, zero-sided, in applying parallel rewriting rules.

The most important L system for our purpose is the ET0L system. Some properties of ET0L languages, including the concept of full AFL, and its relation with Chomsky hierarchy is presented in Section C.2 and shown in Figure C.1.

Theorem 6.2.8 $L = \mathcal{L}(t_\infty)$ is an ET0L language and, consequently, a context-sensitive language.

We begin the discussion by studying several simpler languages.

The first language is

$$K_1 = \{0^m \mid m \geq 0\}.$$

It is easy to see that K_1 is a 0L language, generated by the following 0L system,

$$G_1 = (\{0\}, \{0 \to 0, 0 \to 0^2, 0 \to \varepsilon\}, 0).$$

The second language we need is

$$K_2 = \{u \mid u \text{ is a normal string with } n_{-1} = 0\}.$$

Recalling Definition of normal strings, all words $u \in K_2$ are of the form $t_0^{n_0} t_1^{n_1} \cdots t_l^{n_l}$.

Let an ET0L system be

$$G_2 = (\{S, T, 0, 1\}, H, S, \{0, 1\}),$$

where

$$H = \{h_1, h_2\},$$
$$h_1 = \{S \to \varepsilon, S \to S, S \to ST, T \to T, 0 \to 0, 1 \to 1\},$$
$$h_2 = \{S \to S, T \to 1, 0 \to 11, 1 \to 10\}.$$

Lemma 6.2.9 The language of underlying system $U(G_2)$ is

$$L(U(G_2)) = \{T^n u \mid u \text{ is a normal string with } n_{-1} = 0, n \geq 0\}$$
$$\cup \{ST^n u \mid u \text{ is a normal string with } n_{-1} = 0, n \geq 0\}.$$

Proof. We claim first that the set of the right-hand side is a subset of $L(U(G_2))$. Write $u = t_0^{n_0} \cdots t_l^{n_l}$. If $u = \varepsilon$, then it is obvious that strings T^n and ST^n belong to $L(U(G_2))$ for $n \geq 0$. Otherwise, we proceed inductively on l. For $l = 0$ it is true from the derivations

$$S \xrightarrow{h_1} ST^{n_0} \xrightarrow{h_2} St_0^{n_0} \xrightarrow{h_1} ST^n t_0^{n_0} \xrightarrow{h_1} T^n t_0^{n_0} \text{ for } n \geq 0.$$

Assume inductively that it is established for l, then we have derivations

$$S \Longrightarrow ST^{n_0} t_0^{n_1} \cdots t_l^{n_{l+1}}$$
$$\xrightarrow{h_2} St_0^{n_0} t_1^{n_1} \cdots t_{l+1}^{n_{l+1}}$$
$$\xrightarrow{h_1} ST^n t_0^{n_0} t_1^{n_1} \cdots t_{l+1}^{n_{l+1}}$$
$$\xrightarrow{h_1} ST^n t_0^{n_0} t_1^{n_1} \cdots t_{l+1}^{n_{l+1}}$$

for the case of $l+1$, which completes the induction. (The first derivation is by inductive hypothesis.)

Secondly, we will show that if $z \in L(U(G_2))$, then z is of the form either $T^n u$ or $ST^n u$.

Denote by $S \underset{n}{\Longrightarrow} z$ that starting from S we obtain z after n steps of parallel rewriting.

Now we proceed inductively on n. It's trivial for $n = 0$ or 1. Assume inductively that it is true for the value of n. Consider the case for $n+1$. Assume that $S \underset{n}{\Longrightarrow} z = T^n u$ and $u = t_0^{n_0} \cdots t_l^{n_l}$. We have $h_1(z) = \{z\}, h_2(z) = \{t_0^n t_1^{n_0} \cdots t_{l+1}^{n_l}\}$. The another case is that $S \underset{n}{\Longrightarrow} z = ST^n u$. Then $h_1(z) = \{T^n u, ST^n u, ST^{n+1} u,\}$ and $h_2(z) = \{St_0^n t_1^{n_0} \cdots t_{l+1}^{n_l}\}$.

Combining the discussions above completes our proof. ∎

The third language is

$$K_3 = \{v \mid v = P(t_k) \text{ (a proper prefix of } t_k) \text{ for } k \geq 0\}.$$

Define an E0L system, which is an ET0L system with card $H = 1$, by

$$G_3 = (\{B, 0, 1\}, h, B, \{0, 1\}),$$

where

$$h = \{B \to \varepsilon, B \to B, B \to 1, B \to 1B, 0 \to 11, 1 \to 10\}.$$

Lemma 6.2.10 The language of underlying system $U(G_3)$ is

$$L(U(G_3)) = \{v \mid v = P(t_k), k \geq 0\} \cup \{vB \mid v = P(t_k), k \geq 0\}.$$

Proof. Proceeding inductively on $k \geq 1$ to show that every $P(T_k)$ and $P(T_k)B$ belong to $L(U(G_3))$. It is easier to show that, in excess to this claim, we have $B \underset{k}{\Longrightarrow} P(t_k)$ and $B \underset{k}{\Longrightarrow} P(t_k)B$ exactly in k steps.

For $k = 1$, as $t_1 = 10$, $P(t_1)$ is either ε and 1. From the definition

$$B \underset{1}{\Longrightarrow} \varepsilon, B \underset{1}{\Longrightarrow} 1, B \underset{1}{\Longrightarrow} B, B \underset{1}{\Longrightarrow} 1B$$

the claim is true.

Now assume inductively that it holds for the value k. Consider first $v = P(t_{k+1})$. If $|v| < |t_k| = 2^k$ then it also is a $P(t_k)$. From $B \underset{1}{\Longrightarrow} B \underset{k}{\Longrightarrow} v$ it is also true that $B \underset{k+1}{\Longrightarrow} v$. If $|v| \geq |t_k| = 2^k$, then we can write $v = t_k v'$, where v' is a $P(t_k)$. Here we use an observation that t_k and $\bar{t}_k$ have the same string as their proper prefix. Then it is clear from $B \underset{1}{\Longrightarrow} 1B \underset{k}{\Longrightarrow} t_k v' = v$ that also $B \underset{k+1}{\Longrightarrow} v$.

The discussion about $P(t_k)B$ is quite the same.

Conversely, we will show that every word $z \in L(U(G_3))$ must be of a form either $P(t_k)$ or $P(t_k)B$ for $k \geq 0$.

Let $B \underset{n}{\Longrightarrow} z$ and proceed inductively on n. It is trivial for $n = 1$. Assume it is true also for the value n and consider the situation for $n+1$. If $B \underset{n}{\Longrightarrow} P(t_k)$, then

$h(P(t_k))$ is a $P(t_{k+1})$. In the another case, if $B \underset{n}{\Rightarrow} P(t_k)B$, then we have $h(P(t_k)B) = \{P(t_{k+1}), P(t_{k+1})B, h(P(t_k)1, h(P(t_k)1B\}$. Now since $h(P(t_k))1$ is also a $P(t_{k+1})$, this completes the proof. ∎

Proof of Theorem 6.2.8. From foregoing discussion about K_1, K_2 and K_3, we see that $K_2 = L(G_2)$, $K_3 = L(G_3)$ and reach the conclusion that all of them are ET0L languages. Applying Lemma 6.2.4 about words in $\mathcal{L}(t_\infty)$, it turn out that

$$\begin{aligned} L &= K_1 K_2 K_3, \\ &= \{uvw \mid u \in K_1, v \in K_2, w \in K_3\}. \end{aligned}$$

Since the family L(ET0L) is closed under concatenation, the proof is completed. ∎

Remark. If instead of L we consider another language

$$L_1 = \{w \in \Sigma^* | w \text{ is a finite subword of } t_\infty\},$$

then the conclusions of Theorems 6.2.7 and 6.2.8 still hold for L_1, and their proofs are easier than foregoing ones. It is not strange that since the language L_1 only reflects partial behavior in the attractor itself

6.3 The Approach of Fibonacci Sequences

In order to look for higher degrees of complexity in unimodal maps, an approach of Fibonacci sequences was proposed by Hao (1991). Four numerical examples were reported there. One of them was also discussed in Auerbach and Procaccia (1990).

In this section we will discuss all possibilties provided by the approach of Fibonacci sequences in generating new kneading sequences. It turns out that except in one trivial case this approach is always successful to generate non-regular complexity in unimodal maps.

6.3.1 Fibonacci Sequences and Cyclic Shifts

Definition 6.3.1 Let f_0 and f_1 be two given strings over Σ. A sequence $\{f_n\}_{n\geq 0}$ is said to be a *Fibonacci sequence* generated from f_0 and f_1, if

$$f_n \in \{f_{n-1}f_{n-2}, f_{n-2}f_{n-1}\} \quad \text{for every} n \geq 2, \tag{6.3}$$

that is, if f_n is either $f_{n-1}f_{n-2}$ or $f_{n-2}f_{n-1}$ for every $n \geq 2$.

The motivation of this naming is natural that the lengths of $\{f_n\}_{n\geq 0}$ satisfy the recursion relation $|f_n| = |f_{n-1}| + |f_{n-2}|$. If $|f_0| = |f_1| = 1$, then the sequence $\{|f_n|\}_{n\geq 0}$ is exactly the *Fibonacci numbers*

$$1, 1, 2, 3, 5, 8, 13, 21, 34, 55, \ldots$$

The relation (6.3) includes many special cases used in other works. For example, we see that the following block substitution rules

$$t_n = t_{n-1}t_{n-2} \tag{6.4}$$

and

$$b_{2n} = b_{2(n-1)}b_{2n-1},\ b_{2n+1} = b_{2n}b_{2n-1}$$

were used in Hao (1991), and, starting from two given $\phi^{(0)}$ and $\phi^{(1)}$,

$$\phi^{(n+1)} = \max(\phi^{(n)}, \phi^{(n-1)})\ \min(\phi^{(n)}, \phi^{(n-1)})$$

was proposed in Zheng (1989b).

It is obvious that there are infinitely many way to generate Fibonacci sequences from two given strings f_0 and f_1. But usually we cannot assure that the limit $\lim_{n\to\infty}$ exists and is maximal. In order to obtain kneading sequences of unimodal maps by Fibonacci sequences, an additional operation of cyclic shift (cyclic permutation) on strings was proposed by Hao (1991).

Let $m_n = M(f_n)$ denote the maximal cyclic shift of string f_n for each $n \geq 0$. The following Example by Hao (1991) explains that this operation is necessary.

Example 6.3.2 Let $t_0 = 0$ and $t_1 = 1$. Using the rule (6.4) we have

$$t_2 = 10,\ t_3 = 101,\ t_4 = 10110,\ t_5 = 10110101,\ t_6 = 1011010110110\ldots.$$

Since every t_n has t_{n-1} as its proper prefix, the limit $t_\infty = \lim_{n\to\infty} t_n$ exists. It can be verified that $m_5 = \sigma^5(t_5) = 10110110 > t_5$. Since t_∞ has $t_6 = t_5t_4$ as its prefix, we have $\sigma^5(t_\infty) > t_\infty$, hence t_∞ cannot be a kneading sequence.

Remark. As a fact, here $m_n > t_n$ holds for each $n \geq 5$ (see formulas (6.13) and Proposition 6.3.5.) We will continue the discussion of this Example later (in Example 6.4.3).

The problems considered in this section are whether there exists a limit $m_\infty = \lim_{n\to\infty} m_n$, and whether this limit provides a kneading sequence for unimodal maps, whose grammatical complexity is beyond the regular languages in the Chomsky hierarchy (Hao 1991).

The first step we take is to show that although there are infinitely many possible ways to generate $\{f_n\}$ from two given strings f_0 and f_1, but the sequence $\{m_n\}$ is unique.

Proposition 6.3.3 All Fibonacci sequences generated from two strings f_0 and f_1 have the same $\{m_n\}_{n\geq 0}$, where $m_n = M(f_n)$ for each $n \geq 0$.

Proof. Let $\{t_n\}_{n\geq 0}$ be a particular Fibonacci sequence generated by f_0 and f_1 as follows:

$$t_0 = f_0,\ t_1 = f_1,\ t_n = t_{n-1}t_{n-2} \quad \text{for each } n \geq 2.$$

that is, the sequence $\{t_n\}_{n\geq 0}$ is determined from f_0 and f_1 by rule (6.4). We will prove that

$$m_n = M(f_n) = M(t_n) \quad \text{for each } n \geq 0$$

is true for each Fibonacci sequences $\{f_n\}_{n\geq 0}$ generated from the given strings f_0 and f_1.

Proceed in two steps.

(a) Construct another particular Fibonacci sequence $\{s_n\}_{n\geq 0}$ by

$$s_0 = f_0, s_1 = f_1, s_2 = s_0 s_1 \text{ and } s_n = s_{n-1}s_{n-2} \quad \text{for each } n > 2. \tag{6.5}$$

We will prove $s_n = \sigma^{|t_1|}(t_n)$ for each $n \geq 2$ by induction on n.

It is obviously true for $n = 2$ by $s_2 = t_0 t_1$ and $t_2 = t_1 t_0$. Then suppose that the claim is true for $n \leq k$ and consider the case of $n = k + 1$. Since each t_n $(n \geq 1)$ has t_1 as its prefix, we can calculate as follows

$$\begin{aligned} \sigma^{|t_1|}(t_{k+1}) &= \sigma^{|t_1|}(t_k t_{k-1}) \\ &= (\sigma^{|t_1|}(t_k))(\sigma^{|t_1|}(t_{k-1})) \\ &= s_k s_{k-1} \\ &= s_{k+1} \end{aligned}$$

and complete our induction. It implies that $M(t_n) = M(s_n)$ is true for $n \geq 0$.

(b) Now consider all Fibonacci sequences simultaneously. Proceed again by induction on n. Since for $n = 2$ we have $t_2 = f_1 f_0$ and f_2 is either $f_1 f_0$ or $f_0 f_1$, it is true that $M(f_2) = M(t_2)$. Suppose that for $n = k$ we have $M(f_k) = M(t_k)$ already and consider the case of $n = k + 1$. There are two different kinds of Fibonacci sequences to be discussed.

(b.1) $f_2 = f_1 f_0$. Since f_{k+1} can be seen as the k-th string generated from f_1, f_2 by the rule (6.3) and t_{k+1} the k-th string generated from t_1, t_2 by the rule (6.4), using the facts that $t_1 = f_1$, $t_2 = f_2$ and the inductive hypothesis, we obtain $M(f_{k+1}) = M(t_{k+1})$ for those Fibonacci sequences that have the property of $f_2 = f_1 f_0$.

(b.2) $f_2 = f_0 f_1$. Use the sequence $\{s_n\}_{n\geq 0}$ as defined in (6.5) of step (a), we have $s_1 = f_1$, $s_2 = f_2$ and $s_n = s_{n-1}s_{n-2}$ for $n > 2$, using the inductive hypothesis to $\{s_n\}_{n\geq 1}$ leads to $M(f_{k+1}) = M(s_{k+1})$. Combining it with the results of (a) gives us the required $M(f_{k+1}) = M(t_{k+1})$ for the second kind of Fibonacci sequences and complete our induction. ∎

The consequence of this Proposition is that since we are only interested in asymptotic behaviors of $\{m_n\}_{n\geq 0}$, from now on we can focus our attention to one particular Fibonacci sequence obtained by a fixed rule. In the sequel we will exclusively use the rule (6.4) to calculate every m_n and then to obtain the limit m_∞.

The discussion for the case of $f_0 f_1 = f_1 f_0$ is simple.

Proposition 6.3.4 If $f_0 f_1 = f_1 f_0$, then the limit m_∞ is a maximal and periodic

Proof. Using Proposition 1.1.8, there exists a string u such that both f_0, f_1 are powers of u. This implies (through the rule (6.3)) that every f_n is also a power of u. Denote $M(u)$, the maximal cyclic shift of u, by m, it is obviously true that $m_\infty = m^\infty$. ∎

Since the languages determined by periodic kneading sequences are both regular and finite complement, in this sense the situation discussed in Proposition 6.3.4 is trivial. (For finite complement languages see Theorem 7.2.1 of Chapter 7.)

6.3.2 *Calculation of Cyclic Numbers*

In the remaining part of this section we always assume that $f_0 f_1 \neq f_1 f_0$ for $\{f_n\}_{n\geq 0}$.

By Proposition 6.3.3 we will use the rule (6.4) to calculate $m_n = M(t_n)$. Of course, as we take $t_0 = f_0$ and $t_1 = f_1$, $t_0 t_1 \neq t_1 t_0$ always holds.

In order to calculate $m_n = M(t_n)$ from t_n, we need to calculate the *cyclic number* k_n of t_n, which is defined as an integer k_n such that

$$\sigma^{k_n}(t_n) = M(t_n) \text{ and } 0 \leq k_n < |t_n|.$$

Since the cyclic number of t_n is unique only if t_n is primitive, we need the following result. (see Propositions 1.1.8 and 1.1.11 for details.)

Proposition 6.3.5 If $t_0 t_1 \neq t_1 t_0$, then each t_n is primitive for $n \geq 4$.

Proof. It is easy to see that if $t_0 t_1 \neq t_1 t_0$ then $t_{n-1} t_n \neq t_n t_{n-1}$ holds for each $n > 1$. Write

$$t_n = t_{n-3} t_{n-4} t_{n-3} t_{n-3} t_{n-4}$$

and using Proposition 1.1.10 is sufficient. ∎

Remark. The number 4 given in this Proposition is the best possible one. For example, let $t_0 = 010100$ and $t_1 = 10010010$, then we obtain $t_2 = 10010010010100$ and $t_3 = (10010010010)^2$.

In the calculation of k_n's it turns out that we have to treat two different situations separately.

Definition 6.3.6 Let $\{f_n\}$ be a Fibonacci sequence generated from two given strings f_0 and f_1, and $f_0 f_1 \neq f_1 f_0$ holds. $\{f_n\}$ is called an *even Fibonacci sequence* if both f_0, f_1 are even, and an *odd Fibonacci sequence* if at least one of f_0, f_1 is odd.

Remark. For an even Fibonacci sequence $\{f_n\}$, each member f_n is even. But for an odd Fibonacci sequence $\{f_n\}$, its sequence of parity is (if both f_0, f_1 are odd) odd, odd, even, odd, odd, even, ... Roughly speaking, there are about one third of elements of $\{t_n\}$ being even.

First we establish a formula for calculation of cyclic numbers recursively.

Lemma 6.3.7 If $t_0 t_1 \neq t_1 t_0$, then the cyclic numbers k_n's ($n \geq 4$) satisfy the recursive formula

$$k_n := \max\{\sigma^{k_{n-1}}(t_n), \sigma^{k_{n-2}+|t_{n-1}|}(t_n)\}.$$

Proof. From Proposition 6.3.5 we know that each k_n for $n \geq 4$ is well defined. Use Proposition 1.1.11 in following discussions. Consider $\sigma^i(t_n)$ for $0 \leq i < |t_n|$. When $0 \leq i < |t_{n-1}|$, using the rule (6.4), we have

$$t_n^2 = t_{n-1}^2 t_{n-4} t_{n-3} t_{n-2}$$

and see that $\sigma^i(t_n)$ has $\sigma^i(t_{n-1})$ as its prefix. This means that $i = k_{n-1}$ is the only value to be considered. When $|t_{n-1}| \leq i < |t_n|$, we have

$$t_n^2 = t_{n-1} t_{n-2}^2 t_{n-3} t_{n-2}$$

and find that $\sigma^i(t_n)$ has $\sigma^{i-|t_{n-1}|}(t_{n-2})$ as its prefix. So we need only to consider $i = |t_{n-1}| + k_{n-2}$. Combining these results leads to the formula required. ∎

The next step is to prove that the equality $k_n = k_{n+1}$ must happen for arbitrary big n.

An observation is that, by using Proposition 6.3.5 and Lemma 6.3.7, $k_n \geq k_{n-1}$ for each $n \geq 6$.

Lemma 6.3.8 If $t_0 t_1 \neq t_1 t_0$, then for each integer N there exists an integer $n > N$ such that $k_n = k_{n+1}$.

Proof. Assuming the contrary that the claim is false, then there exists a number N such that $k_n \neq k_{n+1}$ for each $n > N$. From the observation made before, we can select an n such that

$$k_n < k_{n+1} < k_{n+2} < k_{n+3} < k_{n+4} \tag{6.6}$$

hold and the first string t_n being even. We can also assume that the formula of Lemma 6.3.7 holds for all these k_n's. From (6.6) and Lemma 6.3.7 we have four inequalities

$$\sigma^{k_{n+i}}(t_{n+i+1}) < \sigma^{k_{n+i+1}}(t_{n+i+1}) = M(t_{n+i+1}), \quad i = 0, 1, 2, 3 \tag{6.7}$$

For characterizing k_n and k_{n+1} we introduce decompositions $t_n = c_1 c_2$ with $k_n = |c_1|$ and (by Lemma 6.3.7) $t_{n-1} = c_3 c_4$ with $k_{n+1} = |t_n c_3|$.

Through these c_i's the four inequalities (6.7) can be rewritten as follows:

$$\begin{aligned} c_2 t_{n-1} c_1 &< c_4 t_n c_3, \\ c_4 t_n t_n c_3 &< c_2 t_{n+1} c_1, \\ c_2 t_{n+1} t_{n+1} c_1 &< c_4 t_{n+2} t_n c_3, \\ c_4 t_{n+2} t_{n+2} c_3 &< c_2 t_{n+3} t_{n+1} c_1. \end{aligned} \tag{6.8}$$

From the previous two equalities of (6.8) we see that $c_2 t_{n+1}$ is a proper prefix of $c_4 t_{n+2}$, so we can have

$$c_4 t_{n+2} = c_2 t_{n+1} u \quad \text{with } |u| = |c_1| + |c_4|. \tag{6.9}$$

Consider where to embed the substring t_{n+1} of the right-hand side of (6.9) into its left-hand side. From the structure of the string $c_4 t_{n+2} = c_4 t_n c_3 c_4 t_n$ and $t_{n+1} = t_n c_3 c_4$, we see that its every substring of length $|t_{n+1}|$ must be a cyclic shift of t_{n+1}. Combining this fact and Proposition 1.1.11 leads to $c_2 = c_4$.

Using $c_2 = c_4$ to the second inequality of (6.8) and the fact of string t_n being even, we have

$$c_4 t_n c_3 < c_2 c_3 c_4 c_1 = c_2 t_{n-1} c_1, \tag{6.10}$$

which contradicts the first inequality of (6.8). ∎

The rule from $k_n = k_{n+1}$ to $k_{n+1} = k_{n+2}$ is revealed by the following Lemma.

Lemma 6.3.9 If $t_0 t_1 \neq t_1 t_0$, $k_n = k_{n+1}$ holds for some $n \geq 6$, then $k_{n+1} = k_{n+2}$ if and only if t_n is an odd string.

Proof. Calculate k_{n+2} by Lemma 6.3.7 in two cases.

1. $|t_{n-1}| < k_n < |t_n|$. Writing $t_{n-2} = d_1 d_2$ with $|d_1| = k_n - |t_{n-1}|$, we need to compare

$$\begin{aligned}
\sigma^{k_{n+1}}(t_{n+2}) &= \sigma^{k_n}(t_{n-1} t_{n-2} t_{n-1} t_n) \\
&= d_2 t_{n-1} t_n t_{n-1} d_1 \\
&= d_2 t_{n-1} d_1 (d_2 t_{n-3} t_{n-2} d_1) d_2 t_{n-3} d_1,
\end{aligned}$$

and

$$\begin{aligned}
\sigma^{k_n + |t_{n+1}|}(t_{n+2}) &= \sigma^{k_n}(t_n t_{n+1}) \\
&= d_2 t_{n+1} t_{n-1} d_1 \\
&= d_2 t_{n-1} d_1 (d_2 t_{n-1} d_1) d_2 t_{n-3} d_1.
\end{aligned}$$

From $M(t_n) = \sigma^{k_n}(t_n)$ we have

$$M(t_n) = \sigma^{|t_{n-1} d_1|}(t_{n-1} t_{n-2}) = d_2 t_{n-1} d_1 > \sigma^{|d_1|}(t_n) = d_2 t_{n-3} t_{n-2} d_1.$$

Since t_n and $d_2 t_{n-1} d_1$ have the same parity, using the formula (of Lemma 6.3.7) leads to the conclusion: $k_{n+2} = k_{n+1}$ if and only if t_n being odd.

2. $0 \leq k_n \leq |t_{n-1}|$. Writing $t_{n-1} = d_3 d_4$ with $|d_3| = k_n$, we need to compare

$$\begin{aligned}
\sigma^{k_{n+1}}(t_{n+2}) &= \sigma^{k_n}(t_{n-1} t_{n-2} t_{n-1} t_n) \\
&= d_4 t_{n-2} t_{n-1} t_n d_3 \\
&= d_4 t_{n-2} d_3 (d_4 t_n d_3).
\end{aligned}$$

and

$$\begin{aligned}
\sigma^{|t_{n+1}| + k_n}(t_{n+2}) &= \sigma^{k_n}(t_{n-1} t_{n-2} t_{n+1}) \\
&= d_4 t_{n-2} t_{n+1} d_3 \\
&= d_4 t_{n-2} d_3 (d_4 t_{n-2} t_{n-1} d_3).
\end{aligned}$$

The condition $k_{n+1} = k_n$ means that

$$M(t_{n+1}) = \sigma^{k_n}(t_{n+1}) = d_4 t_{n-2} t_{n-1} d_3 > \sigma^{|t_n d_3|}(t_{n+1}) = d_4 t_n d_3.$$

Since t_n and $d_4 t_{n-2} d_3$ have the same parity, it leads to the same conclusion as above and completes our proof. ∎

What happens if $k_n = k_{n+1} < k_{n+2}$ is answered by the following Lemma.

Lemma 6.3.10 If $t_0t_1 \neq t_1t_0$, $n \geq 6$, and $k_n = k_{n+1} < k_{n+2}$, then $k_{n+2} = k_{n+3}$.

Proof. In order to calculate k_{n+3} by Lemma 6.3.7 we need to compare

$$\begin{aligned}\sigma^{k_{n+2}}(t_{n+3}) &= \sigma^{|t_{n+1}|+k_n}(t_{n+3}) \\ &= \sigma^{k_n}(t_nt_{n+1}t_{n+1})\end{aligned}$$

and

$$\begin{aligned}\sigma^{|t_{n+2}|+k_{n+1}}(t_{n+3}) &= \sigma^{|t_{n+2}|+k_n}(t_{n+2}t_{n+1}) \\ &= \sigma^{k_n}(t_{n+2}t_nt_{n-3}t_{n-2}).\end{aligned}$$

From these expressions it can be seen that $\sigma^{k_n}(t_nt_{n+1})$ and $\sigma^{k_n}(t_{n+2})$ are their prefixes respectively.

Using the condition $k_{n+2} > k_{n+1} = k_n$, and Lemma 6.3.7, we have

$$\sigma^{|t_{n+1}|+k_n}(t_{n+2}) = \sigma^{k_n}(t_nt_{n+1}) > \sigma^{k_{n+1}}(t_{n+2}) = \sigma^{k_n}(t_{n+2})$$

and obtain $k_{n+3} = k_{n+2}$. ∎

We need some identities of $\{t_n\}_{n\geq 0}$ for the sequel of this Section. Because all of them are established easily by induction, we list them here and omit their proofs.

$$t_{2l+1} = (\prod_{i=l}^{1} t_{2i})t_1 \qquad \text{for } l \geq 1. \tag{6.11}$$

$$t_{3l+1} = (\prod_{i=l}^{1} t_{3i})t_2(\prod_{i=1}^{l-1} t_{3i}) \qquad \text{for } l > 1. \tag{6.12}$$

Remark. An explanation about the use of notation "$\prod$": since all finite strings are considered as elements of a free monoid $\{0,1\}^*$, in which the binary associative operation is the concatenation of strings, the notation $\prod_{i=n}^{1} t_i$ is simply the product $t_nt_{n-1}\ldots t_1$. But since the concatenation is non-commutative, so in general $\prod_{i=n}^{1} t_i \neq \prod_{i=1}^{n} t_i$.

Proposition 6.3.11 Let $t_0t_1 \neq t_1t_0$ hold. If $\{t_n\}$ is an odd Fibonacci sequence, then $\{k_n\}$, after discarding some finite terms from beginning of $\{t_n\}$ and renaming the remaining terms if necessary, the sequence of cyclic numbers, satisfies the equalities $k_1 = k_2 = k_3$ and

$$k_{3l+1} = k_{3l+2} = k_{3l+3} = k_1 + \sum_{i=1}^{l} |t_{3i}| \qquad \text{for each } l \geq 1. \tag{6.13}$$

Proof. Using Propositions 6.3.8, we can select an integer N such that $k_N = k_{N+1}$ and for all $n \geq (N-1)$ the strings t_n's are primitive and all k_n's satisfy the formula of Lemma 6.3.7.

If t_N is odd and t_{N+1} is even, then from Propositions 6.3.9 and 6.3.10 we will have $k_{N+3i} = k_{N+3i+1} = k_{N+3i+2}$ for $i \geq 0$. If both t_N and t_{N+1} are odd, then for the same reason we have $k_N = k_{N+1} = k_{N+2} = k_{N+3}$ and $k_{N+3i+1} = k_{N+3i+2} = k_{N+3i+3}$ for $i \geq 0$. If t_N is even, then we obtain $k_N = k_{N+1} < k_{N+2}$ and $k_{N+3i+2} = k_{N+3i+3} = k_{N+3i+4}$ for $i \geq 0$.

It is easy to see that in all these cases we can discard some finite terms from beginning of $\{t_n\}_{n\geq 0}$ and rename the remaining terms such that the resulting new sequence $\{t_n\}_{n\geq 0}$ have the following properties: both t_0 and t_1 are odd and

$$k_{3l+1} = k_{3l+2} = k_{3l+3} \quad \text{for } l \geq 0.$$

Because each t_{3l+2} is even, Proposition 6.3.9 implies that $k_{3l+3} < k_{3l+4}$ and that $k_{3l+4} = k_{3l+2} + |t_{3l+3}|$. This leads to the formula of Proposition 6.3.11 inductively. ∎

Using the formula (6.13) and identities (6.12) we obtain the following formulas of m_n's for odd Fibonacci sequence.

If $\{t_n\}$ is an odd Fibonacci sequences, $m_n = M(t_n)$, and the conditions in Proposition 6.3.11 are satisfied, then

$$\begin{aligned} m_{3l+1} &= b\Big(\prod_{i=1}^{l-1} t_{3i}\Big)\Big(\prod_{i=l}^{1} t_{3i}\Big)a, \\ m_{3l+2} &= b\Big(\prod_{i=1}^{l} t_{3i}\Big)\Big(\prod_{i=l}^{1} t_{3i}\Big)a, \\ m_{3l+3} &= b\Big(\prod_{i=1}^{l} t_{3i}\Big)t_{3l+1}\Big(\prod_{i=l}^{1} t_{3i}\Big)a \end{aligned} \tag{6.14}$$

hold for each $l > 0$, where the strings a, b are determined by $t_2 = ab$, $|a| = k_1 = k_2 = k_3$. A direct calculation leads to identities for m_n's:

$$\begin{aligned} m_{3l+2} &= m_{3l+1}m_{3l}, \\ m_{3l+3} &= m_{3l+2}m_{3l+1}, \\ m_{3l+4} &= m_{3l+2}m_{3l+3} \end{aligned} \tag{6.15}$$

for each $l \geq 0$.

The proof of the next Proposition 6.3.12 and formula (6.17) are similar and omitted.

Proposition 6.3.12 Let $t_0t_1 \neq t_1t_0$ hold. If $\{t_n\}$ is an even Fibonacci sequence, then $\{k_n\}$, after discarding some finite terms from beginning of $\{t_n\}$ and renaming the remaining terms if necessary, the sequence of cyclic numbers, satisfies the conditions $k_1 = k_2$ and

$$k_{2l+1} = k_{2l+2} = k_1 + \sum_{i=1}^{l} |t_{2i}| \quad \text{for each } l \geq 1. \tag{6.16}$$

For even Fibonacci sequences $\{t_n\}$ we have the formula for $m_n = M(t_n)$ as follows:

$$\begin{aligned} m_{2l+1} &= b'(\prod_{i=l}^{1} t_{2i})\, a', \\ m_{2l+2} &= b' t_{2l}(\prod_{i=l}^{1} t_{2i})\, a' \end{aligned} \tag{6.17}$$

for each $l \geq 0$, where the strings a', b' are determined by $t_1 = a'b', |a'| = k_1 = k_2$. Similarly we have the identities for $m'_n s$:

$$\begin{aligned} m_{2l+2} &= m_{2l+1} m_{2l}, \\ m_{2l+3} &= m_{2l+1} m_{2l+2} \end{aligned} \tag{6.18}$$

for $l \geq 0$.

From (6.15) and (6.18) we see that the sequence $\{m_n\}_{n \geq 0}$ is also a Fibonacci sequence generated from m_0 and m_1.

6.3.3 *Proof of Theorem 6.3.13*

The main result of this Section is the following theorem, which tells us that except for a trivial case of $f_0 f_1 = f_1 f_0$, the approach of Fibonacci sequences is always successful in finding higher complexity than regularity for unimodal maps, that is to say, the language obtained is non-regular.

Theorem 6.3.13 For all Fibonacci sequences $\{f_n\}_{n \geq 0}$ generated from two given strings f_0 and f_1, the sequence $\{m_n\}_{n \geq 0}$, where $m_n = M(f_n)$, is the same and uniquely determined by f_0 and f_1. The limit

$$m_\infty = \lim_{n \to \infty} m_n, \tag{6.19}$$

exists and is maximal. Moreover, if $f_0 f_1 = f_1 f_0$, then this m_∞ is periodic; otherwise, m_∞ is neither periodic nor eventually periodic.

Proof. The first claim is covered by Proposition 6.3.3. The case of $f_0 f_1 = f_1 f_0$ is treated in Proposition 6.3.4. For the other case of $f_0 f_1 \neq f_1 f_0$ we have to discuss odd and even Fibonacci sequences separately.

(i) Odd Fibonacci sequences. From the formulas of (6.14) for m_n's we see that the string $b \prod_{i=1}^{l-1} t_{3i}$ is their common prefix and obtain

$$m_\infty = \lim_{n \to \infty} m_n = b \prod_{i=1}^{\infty} t_{3i}$$

immediately.

In order to prove that m_∞ is not an eventually periodic sequence, we need only to show that for every pair of two integers i and j, $0 \leq i < j$, the inequality

$$\sigma^i(m_\infty) \neq \sigma^j(m_\infty) \tag{6.20}$$

holds. (This is the method used in the third proof of Lemma 6.1.5.)

From the expression of m_∞ we see that

$$m_\infty = b\prod_{i=1}^{l+1} t_{3_i} \cdots = b(\prod_{i=1}^{l} t_{3_i})t_{3l+3} \cdots = b(\prod_{i=1}^{l} t_{3_i})t_{3l+1}t_{3l}t_{3l+1} \cdots$$

Using the formulas of (6.14) leads to that m_∞ has m_{3l+3} as its prefixes for all $l \geq 0$. Using the identities of (6.15) we can write

$$m_\infty = m_{3l+1}m_{3l}m_{3l+1} \cdots = m_{3l+2}m_{3l+1} \cdots$$

Since m_{3l+2} $(= m_{3l+1}m_{3l})$ is primitive, if we take l big enough such that $0 \leq i < j < |m_{3l+1}|$, then $\sigma^i(m_\infty)$ and $\sigma^j(m_\infty)$ have $\sigma^i(m_{3l+2})$ and $\sigma^j(m_{3l+2})$ as their prefixes (of length $|m_{3l+2}|$) respectively. Using Proposition 1.1.11 leads to the required inequality (6.20).

Since m_{3l+2} is maximal, the discussion above for $i = 0$ already provides the proof about m_∞ being maximal.

(ii) Even Fibonacci sequences. From the formulas of (6.17) for m_n's we see that the string $b't_{2l}$ is their common prefix and obtain

$$m_\infty = \lim_{n\to\infty} m_n = b't_\infty,$$

where $t_\infty = \lim_{n\to\infty} t_n$ is guaranteed by the rule 6.4.

Since t_∞ has each $t_n (n > 0)$ as its prefix and $t_n = t_{n-2}^2 t_{n-5} t_{n-4}$ for each $n > 5$, we have

$$m_\infty = b't_{2l+1}t_{2l+1} \cdots = b't_{2l+1}(\prod_{l}^{1} t_{2_i})t_1 \cdots$$

$$= b'(\prod_{l}^{1} t_{2_i})t_1(\prod_{l}^{1} t_{2_i})a'b' \cdots$$

This leads to

$$m_\infty = m_{2l+1}m_{2l+1} \cdots$$

where the formulas of (6.17) are used again.

Using the fact that m_{2l+1} is primitive and maximal, it is easy to prove that the inequality (6.20) holds for this m_∞ and it is maximal by the same arguments as in the case (i). ∎

6.4 Homomorphisms on Free Submonoids

In previous Section we have seen that some kneading sequences can be generated from Fibonacci sequences. Except for the trivial case of $f_0f_1 = f_1f_0$, such kneading sequences provide higher complexity of unimodal maps. In this section we will show that these infinite strings can also be obtained from two special kinds of homomorphisms on free submonoids, and that it opens a more powerful way to explore grammatical complexity in dynamical systems.

6.4.1 Two Special Kinds of Homomorphisms

In Section 6.1 we have seen that the kneading sequence of the Feigenbaum attractor can be obtained from a homomorphism

$$h = (1 \to 10, 0 \to 11)$$

on the free monoid $\{0,1\}^*$ by taking the limit

$$s = \lim_{n\to\infty} h^n(1) = 1011101010111011\cdots$$

The following discussion shows that it is necessary to extend this idea to homomorphisms on free submonoids which was introduced in Subsection *1.1.2*.

Let α and β are two strings of the free monoid $\{0,1\}^*$. If $\alpha\beta \neq \beta\alpha$, then by Proposition 1.1.6 the submonoid $(\alpha+\beta)^*$ generated from α and β is free. It means that the strings α and β can be treated as if they are indeed letters.

In the following discussion we will call a kneading sequence s an *odd (even) Fibonacci kneading sequence* if s is the limit of m_n's in Theorem 6.3.13 from an odd (even) Fibonacci sequence $\{f_n\}_{n\geq 0}$. The corresponding language $\mathcal{L}(KS)$ is referred to as the *odd (even) Fibonacci language*.

Now we will show that all (either odd or even) Fibonacci kneading sequences can be obtained through homomorphisms on free submonoids.

Proposition 6.4.1 An infinite sequence s is an odd Fibonacci kneading sequence if and only if s is a fixed point of homomorphism h defined by

$$h = (\alpha \to \alpha\beta\alpha\alpha\beta, \beta \to \alpha\alpha\beta)$$

for a pair of strings $\alpha, \beta \in \{0,1\}^*$ which satisfies the conditions:

(i) α is odd and β is even.

(ii) α, $\alpha\beta$ and $\alpha\beta\alpha$ are primitive and maximal.

(iii) $\alpha > \beta$.

Moreover, this s is a global attracting fixed point, that is to say,

$$s = \lim_{n\to\infty} h^n(x)$$

for each nonempty string $x \in (\alpha+\beta)^*$.

Proof. The "Only if" part. If s is an odd Fibonacci kneading sequence, then from Theorem 6.3.13 we have

$$s = m_\infty = \lim_{n\to\infty} m_n,$$

where $m_n = M(f_n)\,(n \geq 0)$ satisfy the identities (6.15). Since we can assume that m_1 has m_0 as its proper prefix, if we take $m_0 = \alpha, m_1 = \alpha\beta$, then it can be verified that each $m_n \in (\alpha+\beta)*$, and $h(m_n) = m_{n+3}$ for $n \geq 0$, where h is defined as desired. This implies that $s = \lim_{n\to\infty} h^n(\alpha)$ and $h(s) = s$.

It remains to verify that the strings α and β satisfy the three conditions as desired. In fact the first two of them are trivial, and the verification of condition (iii) is covered by Proposition 1.2.14.

The "If" part. Let $t_0 = \alpha$ and $t_1 = \beta$. Consider sequence $\{t_n\}_{n\geq 0}$ generated from t_0, t_1 by the rule (6.4). The condition (iii), $\alpha > \beta$, ensures $t_0t_1 \neq t_1t_0$, so the condition (i) means that we have $\{t_n\}_{n\geq 0}$ as an odd Fibonacci kneading sequence. What we need to prove is that the string s is exactly the limit $m_\infty = \lim_{n\to\infty} m_n$, where $m_n = M(t_n)$ for $n \geq 0$.

The condition (ii) means that $t_0, t_1, t_2 (= \alpha\beta\alpha)$ are primitive and maximal strings. By Propositions 1.1.8 and 1.1.10 we know that $t_3 = \alpha\beta\alpha\alpha\beta$ is primitive too. Combining these facts with Proposition 6.3.5 we know that every t_n $(n \geq 0)$ is primitive, and so every k_n $(n \geq 0)$ is well defined. It is easy to verify that t_3 is maximal and leads to $k_0 = k_1 = k_2 = k_3 = 0$. Using Lemmas 6.3.9 and 6.3.10 inductively we obtain

$$k_{3l+1} = k_{3l+2} = k_{3l+3} = \sum_{i=1}^{l} |t_{3i}| \quad \text{for } l \geq 1.$$

Proceed on as in the proof of Theorem 6.3.13, we see that $h(m_n) = m_{n+3}$ holds for each $n \geq 0$ and $m_\infty = \lim_{n\to\infty} m_n$. Since $m_{3n} = h^n(\alpha)$ we have $s = m_\infty$. ∎

The dual result for even Fibonacci kneading sequence is the following Proposition, whose proof is quite similar with Proposition above and thus omitted.

Proposition 6.4.2 An infinite string s is an even Fibonacci kneading sequence if and only if s is a fixed point of homomorphism h defined by

$$h = (\alpha \to \alpha\alpha\beta, \beta \to \alpha\beta)$$

for a pair of strings $\alpha, \beta \in \{0,1\}^*$ which satisfies the conditions:

(i) both α and β are even.

(ii) $\alpha, \alpha\beta$ and $\alpha\alpha\beta$ are primitive and maximal.

(iii) $\alpha > \beta$.

Moreover, this s is a global attracting fixed point, that is to say,

$$s = \lim_{n\to\infty} h^n(x)$$

for each nonempty string $x \in (\alpha+\beta)^*$

Example 6.4.3 (This is a continuation of Example 6.3.2.) Beginning from $t_0 = 0$ and $t_1 = 1$, we can obtain an odd Fibonacci kneading sequence as follows. Calculate their cyclic numbers: $k_0 = k_1 = k_2 = k_3 = k_4 = 0$, $k_5 = k_6 = k_7 = 5$. Compare these with Proposition 6.3.11 we find that the formula 6.13 holds after discarding t_0 and renaming $t'_n = t_{n+1}$ for $n \geq 0$. Using the result in the proof of Theorem 6.3.13, the odd Fibonacci kneading sequence is (Since $t'_2 = 101$ and $k_1 = 0$ the string $b = t'_2$)

$$\begin{aligned} m_\infty &= t'_2 t'_3 t'_6 t'_9 \cdots \\ &= 10110,11010,11010,11011,01011,01011\cdots \end{aligned}$$

On the other hand, from the proof of Proposition 6.4.1, we obtain a homomorphism

$$h = (1 \to 10110, 0 \to 110)$$

and $m_\infty = h(m_\infty) = \lim_{n\to\infty} h^n(1)$.

Example 6.4.4 Let $t_0 = 0$ and $t_1 = 11$, we can obtain an even Fibonacci kneading sequence. Calculate $t_2 = 110$, $t_3 = 11011$, $t_4 = 11011110$, $t_5 = 1101111011011$, ... From their cyclic numbers $k_2 = k_3 = 1$, $k_4 = k_5 = 6$, $k_6 = 19$ we see that only one term t_0 need to be discarded. In the new sequence $\{t'_n\}_{n\geq 0}$ we have $t'_0 = 11$ and $t'_1 = 110$. From the proof of Theorem 6.3.13 and $k_1 = 1$, it is easy to obtain

$$\begin{aligned} m_\infty &= 10t_\infty \\ &= 10110,11110,11011,11011,11011,01111\cdots \end{aligned}$$

From the proof of Proposition 6.4.2 and $m_0 = 11$, $m_1 = 101$ we can define homomorphism

$$h = (101 \to 10110111, 11 \to 10111)$$

and know that $m_\infty = h(m_\infty) = \lim_{n\to\infty} h^n(101)$. Here h is a homomorphism on submonoid $(101 + 11)^*$

This example also appeared in Auerbach and Procaccia (1990). Since the kneading sequence defined there is $\sigma(m_\infty)$, a different homomorphism

$$h' = (011 \to 01101111, 01111 \to 0110111101111)$$

was used. Using the cyclic shift operator σ to 011 and 01111 we can easily obtain h from h'.

Example 6.4.5 Let $t_0 = 0$, $t_1 = 10$ and we obtain again an odd Fibonacci kneading sequence. Calculate $t_2 = 100$, $t_3 = 10010$, $t_4 = 10010100$, ..., and cyclic numbers $k_0 = k_1 = k_2 = k_3 = k_4 = 0$ and $k_5 = k_6 = 8$. Discarding t_0 and renaming $t'_0 = 10$ and $t'_1 = 100$, we can use Theorem 6.3.13 and obtain

$$\begin{aligned} m_\infty &= t'_2 t'_3 t'_6 t'_9 \cdots \\ &= 10010,10010,10010,01010,01001,01001\cdots \end{aligned}$$

Calculating homomorphism h according Proposition 6.4.1, we have

$$h = (10 \to 10010100, 0 \to 10100).$$

But in this example α has β as its suffix, this homomorphism h can be replaced by

$$h' = (1 \to 100, 0 \to 10100).$$

Example 6.4.6 Let $t_0 = 1$ and $t_1 = 00$ and we will also obtain an odd Fibonacci kneading sequence. Calculate $t_2 = 001$, $t_3 = 00100$, $t_4 = 00100001, \ldots$, and cyclic numbers $k_2 = k_3 = k_4 = k_5 = 2$ and $k_6 = 15$. Discarding first two terms and renaming $t'_0 = 001$ and $t'_1 = 00100$, we have

$$\begin{aligned} m_\infty &= 100001\, t'_3 t'_6 t'_9 \cdots \\ &= 10000, 10010, 00010, 01000, 01000, 01001 \cdots \end{aligned}$$

The homomorphism is

$$h = (100 \to 1000010010000, 00 \to 10010000),$$

which can also be simplified as in the previous example to

$$h' = (1 \to 10000, 00 \to 10010000).$$

The Fibonacci kneading sequences obtained from these four Examples include all those kneading sequences that can be generated under condition of $|t_0| + |t_1| = 3$, except for the trivial results of 1^∞ and 0^∞.

6.4.2 A More General Route

The previous Subsection shows that all Fibonacci kneading sequence can be obtained from two special kinds of homomorphisms on free submonoids. In this Subsection we will propose a much more general result which includes results of previous Subsection and all those kneading sequence which can be obtained through applying either the $*$-composition law or the generalized composition law infinitely many times as well.

Let α and β be two strings in the free monoid $\{0, 1\}^*$ and the condition $\alpha\beta \neq \beta\alpha$ hold. Thus the submonoid $(\alpha + \beta)^*$ is free.

Let ρ and λ be two strings in the same free submonoid $(\alpha + \beta)^*$ and satisfy the following conditions:

$$\rho \text{ has } \alpha \text{ as its proper prefix,} \tag{H1}$$

$$\alpha, \rho \text{ have the same parity and } \beta, \lambda \text{ have the same parity,} \tag{H2}$$

$$\alpha > \beta \text{ and } \rho > \lambda, \tag{H3}$$

$$\text{each } \rho\lambda^n \text{ is primitive and maximal for } n \geq 0. \tag{H4}$$

The main result in this Subsection is

Theorem 6.4.7 Let α, β be two strings of $\{0,1\}^*$ and $\alpha\beta \neq \beta\alpha$ hold. If a homomorphism

$$h = (\alpha \to \rho, \beta \to \lambda)$$

satisfy the conditions (H1)–(H4), then the limit

$$s = \lim_{n\to\infty} h^n(\alpha)$$

exists and maximal. If s is not $\alpha\beta^\infty$, then it is neither periodic nor eventually periodic.

First we will explain how general this result is and then proceed to prove it.

Case 1. It is easy to verify that the homomorphism defined in Proposition 6.4.1 by

$$\rho = \alpha\beta\alpha\alpha\beta, \quad \lambda = \alpha\alpha\beta$$

is a special case of this Theorem.

Case 2. It is also easy to verify that the homomorphism of Proposition 6.4.2 by

$$\rho = \alpha\alpha\beta, \lambda = \alpha\beta,$$

is also a special case of this Theorem.

Case 3. The $*$-composition law of Derrida, Gervois and Pomeau (1978) is discussed in Section 4.5. It is a homomorphism on $\{0,1\}^*$ defined by

$$h = (1 \to \overline{x}, 0 \to x),$$

where x is an even string satisfying the condition (P3) of Definition 1.2.7. It is easy to find that this h is a special case of this Theorem.

For instance, let $x = 101$, then $h = (1 \to 100, 0 \to 101)$ represents a self-similar map from the set of all kneading sequences (of a full family of unimodal maps) to those of its period 3 window. The limit $s = \lim_{n\to\infty} h^n(1)$ is the kneading sequence of the limit of period-n-tupling sequences.

Case 4. The generalized composition law in Zheng (1989b), Hao (1991) is also discussed in Section 4.5. It is a homomorphism on $\{0,1\}^*$ defined by $h = (1 \to \rho, 0 \to \lambda)$, where ρ and λ satisfy conditions: (1) ρ is odd and λ is even, (2) $\rho > \lambda$, (3) $\rho|_c$, $\rho\lambda|_c$ and $\rho\lambda^\infty$ are maximal. Here the notation $x|_c$ is a string obtained from x by changing the last symbol of x by c.

It is not difficult to prove that these conditions (1)–(3) are equivalent to conditions (1)–(2) and (3)$'$ that ρ and $\rho\lambda$ are primitive and maximal, and $\rho\lambda^\infty$ is maximal. This fact shows clearly that the generalized composition law may be obtained as a special case of Theorem 6.4.7.

In Subsection *4.5.3* it has been proved in Theorem 4.5.9 that all generalized composition laws can be generated from either (i) eventually periodic kneading sequences, or (ii) periodic kneading sequences x^∞, where x is even maximal primitive. An example is Example 6.4.3 of the previous Section:

$$h = (1 \to 10110, 0 \to 110).$$

This is a generalized composition law. Since $\rho\lambda^\infty = (101)^\infty$, it can be generated from periodic kneading sequence.

Now we proceed to prove Theorem 6.4.7. Here some preparation is in need.

Lemma 6.4.8 Let $x, y \in \{0,1\}^*$ and $x > y$. If for some integer $n > 0$, xy^n is maximal and primitive, xy^{n+1} is maximal, and v is a nonempty suffix of y, then $vy^{n-1}x < xy^n$.

Proof. Since xy^n is maximal, we have $vy^{n-1}x \le xy^n$. If vy^{n-1} is not a prefix of xy^n, then the claim holds already. Otherwise, vy^{n-1} is a prefix xy^n, and also its suffix. Since xy^n is maximal and primitive, using Corollary 1.2.11 leads to that vy^{n-1} must be an odd string.

Combining this fact with $x > y$ leads to $vy^{n-1}x < vy^n$. Since vy^n is a suffix of xy^{n+1}, and the latter is maximal, we have $vy^{n-1}x < vy^n \le xy^{n+1}$. But from $|vy^{n-1}x| < |xy^n|$, this inequality implies already the required conclusion. ∎

Remark. The condition of xy^{n+1} being maximal cannot be removed. For instance, if $x = 1$, $y = 0110$, $v = 10$ and $n = 1$, then the conclusion of Lemma is false. The reason is that $xy^2 = 101100110$ is not maximal.

From conditions (H2) and (H3) we see that the homomorphism h of Theorem 6.4.7 preserve the order relation in $(0+1)^*$. That is to say, if $w_1, w_2 \in (\alpha+\beta)^*$ and $w_1 < w_2$, then $h(w_1) < h(w_2)$.

The next step is to prove that h preserve the primitivity of strings under some conditions. As a matter of fact, we can establish a stronger result.

Proposition 6.4.9 If string $w \in (\alpha+\beta)^*$ begins from α and is maximal primitive in the free submonoid $\{\alpha,\beta\}^*$, then $h(w)$ is maximal primitive in the free monoid $\{0,1\}^*$.

Proof. Let $w = r_1 r_2 \cdots r_n$, in which $r_1 = \alpha$ and each r_i else (for $i \in \{2,3,\ldots,n\}$) is either α or β. Consider $h(w) = h(r_1)h(r_2)\cdots h(r_n)$. What we need to prove is

$$\sigma^i(h(w)) < h(w) \quad \text{for } 1 \le i < |h(w)|. \tag{6.21}$$

Since $r_l \cdots r_n r_1 \cdots r_{l-1} < w$ for $1 < l \le n$, so $h(r_l \cdots r_n r_1 \cdots r_{l-1}) < h(w)$ holds. It implies that if $i = \sum_{k=1}^{l-1} |h(r_k)|$ then the inequality (6.21) is true.

Consider more general cases. If $0 < i < |h(r_1)|$, then let $w' = w$ and $j = i$. On the other hand, if there is some $l > 1$ such that $|h(r_1 \cdots r_{l-1})| < i < |h(r_1 \cdots r_l)|$, let $w' = r_l \cdots r_n r_1 \cdots r_{l-1}$ and $j = i - |h(r_1 \cdots r_{l-1})|$. After such selection of w' and j we have

$$\sigma^i(h(w)) = \sigma^j(h(w')).$$

There are several possibilities for string w'.

Case (1). $w' = \alpha^2\cdots$. Since $h(w') = \rho^2\cdots$, $0 < j < |\rho|$ and ρ being maximal and primitive, we have

$$\sigma^i(h(w)) = \sigma^j(h(w')) = \sigma^j(\rho)\cdots < h(w') \le h(w).$$

Case (2). $w' = w = \alpha\beta^p$ with some $p > 0$. Using the condition that $h(w) = \rho\lambda^p$ being maximal and primitive, (6.21) holds.

Case (3). $w' = \alpha\beta^q\alpha\cdots$ with some $q > 0$. Since $h(w') = \rho\lambda^q\rho\cdots$, $0 < j < |\rho|$ and $\rho\lambda^q$ being maximal and primitive, we have

$$\sigma^i(h(w)) = \sigma^j(h(w')) = \sigma^j(\rho\lambda^q)\cdots < h(w') \leq h(w).$$

Case (4). $w' = \beta^q\alpha\cdots$ with some $q > 0$. Here $h(w') = \lambda^q\rho\cdots$ and $0 < j < |\lambda|$. Taking $\lambda = l_1 l_2$ with $|l_1| = j$, we have $\sigma^i(h(w)) = \sigma^j(h(w')) = l_2\lambda^{q-1}\rho\cdots$. Using Lemma 6.4.8 we have $l_2\lambda^{q-1}\rho < \rho\lambda^q$. Since w' is a cyclic shift of w and the latter is maximal, w must have $\alpha\beta^q$ as its prefix. It implies that $h(w)$ has $\rho\lambda^q$ as its prefix, this leads to

$$\sigma^i(h(w)) = \sigma^j(h(w')) < \rho\lambda^q\cdots = h(w)$$

and finishes our proof. ∎

Proof of Theorem 6.4.7

The condition (H1) guarantees the existence of limit of $\{h^n(\alpha)\}_{n\geq 0}$ when $n \to \infty$ (cf. Lemma C.1.3). Denote this limit by s, then s has each $h^n(\alpha)$ as its prefix. From Proposition 6.4.9 each $h^n(\alpha)$ is maximal and hence s is also maximal.

Consider the structure of s. It depends of string ρ (and λ). There are three cases to be discussed.

(i) $\rho\lambda = \alpha\beta^m$ with $m > 0$. It is easy to see in this (and only in this) case $s = \alpha\beta^\infty$.

(ii) $\rho\lambda = \alpha\beta^m\alpha\cdots$ with $m > 0$. Consider the prefix of s

$$\begin{aligned} h^n(\alpha) &= h^{n-1}(\rho) = h^{n-1}(\alpha\beta)\cdots = h^{n-2}(\rho\lambda)\cdots \\ &= h^{n-3}(\rho\lambda^m)h^{n-3}(\rho)\cdots \end{aligned}$$

From (H4) and Proposition 6.4.9 we know that $h^{n-3}(\rho\lambda^m)$ is maximal and primitive, using the same argument as in Theorem 1 is enough to conclude that s is not an eventually periodic sequence.

(iii) $\rho = \alpha\alpha\cdots$. (This can only happen for even α and ρ.) Similarly we have

$$h^n(\alpha) = h^{n-1}(\rho) = h^{n-2}(\rho)h^{n-2}(\rho)\cdots$$

Using (H4) and Proposition 6.4.9 and the same argument in Theorem 1 finish our proof in this case. ∎

Remark The verification of condition (H4) for two given strings ρ and λ can be done in finite steps as shown by the following Lemma.

Lemma 6.4.10 If both strings xy^{k+2} and xy^{k+3} are primitive and maximal, where $k = [|x|/|y|]$, then each xy^n for $n > k+3$ is primitive and maximal.

Proof. It is enough to prove that xy^∞ is maximal (see Zheng 1989b). The value of k means that $k|y| \leq |x| < (k+1)|y|$. Assume the contrary that there is an integer i such that

$$\sigma^i(xy^\infty) > xy^\infty$$

Consider the longest common prefix of both hand sides, and denote it by z.

(i) $0 < i \leq |x|$. We have $\sigma^i(xy^\infty) = vy^\infty$, where v is a proper suffix of x. If $|z| \geq |vy^{k+2}| \geq (k+2)|y|$ then z contains xy as its proper prefix. Let $y = u^m$, where u is primitive, then we see that $\sigma^i(xy^\infty) = vu^\infty = xy^\infty$, a contradiction to the selection of i. Otherwise, if $|z| < |vy^{k+2}|$, then $\sigma^i(xy^{k+2}) = vy^{k+2}\cdots > xy^{k+2}$, which contradicts the primitivity of xy^{k+2}.

(ii) $i > |x|$. We have $\sigma^i(xy^\infty) = v'y^\infty > xy^\infty$, where v' is a proper suffix of y. If $|z| \geq |y^{k+2}|$ then arguing as in (i) is enough. Otherwise, no matter how large the integer i is, we can find string z from xy^{k+3}. Using the primitivity of the latter leads to a contradiction again. ∎

CHAPTER 7

DISTINCT EXCLUDED BLOCKS OF UNIMODAL MAPS

In Section 2.2 we have introduced the concept of distinct excluded block (DEB) for dynamical languages. In this chapter we will see its role in complexity study of languages $\mathcal{L}(KS)$ generated from unimodal maps. Here the notations are the same as in Section 2.2. Denote by L' the complement language $\Sigma^* \setminus L$ of L, and by L'' the set of all DEB's of L. We also need the notations introduced in Subsection *1.1.5*.

In Section 7.1 two questions are discussed: that is, how to decide if a string is a DEB for a given kneading sequence, and if a DEB is given, how to find the next one.

The structure of DEB's for regular languages of unimodal maps is discussed in Section 7.2. It is found that, if a given kneading sequence KS is either periodic or 10^∞, then $\mathcal{L}(KS)$ is finite complement, which corresponds the subshift of finite type in the theory of symbolic flow. Otherwise, the regular $\mathcal{L}(KS)$ is infinite complement, which corresponds the sofic systems, and L'', the set of all DEB's, has a semilinear structure.

Section 7.3 is a continuation of discussion in Section 2.2 for dynamical languages generated from unimodal maps. Two new Conjecture are proposed.

In Section 7.4, non-regular L'' are discussed for those languages which have been discussed in Chapter 6. Some distribution functions about DEB's are studied.

7.1 A General Discussion

7.1.1 Criterion of being DEB

We begin the discussion by providing a criterion of finding DEB's of $L = \mathcal{L}(KS)$ for a giving kneading sequence KS.

Proposition 7.1.1 If KS is a given kneading sequence, then a string x is a DEB of $L = \mathcal{L}(KS)$ if and only if its dual string $\bar{x}$ satisfies conditions:

1. $\bar{x}$ is an even prefix of KS,
2. $\bar{x}$ has no even proper PS with respect to KS.

Proof. The "only if" part. If x is a DEB, then from Definition of the DEB every proper suffix of x belongs to L. Using Proposition 3.4.2 we must have

$$x > KS.$$

Combining this inequality with $x\bar{x} \le KS$ leads to the conclusion that $\bar{x}$ is a prefix of KS. Since $x > \bar{x}$, $\bar{x}$ is even.

If $\bar{x}$ has a proper even PS, that is, a proper suffix, say $\bar{v}$, which is also an even prefix of $\bar{x}$ (and of KS), then v is a suffix of the DEB x. Since v is odd and $v > \bar{v}$, we know that $v \in L'$, a contradiction to the definition of DEB.

The "if" part. Since x is odd and $x > \bar{x}$, we have $x > KS$ and $x \in L'$. In order to prove that x is a DEB of L, it is enough to consider if every proper prefix and proper suffix of x is a word of L.

From the fact that $\bar{x}$ is a prefix of KS, its every proper prefix belongs to L. As to suffixes, assuming the contrary that x has a proper suffix $v \in L'$, then we have

$$v\bar{x} \in L \text{ and } v > KS.$$

This implies that $\bar{v}$ is an even prefix of KS, and hence $\bar{x}$ has an even PS $\bar{v}$ with respect to KS, a contradiction. ∎

Using Theorem 1.2.10 we obtain the following Corollary. It characterize those strings which can be DEB's for some kneading sequences of unimodal maps.

Corollary 7.1.2 A string x is a DEB for a kneading sequences KS if and only if its dual string $\bar{x}$ is KS's even prefix that satisfies property (P3) of Definition 1.2.7, that is to say, $\bar{x}$ is either maximal primitive or $\bar{x} = yy$ where y is an odd maximal primitive string.

Remark. As pointed out in the proof of Proposition 5.1.1, from Proposition 1.2.4 we know that if a prefix of KS is of the form yy where y is odd, then we can only have $KS = y^\infty$, and $y\bar{y}$ is the longest DEB for $\mathcal{L}(y^\infty)$. Thus except this special case each DEB is dual to an even maximal primitive prefix of KS and vice versa.

7.1.2 Finding the Next DEB

Here we present a result that allows us to find the next longer DEB if a DEB is already given for a kneading sequence KS.

Proposition 7.1.3 Let KS be a given kneading sequence, $\bar{x}$ a DEB of the $L = \mathcal{L}(KS)$, and xy ($|y| > 0$) a prefix of KS. $x\bar{y}$ is the next longer DEB after $\bar{x}$ if and only if $\bar{y}$ is a prefix of KS.

Proof. The "if" part is already contained in the proof of Proposition 5.1.1.

The "only if" part is similar to the proof of Proposition 5.1.4. Suppose $x\bar{y}$ is the next longer DEB after $\bar{x}$. Thus we have at least two DEB's. Since we have $L'' = \emptyset$ for $L = \mathcal{L}(10^\infty)$, $L'' = \{1\}$ for $L = \mathcal{L}(0^\infty)$, and $L'' = \{10\}$ for $L = \mathcal{L}(1^\infty)$, hence the kneading sequence KS must begin from 10^n1 for some $n > 0$. Observe the suffix s after x in $KS = xs$. If s begins from 0, then $\overline{x0} = x1$ is the next DEB as desired. Otherwise, s begins from 1. Search for a prefix z of s such that $\bar{z}$ is a prefix of KS. If there exists no such z, then we have $KS = x^\infty$, and $\bar{x}$ is the longest DEB of L which contradicts the existence of a longer DEB $x\bar{y}$. Use the established sufficient part, $x\bar{z}$ is the next longer DEB after $\bar{x}$, and we have $y = z$. ∎

Using this result it is very easy to find DEB's for a given KS. If a KS begins from 10^n1, then the shortest DEB is 10^{n+1}, and then using Proposition 7.1.3 to find others. For example, if $KS = 10111111\cdots$ then

$$L'' = \{100, 10110, 1011110, 101111110, \ldots\};$$

if $KS = 100100\cdots$ then

$$L'' = \{1000, 10011, 100101\}.$$

Finally, we note that the lengths of DEB's determine the kneading sequence KS and the language $\mathcal{L}(KS)$ uniquely. Let N_L be an integer sequence $\{n_1, \ldots, n_k, \ldots\}$, where n_k is the length of the k-th DEB of L. We call N_L the *length set* of DEB's. By Proposition 7.1.3 the following claim is obviously true.

Proposition 7.1.4 For $L = \mathcal{L}(KS)$ the sets L'' and L are completely determined by N_L, the length set of DEB's.

7.2 Structure of Regular L''

7.2.1 Finite Complement Languages $\mathcal{L}(KS)$

Recall the following facts about finite complement languages.

A dynamical language L is finite complement if L'', the set of all DEB's of L, is finite (see Definition 2.2.2). We know that each finite complement language is regular (see Remark after Lemma 2.2.6). From Definition 2.3.6 a symbolic flow is a subshift of finite type if its associate language is finite complement. From Theorem 3.3.7 each $L = \mathcal{L}(KS)$ is dynamical and Y, the set of all admissible sequences with respect to KS, is the symbolic flow corresponding to L.

In this subsection we discuss how to characterize a finite complement language $\mathcal{L}(KS)$ through the kneading sequence KS. The following result answers the question completely.

Theorem 7.2.1 The language $L = \mathcal{L}(KS)$ is finite complement if and only if KS is either periodic or 10^∞.

Proof. The "only if" part is a simple consequence of Proposition 7.1.3. As a fact, if the longest DEB is denoted by $\bar{x}$, then x is a prefix of KS, and $\sigma^{|x|}(KS) = KS$. Thus $KS = x^\infty$.

Conversely, if KS is 10^∞, then $L = \Sigma^*$ and $L' = \emptyset$. If $KS = x^\infty$, without loss of generality we can assume that x is maximal and primitive. If x is even, then by Corollary 1.2.11 $\bar{x}$ is a DEB. Using Proposition 7.1.3 we know that there exists no longer DEB. If x is odd, then we can use Theorem 1.2.10 instead and obtain the conclusion that $x\bar{x}$ is the longest DEB. ∎

Remark. The Theorem 7.2.1 was proved in Xie (1995a) without using Proposition 7.1.3.

7.2.2 *Semilinear Structure of Regular L''*

From results of Chapter 4 and Lemma 2.2.6 we know that for $L = \mathcal{L}(KS)$, the set L'' is regular if and only if the kneading sequence KS is eventually periodic. From Theorem 7.2.1 we know that if KS is eventually periodic, but neither 10^∞ nor periodic, then its L'' is infinite.

Here we show that for such regular languages the set L'' has a special structure.

Recall the concept of linear and semilinear integer sets introduced in Definition B.2.14. A set of integers is said to be linear if it is of the form $\{c+pi \mid i = 0, 1, \ldots\}$, and a set of integers is said to be semilinear if it is a finite union of linear sets. Note that every finite set of integer can be seen as a semilinear set. For convenience we also say the empty set $\emptyset$ is semilinear.

Lemma 7.2.2 If $KS = \rho\lambda^\infty$ is a eventually periodic kneading sequence, where λ is even, and $x = \rho\overline{\lambda\mu}$ is a DEB of $\mathcal{L}(KS)$, where μ is a proper prefix of λ, then $\rho\overline{\lambda^2\mu}$ is also a DEB.

Proof. Since the form of $KS = \rho\lambda^\infty$ is not unique, it suffices to discuss the special case of $\mu = \varepsilon$. As a matter of fact, for general case we can write

$$KS = \rho\mu(\sigma^{|\mu|}(\lambda))^\infty = \rho'\lambda'^\infty,$$

where $\rho' = \rho\mu$ and $\lambda = \sigma^{|\mu|}(\lambda)$. Rewriting ρ' and λ' by ρ and λ, we need only to show that if $\rho\overline{\lambda}$ is a DEB, then $\rho\lambda\overline{\lambda}$ is also a DEB. Then it is a simple consequence by applying Proposition 7.1.1 and the condition of λ being even. ∎

Theorem 7.2.3 The language $L = \mathcal{L}(KS)$ is regular if and only if N_l, the length set of DEB's, is semilinear.

Proof. The "only of" part. If KS is either periodic or 10^∞, then by Theorem 7.2.1 the set L'' is finite, and hence semilinear.

Now suppose $KS = \rho\lambda^\infty$ is eventually periodic but neither 10^∞ nor periodic. Without loss of generality, assume λ is an even string. By Theorem 7.2.1 the set L'' is infinite, hence we can consider a DEB x that satisfies the condition $|x| \geq |\rho\lambda|$.

The conclusion of Lemma 7.2.2 implies that each DEB x with the condition of length $|x| \geq |\rho\lambda|$ will generate a family of DEB's whose length set is linear. Thus our claim is true.

The "if" part. Let N_l be semilinear, and n_k the length of the n-th DEB of L. Consider another sequence $\{n_k - n_{k-1}\}$. Using Proposition 7.1.3 we can use this sequence $\{n_k - n_{k-1}\}$ to reproduce the kneading sequence KS. By the structure of semilinearity, this sequence of integers is eventually periodic, and hence KS is an eventually periodic kneading sequence. By Theorem 4.2.1 $L = \mathcal{L}(KS)$ is regular. ∎

Example 7.2.4 We list the results for period 3 kneading sequences. There are three cases.

If $KS = (101)^\infty$ or $(10c)^\infty$, then $x = 101$, and the unique DEB is 100. As a fact, for $L = \mathcal{L}((101)^\infty)$, its $L'' = \{100\}$..

If $KS = (100)^\infty$, then we have to take $x = 100100$, and the longest DEB is 100101. For $L = \mathcal{L}((100)^\infty)$, we have $L'' = \{1000, 10011, 100101\}$.

7.2.3 *Calculation of Regular L''*

As a matter of fact, we can easily determine all DEB's for regular languages generated from unimodal maps. If L'' is finite, then we can use Proposition 7.1.3 to find all DEB's one by one. From Theorem 7.2.1 we know that if $KS = x^\infty$ with x satisfying the property (P3) (of Definition 1.2.7), then $\bar{x}$ is the longest DEB.

In this subsection we will calculate those L'' that are regular and infinite. Here we call A, a set of strings, a *linear set of strings* if the corresponding length set of integers is linear; and a set of strings *semilinear* if it is the finite union of linear sets. Evidently, each finite set of strings is semilinear. Of course, what interests us is the case of $p \neq 0$, that is, the semilinear set is infinite.

Theorem 7.2.5 If a kneading sequence is eventually periodic but neither periodic nor 10^∞, then L'', the set of all distinct excluded blocks, has a special semilinear infinite set. More specifically, if this KS is written in the form of

$$KS = \rho\lambda^\infty,$$

where λ is even and minimal (in length), then

$$L'' = F \cup G,$$

where F is a finite set, and

$$G = \cup_{j=1}^{q} \{\overline{\rho\lambda^i\lambda'_j}\}_{i\geq n}$$

for an integer n and some proper prefixes of λ, $\lambda'_1, \cdots, \lambda'_q$ with the following properties:

(1) $0 \leq |\lambda'_1| < |\lambda'_2| < \cdots < |\lambda'_q| < |\lambda|$,

(2) if $n > 0$, then any $\overline{\rho\lambda^{n-1}\lambda'_j}$ for $1 \leq j \leq q$ is not a distinct excluded block,

(3) $|x| < \min\{|y|, |\rho\lambda|\}$ for all $x \in F$ and $y \in G$.

By the conclusion of this Theorem, we call every set of the form $\{\overline{p\lambda^i\lambda'}\}_{i\geq n}$, where λ' is a proper prefix of λ and $n \geq 0$, a *linear set* (or *linear family*) of DEB's. On the other hand, every element of F is called an *isolated* DEB.

Remark. Here we explain the reason that why in Lemma 7.2.2 (and Theorem 7.2.5) we have to take the string λ (in $KS = \rho\lambda^\infty$) being even.

As a fact, if λ is odd and $n \geq k$ $(= [|\rho|/|\lambda|]+2)$, then $\overline{\rho\lambda^n\lambda'}$ is a DEB $\Rightarrow$ $\overline{\rho\lambda^{n+1}\lambda'} \in L$. Its proof is simple. Since $\rho\lambda^{n+1}\lambda'$ is odd, if $\overline{\rho\lambda^{n+1}\lambda'} \notin L$, then it must have a proper suffix being a DEB. Using the condition that $\rho\lambda^n\lambda'$ has no even PS and Lemmas 4.4.1 and 4.4.2 $\rho\lambda^{n+1}\lambda'$ also cannot have even PS. This contradicts the Proposition 7.1.1.

Lemma 7.2.6 If $KS = \rho\lambda^\infty$, where λ is even, and λ' is a proper prefix of λ, $\lambda = \lambda'\lambda''$, then $\overline{\rho\lambda^n\lambda'}$ is a DEB for each $n \geq k = [|\rho|/|\lambda|] + 2$ if and only if λ' satisfies the conditions:

(1) ρ and λ' are of the same parity,

(2) $\lambda''\lambda'$ has no even PS with respect to KS.

Proof. The "only if" part. If $\overline{\rho\lambda^k\lambda'}$ is a DEB, then $\rho\lambda'$ is obviously even. Since $\rho\lambda^k\lambda'$ has no even PS, so it is also true for its suffix $\lambda''\lambda'$.

The "if" part. Using the conditions (1), (2) and Lemmas 4.4.1 and 4.4.2, the conclusion is true by Proposition 7.1.1. ∎

Proof of Theorem 7.2.5. At first we write KS in the form $KS = \rho\lambda^\infty$, and make the string λ even and minimal in length, that is, either λ is primitive and even or $\lambda = yy$ where y is primitive and odd. We can also suppose the strings ρ and λ having different last symbols. Since KS is not periodic, this is always possible.

Now taking all proper prefix of λ and using Lemma 7.2.6 as the rule to obtain a set

$$\Lambda = \{\lambda'_1, \cdots, \lambda'_q\},$$

where $0 \le |\lambda'_1| < |\lambda'_2| < \cdots < |\lambda'_q| < |\lambda|$. By Lemma 7.2.2 we can determine integers $n_1, \cdots, n_q$, such that every family $\{\overline{\rho\lambda^i\lambda'_j}\}_{i \ge n_j}$ for $j = 1, \cdots, q$ is a linear set of DEB's. We can require that every n_j is the least possible value, that is to say, if $n_j > 0$, then $\overline{\rho\lambda^{n_j-1}\lambda'_j}$ is not a DEB.

Using these considerations and Lemma 7.2.2 and 7.2.6, L'', the set of all DEB's of L, can be written as $L'' = F \cup G$, where F consists of all isolated DEB's which is of length less than $|\rho\lambda|$, and G consists of q linear sets. It is important to note that this decomposition of L'' is independent of any special way of writing KS as $\rho\lambda^\infty$.

Using Proposition 7.1.1, it is easy to show that $0 \le n_q \le n_{q-1} \le \cdots \le n_1 (\le k)$ and $0 \le n_1 - n_q \le 1$.

For instance, using Proposition 7.1.1 to $\overline{\rho\lambda^{n_1}\lambda'_1}$ and $\overline{\rho\lambda^{n_2}\lambda'_2}$ and the fact of $|\lambda'_1| < |\lambda'_2|$, the string $\rho\lambda^{n_1}\lambda'_2$ is even and has no even PS with respect to KS, and this shows that $\overline{\rho\lambda^{n_1}\lambda'_2}$ is a DEB. Thus we have $n_1 \ge n_2$. As for the inequality $n_1 \le n_q + 1$, we can use the inequality $|\lambda'_q| < |\lambda\lambda'_1|$ instead.

These inequalities mean that the difference between lengths of any two members of the first elements of linear sets, $\rho\lambda^{n_1}\lambda'_1, \cdots, \rho\lambda^{n_q}\lambda'_q$, is less than $|\lambda|$. Therefore, if we select a proper way to rewrite KS, for instance to let $\bar{\rho}$ equal to the shortest element of G, then the conditions (1) and (2) in Theorem 7.2.5 can be satisfied for some integer n. (We can then make $\lambda'_1 = \varepsilon$ and, consequently, ρ is even.)

Finally, if $x \in F, y \in G$ and $|x| > |y|$ happens, then we can write $y = \overline{\rho\lambda'}, x = \overline{\rho\lambda''}$, where λ' and λ'' are proper prefixes of λ and $|\lambda'| < |\lambda''|$. Since $y \in G$ means that $\overline{\rho\lambda^i\lambda'} \in G$ for each $i \ge 0$, using Proposition 7.1.1 again we see that $\overline{\rho\lambda^i\lambda''}$ are also DEB's for each $i \ge 0$. This contradicts the fact that $x \in F$ is an isolated DEB. ∎

Example 7.2.7 $KS = 10110(11011)^\infty$

Using Lemma 7.2.6, there are two odd prefixes of λ, $\lambda'_1 = 1$ and $\lambda'_2 = 1101$, but since $11101(= \lambda''_2\lambda'_2)$ has an even PS 101, so that we obtain $\Lambda = \{1\}$. It is easy to obtain F by Proposition 7.1.1 and

$$L'' = \{100\} \cup \{10110(11011)^n 0\}_{n \ge 1}.$$

Example 7.2.8 $KS = 10110(10111)^\infty$

Using Proposition 7.1.1 and Lemma 7.2.6, we obtain

$$\begin{aligned} L'' = \{100\} &\cup \{10110(10111)^n0\}_{n\geq 1} \cup \{10110(10111)^n11\}_{n\geq 0} \\ &\cup \{10110(10111)^n1010\}_{n\geq 0}. \end{aligned}$$

We can also rewrite it as that:

$$\begin{aligned} L'' = \{100\} &\cup \{101101(01111)^n1\}_{n\geq 0} \cup \{101101(01111)^n010\}_{n\geq 0} \\ &\cup \{101101(01111)^n01110\}_{n\geq 0}. \end{aligned}$$

Example 7.2.9 $KS = 10^31(101)^m10^2110^2(011)^\infty$, where m is a fixed integer.

From Proposition 7.1.1 we obtain

$$F = \{10^4, 10^31100, 10^31(101)100, \cdots, 10^31(101)^{m-1}100, 10^31(101)^m10^3\},$$

where card $F = m + 2$. Using Lemma 7.2.6 we obtain

$$G = \{10^31(101)^m10^2110^2(011)^i00\}_{i\geq m+1}.$$

7.3 Complexity Levels of L and L''

As the results obtained above are with $\mathcal{L}$(RGL), that is, the class of regular languages, in this section we will study the higher levels of L and L'' in Chomsky hierarchy for $L = \mathcal{L}(KS)$ and continue the discussion in Section 2.2.

For the next level of CFL (the class of context-free language), however, we have proposed Conjecture 1 in Section 2.2 that L is context-free if and only if L'' is context-free. We do not know if this conjecture is correct or not. But for languages generated from unimodal maps we have a partial answer.

Proposition 7.3.1 Let $L = \mathcal{L}(KS)$. If L'' is a context-free language, then L'' is a regular language.

Proof. Let $L'' = L(G)$, where G is a context-free grammar (CFG) (see Definition B.2.1 of Subsection *B.2.1*):

$$G = (V, S, P, s_0),$$

where V is an alphabet of variables (nonterminals), S is an alphabet of terminals, P is a set of grammar rules, and s_0 is an initial symbol. Without loss of generality, we can assume that the grammar is reduced: (i) for each variable $x \neq s_0$, s_0 generates a string containing x; and (ii) x generates a string ($\in L(G)$) (see, e.g., Salomaa 1973).

Here we need the Theorem B.2.13: a context-free language $L(G)$ is regular if and only if it is not self-embedding.

First recall from Definition B.2.12 that a CFG is self-embedding if and only if $A \stackrel{*}{\Rightarrow} pAq$ for some $A \in V$ and $p, q \in (V \cup S)^*$ such that $p \neq \varepsilon$ and $q \neq \varepsilon$. A CFL L is self-embedding if and only if all CFG generating L are self-embedding.

If the grammar G for $L'' = L(G)$ is not self-embedding then our proof is completed. Assume the contrary, that this G is self-embedding. Then there is a variable A such that

$$A \stackrel{*}{\Rightarrow} pAq,$$

and since G is reduced, and $\varepsilon \notin L''$ we have

$$p \neq \varepsilon, q \neq \varepsilon, \text{ and } p, q \in S^*$$

Again using the fact that G is reduced, we obtain

$$s_0 \stackrel{*}{\Rightarrow} wAx$$

and these strings w and x have the same properties as p, q. Now we can write

$$s_0 \stackrel{*}{\Rightarrow} wp^nAq^nx \in L'' \quad \forall n \geq 0.$$

Using Proposition 7.1.1 we have

$$KS = wp^\infty$$

Finally, using Theorem 4.2.1 of Chapter 4 completes our proof. ▮

Using Theorem B.2.15 and Proposition 7.1.3 we can give another proof for Proposition 7.3.1.

Second proof of Proposition 7.3.1. Let a homomorphism h on $\{0, 1\}^*$ is defined by

$$h = (1 \to 0, 0 \to 0).$$

If L'' is a CFL, then $h(L'')$ is also a CFL. Using Theorem B.2.15, if a language over one symbol is a CFL, then the length set of all words of this language is semilinear. By Theorem 7.2.3 the length set of all DEB's of L is semilinear, and hence $h(L'')$ is regular. Since $\mathcal{L}$(CFL) is closed under inverse homomorphism (see Subsection C.2.2), the proof is completed. ▮

Remark. Thus for languages L generated from unimodal maps,

$$L'' \text{ is CFL} \Longrightarrow L \text{ is CFL}$$

is true.

Another Conjecture that is weaker than Conjecture 1 is proposed as follows. If it is true, then the Conjecture 1 is true for languages generated from unimodal maps, and vice versa.

Conjecture 2. If $L = \mathcal{L}(KS)$ is a context-free language, then it is also a regular language.

Proposition 7.3.2 Let $L = \mathcal{L}(KS)$. If its L'' is a recursively enumerable language, then both L and L'' are recursive languages.

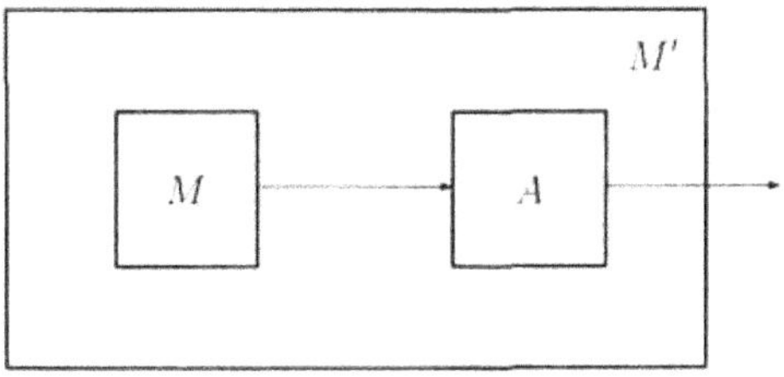

Figure 7.1: The Turing machine generating L'' in Proposition 7.3.2

Proof. Recalling the contents of Theorems B.1.3 and B.1.6, we know that there exists a Turing machine (TM) M as an enumerator or generator of L''. What we have to do is to construct a new TM M' for L'' such that its strings can be generated in order of increasing size.

This TM can be designed as follows (see Figure 7.1).

For each DEB x generated from M, the processor A can calculate all DEB's whose length is less than x and print out all new DEB's (if there is any) in the order of increasing size. This is possible by Proposition 7.1.1. At the same time the length of x is recorded and used to remember that all DEBs of length up to this number have been output already. It is obvious that the TM M', combined from M and A, is the required one. Using Proposition 2.2.8 completes our proof. ∎

A consequence of this result is: if $L = \mathcal{L}(KS)$ is a nonREL, then its L'' is also a nonREL.

Here is another conjecture.

Conjecture 3. If $L = \mathcal{L}(KS)$ is a recursively enumerable language, then it is also a recursive language.

Now we discuss what is meant when we say that a kneading sequence KS is given. From the viewpoint of the theory of computability (see, e.g., Pour-El and Richards 1989), we give the following Definition. Here we need an intuitive understanding about the notion of recursive function. Its rigorous definition can be found in Hopcroft and Ullman (1979).

Definition 7.3.3 A kneading sequence $KS = e_1 \ldots e_n \ldots$ is *computable* if the function

$$e : \mathbf{N} \longrightarrow S = \{0, 1\}$$
$$n \longmapsto e(n) = e_n$$

is recursive, that is to say, it can be computed by a Turing machine for all values of its arguments.

Using the same idea as in the proof of Proposition 7.3.2, it is easy to obtain the following Proposition.

Proposition 7.3.4 The language $L = \mathcal{L}(KS)$ is recursive if and only if its KS is computable.

7.4 DEB of Non-Regular Languages

If KS is a given non-periodic kneading sequence, and $L = \mathcal{L}(KS)$ the language decided by KS, then by applying Propositions 7.1.2 and 5.1.2 to this KS, the non-regular language L'', the set of all DEB's of L, can be directly obtained from all even maximal primitive prefixes of KS. From Theorem 7.2.3 we know that L'' is not semilinear. Then what is the structure of non-regular L''?

Until now there is not many results known about L'' for non-regular languages $L = \mathcal{L}(KS)$ generated from unimodal maps. In this section we will determine L'' for language of Feigenbaum attractor of Section 6.1, and for even and odd Fibonacci languages of Section 6.4. Finally, we will calculate the distribution functions of DEB for languages discussed already.

7.4.1 DEB of Feigenbaum Attractor

This calculation is not difficult. The tools used are Proposition 7.1.1 and properties of t_n in Lemma 6.1.1. Here the homomorphism h is defined by $h = (1 \to 10, 0 \to 11)$ as in (6.1).

Proposition 7.4.1 The set of all DEB's of the language $L = \mathcal{L}(t_\infty)$ is given by

$$L'' = \{h^n(100)\}_{n\geq 0} \cup \{h^n(10110)\}_{n\geq 0}.$$

Proof It is easy to verify that each element of either $\{h^n(100)\}$ or $\{h^n(10110)\}$ is a DEB of L. For example, consider an element of the first family $h^n(100) = t_n\bar{t}_n\bar{t}_n$. By Proposition 7.1.1 we need to verify that $t_n\bar{t}_nt_n$ is an even prefix of t_∞, and has no even proper PS with respect to t_∞. Since t_∞ has $t_{n+2} = t_n\bar{t}_nt_nt_n$ as its prefix, the former point is obviously true. As to the latter point, using property (5) of t_n in Lemma 6.1.1 that t_n is maximal and primitive, it is also an easy exercise to obtain this conclusion. The discussion for the second family is quite the same and omitted.

Now suppose $x \in L''$. If the length of x is an odd number, then since $\bar{x}$ is a prefix of t_∞, and each e_{2k-1} in $t_\infty = e_1 \cdots e_n \cdots$ is 1, we can write this DEB into the form

$$x = t_n\alpha 0,$$

where the integer n is determined by $2^n = |t_n| < |x| < 2^{n+1} = |t_{n+1}|$. Note that since t_n is odd, the length of a DEB cannot be a power of 2.

If $|\alpha 0| > 1$, we can find an integer i such that $\alpha = t^i\alpha'$ and $|\alpha'| < 2^i$. Since all x, t_n, t_i are odd, $\alpha'0$ is odd. On the other hand, α and α' are prefixes of t_∞. Since $|\alpha'|$ is even, $\alpha'1$ is also a prefix of t_∞. From $\alpha'0 > \alpha'1$ we have $\alpha'0 \notin L$. This contradicts the assumption of x being a DEB.

If $\alpha = \epsilon$, then $x = t_n0$. If n is odd, then it can be shown that each t_n with odd n has 10 as suffix, hence x has 100 as suffix. From $100 \in L''$ we can only have $x = 100$. If n is even, then it can also be shown that each t_n with even $n \geq 2$ has 1011 as suffix, and similarly we obtain $x = 10110$. Thus for the case of $|x|$ being odd we obtain only two DEB's: 100 and 10110.

If $x \in L''$ and $|x|$ is even, then $\bar{x}$ is a prefix of t_∞ and $h^{-1}(x)$ is well defined. Since the homomorphism h preserves both the parity and order of strings, by Proposition 7.1.1 it is evident that $h^{-1}(x)$ is also a DEB. If its length is even, proceed on similarly, until we find an integer n such that $h^{-n}(x)$ is a DEB and of odd length, then we know that it is either 100 or 10110, and this is the conclusion desired. ∎

Therefore, the set L'' for the language of Feigenbaum attractor consists of two families, each of them is generated by the homomorphism h. Since the lengths of strings of each family is a geometric progression, by Theorem 7.2.5 and Lemma 2.2.6 we know that both L and L'' are non-regular. By Proposition 7.3.1 or Theorem B.2.15 this L'' is not a context-free language, but a context-sensitive language.

7.4.2 DEB of Even Fibonacci Languages

Let α and β satisfy the conditions in Theorem 6.4.2 that both α, β are even, $\alpha, \alpha\beta$ and $\alpha\alpha\beta$ are maximal and primitive, and $\alpha > \beta$. By using the homomorphism

$$h = (\alpha \to \alpha\alpha\beta, \beta \to \alpha\beta)$$

we obtain an even Fibonacci kneading sequence

$$KS = \lim_{n\to\infty} h^n(\alpha).$$

We also know that all strings $h^n(\alpha)$ are maximal and primitive.

By Proposition 7.1.2 it is easy to know that each $h^n(\alpha)$ is dual to a DEB.

Let l_n be the number of DEB's whose lengths does not exceed $h^n(\alpha)$, that is,

$$l_n = \text{card}\{y \mid y \in L'', |y| \le |h^n(\alpha)|\}.$$

Proposition 7.4.2 l_n is a linear function of n.

Proof. Observe the expression

$$\begin{aligned} h^{n+1}(\alpha) &= h^n(\alpha)h^n(\alpha)h^n(\beta) \\ &= h^n(\alpha)h^n(\alpha)h^{n-1}(\alpha)\cdots h(\alpha)\alpha\beta, \end{aligned}$$

and compare with

$$\begin{aligned} KS &= h^{n+1}(\alpha)\cdots \\ &= h^n(\alpha)h^n(\alpha)\cdots \\ &= h^n(\alpha)h^{n-1}(\alpha)\cdots h(\alpha)\alpha\alpha\beta. \end{aligned}$$

By Proposition 7.1.3 and the fact that $\alpha \ne \beta$, it is clear that each dual string of DEB whose length lying between $|h^n(\alpha)|$ and $|h^{n+1}(\alpha)|$ can only be of the form

$$x = h^n(\alpha)h^n(\alpha)h^{n-1}(\alpha)\cdots h(\alpha)\alpha u$$

where u is an even prefix of β. From Proposition 7.1.1 we see that $\bar{x}$ is a DEB if and only if u has no even PS to $KS = \alpha\alpha\beta\cdots$. If we denote the number of such strings u by c, a constant, then it is determined completely by β. Therefore, we have proved already

$$l_{n+1} - l_n = c$$

and it follows our conclusion. ∎

Corollary 7.4.3 For even Fibonacci language L the set $L'' = F \cup G$ where

$$F = \{x \mid \bar{x} \text{ is an even maximal primitive prefix of } \alpha\},$$
$$G = \{h^n(\alpha)h^n(\alpha)h^{n-1}(\alpha)\cdots h(\alpha)\alpha\bar{u} \mid n > 0, u \text{ is an even prefix of } \beta \text{ and has no even PS to } \alpha\alpha\beta\}.$$

A simple example is as follows.

Example 7.4.4 Recalling Example 6.4.4, denote the strings $m_0 = 11$, $m_1 = 101$ by

$$\alpha = 101, \ \beta = 11,$$

then the homomorphism

$$h = (101 \to 10110111, 11 \to 10111)$$

generate sequence $\{s_n\} = h^n(\alpha)$ and hence the even Fibonacci kneading sequence

$$s = \lim_{n\to\infty} s_n = 10110111, 10110111, 10111, \cdots$$

By the Proposition above the string $u = 11$ is unique. Thus the set of all DEB's of the language $L = \mathcal{L}(s_\infty)$ is given by

$$L'' = \{\bar{s}_n\}_{n\geq 0} = \{100, 10110110, \ldots\}.$$

7.4.3 DEB of Odd Fibonacci Languages

Let α and β satisfy the conditions in Theorem 6.4.1 that α is odd, β is even, $\alpha, \alpha\beta$ and $\alpha\beta\alpha$ are maximal and primitive, and $\alpha > \beta$. By using a homomorphism

$$h = (\alpha \to \alpha\beta\alpha\alpha\beta, \beta \to \alpha\alpha\beta),$$

we obtain an odd Fibonacci kneading sequence

$$KS = \lim_{n\to\infty} h^n(\alpha).$$

We also know that all strings $h^n(\alpha)$, $h^n(\alpha\beta)$, and $h^n(\alpha\beta\alpha)$ are maximal and primitive.

Let l_n be the number of DEB's whose lengths does not exceed $h^n(\alpha)$, that is,

$$l_n = \text{card}\{y \mid y \in L'', |y| \leq |h^n(\alpha)|\}.$$

We will prove that l_n satisfies a difference equation of order 3 and has a same asymptotic behavior for all odd Fibonacci languages.

Proposition 7.4.5 $l_n \sim An^\lambda$ for $\lambda = 1 + \sqrt{2}$.

In the proof of this result we need two sets M_n and P_n defined by

$M_n = \{x \mid \overline{x} \in L'', x = h^n(\alpha\beta)u$ where u is an odd prefix of $h^n(\alpha)\}$,
$P_n = \{v \mid v$ is an odd prefix of $h^n(\alpha\beta)$, and $h^n(\alpha)v$ has no even PS to $KS\}$,

and their cardinal numbers

$m_n = \text{card} M_n$ and $p_n = \text{card} P_n$.

Lemma 7.4.6 The relation

$$l_{n+1} - l_n = m_n \text{ holds for every } n \geq 0. \tag{7.1}$$

Proof. We need to show that if $\overline{x} \in L''$ and $|h^n(\alpha)| < |x| \leq |h^{n+1}(\alpha)|$, then $x \in M_n$, which is equivalent to

$$|h^n(\alpha\beta)| < |x| \leq |h^n(\alpha\beta\alpha)| \tag{7.2}$$

Assume the first inequality of (7.2 does not hold, that is, $|x| \leq |h^n(\alpha\beta)|$. Writing

$$h^n(\alpha\beta) = h^{n-1}(\alpha\beta\alpha)h^n(\alpha),$$

and using Propositions 7.1.1 and 7.1.3, we see that $h^{n-1}(\alpha\beta\alpha)$ is dual to an DEB, and there exists no longer DEB that can be obtain from this expression. Thus $|x| \leq |h^{n-1}(\alpha\beta\alpha)| < |h^n(\alpha)|$, a contradiction.

The second inequality of (7.2) can be obtained by writing

$$h^{n+1}(\alpha) = h^n(\alpha\beta\alpha)h^n(\alpha\beta),$$

and using the same argument above. ∎

Lemma 7.4.7 The relation

$$m_n = 3p_{n-2} + p_{n-3} \text{ holds for every } n \geq 3. \tag{7.3}$$

Proof. We will establish the following equality

$$M_n = M_n^1 \cup M_n^2 \cup M_n^3 \cup M_n^4$$

for every $n \geq 3$, where

$$M_n^1 = \{h^n(\alpha\beta)h^{n-1}(\alpha)h^{n-2}(\alpha)v \mid v \in P_{n-2}\},$$
$$M_n^2 = \{h^n(\alpha\beta)h^{n-1}(\alpha\beta\alpha)h^{n-2}(\alpha)h^{n-3}(\alpha)v' \mid v' \in P_{n-3}\},$$
$$M_n^3 = \{h^n(\alpha\beta)h^{n-1}(\alpha\beta\alpha)h^{n-2}(\alpha\beta\alpha)v \mid v \in P_{n-2}\},$$
$$M_n^4 = \{h^n(\alpha\beta)h^{n-1}(\alpha\beta\alpha\alpha)h^{n-2}(\alpha)v \mid v \in P_{n-2}\},$$

and hence obtain the conclusion desired.

Here we only give the detail for M_n^1. The proof for others are quite similar and omitted. Observe the expression

$$h^n(\alpha\beta\alpha) = h^{n-1}(\alpha\beta\alpha)h^n(\alpha)h^{n-1}(\alpha)h^{n-2}(\alpha)h^{n-2}(\alpha\beta)\cdots,$$

and compare it with

$$\begin{aligned} KS &= h^{n+1}(\alpha) \\ &= h^n(\alpha)h^{n-1}(\alpha)h^{n-2}(\alpha)h^{n-2}(\beta\alpha)\cdots \end{aligned}$$

Here both $h^{n-1}(\alpha\beta\alpha)$ and $h^n(\alpha\beta\alpha)$ are dual to DEB's. By $\alpha\beta \neq \beta\alpha$ and Proposition 7.1.3 there must exist some dual strings of DEB's whose lengths between those of $h^{n-1}(\alpha\beta\alpha)$ and $h^n(\alpha\beta\alpha)$ and of the form $x = h^n(\alpha\beta)h^{n-1}(\alpha)h^{n-2}(\alpha)v$ where v is an odd prefix of $h^{n-2}(\alpha\beta)$.

We need to establish

$$x = h^n(\alpha\beta)h^{n-1}(\alpha)h^{n-2}(\alpha)v \in M_n \iff v \in P_{n-2}.$$

By Proposition 7.1.1 the proof of "$\Rightarrow$" is obvious. Consider its converse. If $v \in P_{n-2}$, then we should show there exists no even proper PS of x to KS. Rewrite x into the form

$$x = y^3 h^{n-2}(\alpha)v, \text{ where } y = h^{n-1}(\alpha\beta\alpha)$$

and observe any even proper PS of x. Since $v \in P_{n-2}$ and y is even maximal primitive, we need only to examine two suffixes of x as follows:

$$yh^{n-2}(\alpha)v \text{ and } yyh^{n-2}(\alpha)v.$$

Since $KS = y^3\cdots$, if any of them is a prefix of KS, then $h^{n-2}(\alpha)v$ itself would become an even prefix of KS, a contradiction with $v \in P_{n-2}$. This finishes the proof for M_n^1.

Taking $v = h^n(\alpha)$ in M_n^1, we obtain the longest dual string in M_n^1 for DEB's. Thus we can proceed on by using Proposition 7.1.3 to discuss M_n^2 etc. ∎

Lemma 7.4.8 The relation

$$p_n - p_{n-1} = m_n \text{ holds for every } n > 0. \tag{7.4}$$

Proof. If $v \in P_n$, then there are two cases to be discussed.

(1) $|v| \le |h^n(\alpha)|$. Since then we can write $h^n(\alpha\beta)v = h^{n-1}(\alpha\beta\alpha)h^n(\alpha)v$. It is simple to show that it is an even maximal primitive prefix of KS, thus a dual string of DEB by Proposition 7.1.2. Therefore, each v of this case corresponds an element of M_n. The converse holds too.

(2) $|v| > |h^n(\alpha)|$. By the identity

$$h^n(\alpha\beta) = h^{n-1}(\alpha)h^{n-1}(\alpha)h^{n-1}(\alpha\beta),$$

in this case we must have $|v| > |h^n(\alpha)h^{n-1}(\alpha)|$ if $v \in P_n$. Writing $v = h^n(\alpha)h^{n-1}(\alpha)v'$, it is trivial to verify that $v \in P_n \Leftrightarrow v' \in P_{n-1}$. ∎

Proof of Proposition 7.4.5. Combining the equalities (7.3) and (7.4) together, we obtain a homogeneous difference equation of third order for p_n.

$$p_n - p_{n-1} - 3p_{n-2} - p_{n-3} = 0.$$

From the equalities (7.1) and (7.4) we see that $l_{n+1} - p_n = \text{const}$, and hence l_n satisfies the corresponding nonhomogeneous equation with a constant at the right-hand side. From their characteristic equation

$$\lambda^3 - \lambda^2 - 3\lambda - 1 = (\lambda + 1)(\lambda^2 - 2\lambda - 1)$$

we obtain the required value $\lambda = 1 + \sqrt{2}$ for each odd Fibonacci languages. ∎

7.4.4 Distribution of DEB

If we use $\pi(n)$ to denote the number of DEB's whose lengths do not exceed n, then we can compare the different situations of distribution of DEB for several languages.

For the regular languages generated brom unimodal maps, there are three cases. For $KS = 10^\infty$ we have $\pi(n) \equiv 0$. For periodic kneading sequences, $\pi(n) = 0$ holds for every large enough n. For the third regular case, that is, kneading sequences being eventually periodic but neither periodic nor 10^∞, by Theorem 7.2.5 we have

$$\pi(n) \sim cn \text{ for some constant } c \in (0, 1). \tag{7.5}$$

For the language of Feigenbaum attractor, by the result of Proposition 7.4.1 it is easy to obtain

$$\pi(n) \sim 2 \log n, \tag{7.6}$$

where the base of logarithm is 2.

In order to do the same analysis for even Fibonacci languages, we need to estimate $|h^n(\alpha)|$ for them. Denote $|h^n(\alpha)| = a_n$ and $|h^n(\beta)| = b_n$ and use the homomorphism $h = (\alpha \to \alpha\alpha\beta, \beta \to \alpha\beta)$, we obtain equations for them as follows.

$$a_{n+1} = 2a_n + b_n, \quad b_{n+1} = a_n + b_n.$$

Solve the characteristic equation $\lambda^2 - 3\lambda + 1 = 0$, we obtain

$$a_n \sim An^\lambda \text{ with } \lambda = \frac{3 + \sqrt{5}}{2} \approx 2.618$$

Combining this result with Proposition 7.4.2 we have similar asymptotic behavior:

$$\pi(n) \sim c \log n \tag{7.7}$$

for some constant c.

For the odd Fibonacci languages, the result is different. First we estimate the asymptotic behavior of $a_n = |h^n(\alpha)|$ by the same method as above. Denote $b_n =$

$|h^n(\beta)|$ and use the homomorphism $h = (\alpha \to \alpha\beta\alpha\alpha\beta, \beta \to \alpha\alpha\beta)$, we have the equations for a_n and b_n as follows:

$$a_{n+1} = 3a_n + 2b_n, \quad b_{n+1} = 2a_n + b_n.$$

Thus we obtain the characteristic equation $\lambda^2 - 4\lambda - 1 = 0$ and

$$a_n \sim An^\lambda \text{ with } \lambda = 2 + \sqrt{5} \approx 4.23607.$$

Combining this result with Proposition 7.4.5 we obtain

$$\pi(n) \sim cn^\mu \tag{7.8}$$

for some constant c and $\mu = \log(1 + \sqrt{2}) / \log(2 + \sqrt{5}) \approx 0.61052$.

CHAPTER 8
TOPOLOGICAL ENTROPY OF UNIMODAL MAPS

In Section 2.5 the concept of topological entropy has been introduced for dynamical languages (and symbolic flows). In this chapter we discuss topological entropy for unimodal maps (see, e.g., Milnor and Thurston 1988, Douady 1995). The results presented below are largely well-known, but the approach is consistent with other Chapters of this book, and new proofs for some results are included in it. Here all calculations depends on counting words of languages generated by unimodal maps.

In Section 8.1 two sequences $\{N_n\}$ and $\{S_n\}$ are defined, which will be used to calculate the growth number and the topological entropy. Some simple examples are presented.

In Section 8.2 is established a general recursive relations for the sequence $\{S_n\}$, which leads directly to difference equations for periodic and eventually periodic kneading sequences. The kneading invariant is also obtained in Section 8.3 through the recursive relations of $\{S_n\}$. The dynamical meaning of $\{N_n\}$ and $\{S_n\}$ is explained.

The final Section is devoted to applications, which include a Theorem of Bowen and Frank, positivity of entropy, periodic orbits with odd primitive period, and constant entropy in windows.

8.1 Introduction

8.1.1 Growth Number and Entropy

Let KS be the kneading sequence of a unimodal map f, and $L = \mathcal{L}(KS)$ be the language generated by KS. Assume as before that there is no symbol c appearing in KS. Since the topological entropy $h(f)$ is completely determined by L or KS, the notations $h(L)$ or $h(KS)$ are also used.

For each integer $n \geq 0$ we denote by

$$N_n = \text{card}\{x \in L \mid |x| = n\}$$

the number of words of L which are of length n (see Definition 2.5.1 in Chapter 2). Since we always have $\varepsilon \in L$ by Lemma 2.1.2, $N_0 = 1$ holds. The mapping $n \mapsto N_n$ is called the structure function in Mandelbrot (1954), and the complexity function of L in Queffélec (1987).

It has been proved in Section 2.5 that the limit

$$\lim_{n\to\infty} \frac{\log N_n}{n} = \inf_n \frac{\log N_n}{n}$$

always exists and is called the topological entropy of language L, and is denoted by $h(L)$ (or simply h). For $L = \mathcal{L}(KS)$ generated by a given kneading sequence KS, the notation $h(KS)$ is also used.

By the same method it can be shown that the following limit

$$s = \lim_{n\to+\infty} \sqrt[n]{N_n} = \inf_n \sqrt[n]{N_n}$$

exists and is called the *growth number* of $\{N_n\}$. It is obvious that $h = \log s$.

From the obvious fact $N_n \leq N_{n+1} \leq 2N_n$, we have $1 \leq s \leq 2$ for unimodal maps. For the surjective case or any unimodal map with $KS = 10^\infty$, we have $s = 2$. Thus in this case the topological entropy reaches the maximum value $h = \log 2$.

8.1.2 The Sequence $\{S_n\}$

Besides the sequence $\{N_n\}$, we need another sequence of integers $\{S_n\}$ defined by

$$S_n = \text{card}\{x \in L \mid |x| = n \text{ and } x \text{ begins from } 1\}$$

It is straightforward to relate $\{N_n\}$ with $\{S_n\}$ as follows:

$$\begin{aligned} N_n &= S_n + S_{n-1} + \cdots + S_1 + 1, \\ S_n &= N_n - N_{n-1}. \end{aligned} \tag{8.1}$$

As a matter of fact, if a word $z \in L$ is of length n, then either z has 0^i1 $(0 \leq i < n)$ as its prefix or $z = 0^n$. By Proposition 3.4.2 it follows that

$$0^i1x \in L \Longleftrightarrow 1x \in L,$$

thus obtain the first equality of (8.1). The second one is its consequence. By the basic property (D2) in Definition 2.1.1 both $\{N_n\}$ and $\{S_n\}$ are nondecreasing, thus we have the inequalities

$$S_n < N_n \leq nS_n + 1.$$

If $S_n > 0$, we have

$$\sqrt[n]{N_n/(2n)} < \sqrt[n]{S_n} < \sqrt[n]{N_n},$$

and thus obtain

$$\lim_{n\to+\infty} \sqrt[n]{S_n} = \lim_{n\to+\infty} \sqrt[n]{N_n}.$$

Here the unique exception is $KS = 0^\infty$, where each $S_n = 0$, since the language $L = \mathcal{L}(0^\infty) = \{0^i \mid i \geq 0\}$.

Therefore, with the exception of this trivial case, we can always use $\{S_n\}$ to compute growth number and topological entropy of languages generated from unimodal maps. In one word, in order to compute topological entropy we need only to count words beginning from the symbol 1. As shown in the sequel this is often an easier task.

8.1.3 Kneading Sequences and Entropy

By Proposition 3.4.3 we see that in computation of topological entropy it is not necessary to consider those kneading sequences that contain the symbol c.

From the definition of topological entropy it is easy to know that as a function of kneading sequence the entropy is monotone increasing.

Lemma 8.1.1 Let KS_1 and KS_2 be two kneading sequences, and h_1 and h_2 their corresponding topological entropies. If $KS_1 < KS_2$, then $h_1 \leq h_2$.

Proof. From Proposition 3.4.2 it is obvious that

$$\mathcal{L}(KS_1) \subset \mathcal{L}(KS_2)$$

and it follows the conclusion. ∎

8.1.4 Some Examples

Here we give some examples of calculation of topological entropy.

Example 8.1.2 Compute the topological entropy for $KS = (101)^\infty$

Solution 1. By Theorem 2.5.3 we need only to construct the associate matrix for this language. From the minDFA shown in Figure 4.4 we obtain directly

$$\begin{pmatrix} 1 & 1 & \\ & 1 & 1 \\ & 1 & \end{pmatrix}$$

Its largest eigenvalue β is $(1+\sqrt{5})/2 \approx 1.618$. Thus we have the topological entropy

$$h = \log \beta \approx 0.6942.$$

(Here the base of logarithm is 2.)

Solution 2. First compute the growth number s. It is easy to use the set of DEB to obtain a recursive relation for $\{N_n\}$. Here we have $L'' = \{100\}$ (cf. Example 7.2.4).

Consider any word $z \in L$ which is of length $n+2$. If $z = x1$, then by the knowledge of $L'' = \{100\}$ it is easy to find

$$z = x1 \in L \Longleftrightarrow x \in L.$$

Similarly, if the last symbol of z is 0, then either $z = 0^{n+2}$ or $z = y10$. We can also establish

$$z = y10 \in L \Longleftrightarrow y \in L.$$

Therefore, we obtain a recursive relation

$$N_{n+2} = N_{n+1} + N_n + 1.$$

Its characteristic equation is $\lambda^2 - \lambda - 1 = 0$. By the well-known structure of solutions of difference equations, its largest positive root $(1 + \sqrt{5})/2$ is exactly the required growth number s.

Solution 3. It is often easier to consider $\{S_n\}$. If $z \in L$ and begins from symbol 1, then it can be either $z = 11x$ of $z = 10y$. It is easy to find

$$z = 11x \in L \Longleftrightarrow 1x \in L,$$
$$z = 10y \in L \Longleftrightarrow y \in L \text{ and } y \text{ begins from } 1.$$

Therefore, we obtain a new recursive relation

$$S_{n+2} = S_{n+1} + S_n$$

and the same growth number.

A comment. Although for regular languages $L = \mathcal{L}(KS)$ we can use results in Chapter 4 to obtain the associate matrices for them, but it naturally leads to calculations of characteristic determinant for the matrix. If its order is big, then this method is not a good choice. In the next example we only give solution by counting S_n.

Example 8.1.3 Compute the topological entropy for $KS = (101^{k-2})^\infty$, in which $k \geq 3$ is odd. (Example 8.1.2 is its special case of $k = 3$.)

Solution. By applying Proposition 7.1.3 it is easy to obtain the set of DEB

$$L'' = \{100, 10110, \ldots, 101^{k-3}0\}.$$

Consider S_n, the number of words which begin from 1 and of length n. Assume $n > k$. If z is a such word, then it begins from either 11 of 101. From $11x \in L \Leftrightarrow 1x \in L$, we have $\text{card}\{11x \in L \mid |1x| = n - 1\} = S_{n-1}$. As a fact used below, $\text{card}\{10x \in L \mid |10x| = n\} = S_n - S_{n-1}$ holds for each $n > 0$.

If z begins from 101, then it has either 1010 or 10111 as its prefix. Arguing as above we find $\text{card}\{1010x \in L\} = \text{card}\{10x \in L\} = S_{n-2} - S_{n-3}$. Proceed on in this way, the last kind of words are $\{101^{k-3}x \in L\}$. Since the longest DEB is $101^{k-3}0$, $x \in L$ have to begin from 1, and vice versa. Hence the cardinal number of this set is exactly S_{n-k+1}. Therefore, we obtain the desired relation

$$S_n = S_{n-1} + (S_{n-2} - S_{n-3}) + \cdots + (S_{n-k+3} - S_{n-k+2}) + S_{n-k+1}.$$

Writing out its characteristic equation

$$\lambda^{k-1} - \lambda^{k-2} - \lambda^{k-3} + \lambda^{k-4} - \cdots - \lambda^2 + \lambda - 1 = 0,$$

and multiplying the factor $(\lambda + 1)$ to it, we obtain the equation

$$f(\lambda) = \lambda^k - 2\lambda^{k-2} - 1 = 0.$$

It is easy to find that this equation has only one positive root. Denote it by ω. From $f(\sqrt{2}) = -1$ it is obvious that

$$\omega > \sqrt{2}.$$

Thus we obtain the growth number $s = \omega$ and the topological entropy $h = \log \omega$.

Example 8.1.4 Compute the topological entropy for $KS = 100(101)^\infty$

Solution 1. Using the minDFA in Figure 4.7 we obtain directly the associate matrix as follows (Here we do not consider the starting state, that is to say, omit all words beginning from 0):

$$\begin{pmatrix} 1 & 1 & & & \\ 1 & & 1 & & \\ & & & 1 & \\ & & & & 1 \\ & & 2 & & \end{pmatrix}$$

Its characteristic equation is

$$(\lambda^2 - \lambda - 1)(\lambda^3 - 2) = 0.$$

Thus the growth number is the same as in Example 8.1.2.

Solution 2. Consider the sequence $\{S_n\}$. Since 100 is a DEB, it is easy to find that for $n > 3$,

$$\mathrm{card}\{z \in L \mid |z| = n, z = 100x\} = S_n - S_{n-1} - S_{n-2}.$$

Using the fact that if z begins from 100, then it is followed by any combination of 100 and 101, hence we can obtain recursive relations as follows.

$$\begin{aligned} S_{3n+2} - S_{3n+1} - S_{3n} &= 2^{n-1}, \\ S_{3n+1} - S_{3n} - S_{3n-1} &= 2^{n-1}, \\ S_{3n} - S_{3n-1} - S_{3n-2} &= 2^{n-1} \end{aligned}$$

From them we have the desired difference equation

$$S_{n+3} - 2S_{n+2} + S_n = 0.$$

Its characteristic equation is

$$\lambda^3 - 2\lambda^2 + 1 = (\lambda - 1)(\lambda^2 - \lambda - 1)$$

and provides the same result as in Solution 1.

Remark. Since the period 3 window in Figure 3.1 begins from $KS = (101)^\infty$ and ends at $KS = 100(101)^\infty$, and the monotone increasing property of kneading sequences in this window, the result of Examples 8.1.2 and 8.1.4 means that the topological entropy in period 3 window is constant. Later we will see that this is always true for windows of kneading sequences (cf. Figure 3.1).

8.2 General Recursive Relations of $\{S_n\}$

Instead of using tricks in computing topological entropy in this section we present the general result for such computing.

8.2.1 The Main Theorem

Let $KS = e_1e_2 \cdots e_n \cdots$ be a given kneading sequence and $L = \mathcal{L}(KS)$. Without loss of generality, assume that there is no symbol c appearing in it. Use $\varepsilon(0) = +1$ and $\varepsilon(1) = -1$ to denote the degree of branches of unimodal maps, that is, the property of monotone increasing and decreasing. Introduce notations

$$\varepsilon_i = \varepsilon(e_i) \text{ and } \varepsilon'_i = \varepsilon_1 \cdots \varepsilon_i \text{ for } i \geq 1.$$

The general recursive relations for the sequence $\{S_n\}$ are given as follows.

Theorem 8.2.1 The following equality holds for each $n > 0$:

$$S_n + \varepsilon'_1 S_{n-1} + \varepsilon'_2 S_{n-2} + \cdots + \varepsilon'_{n-1} S_1 + \frac{1}{2}\varepsilon'_n = \frac{1}{2}. \tag{8.2}$$

Proof. If $KS = 0^\infty$ then $S_i = 0$ and $\varepsilon_i = 1$ for each $i > 0$, hence the claim is trivial. If $KS = 1^\infty$ then $S_i = 1$ and $\varepsilon_i = -1$ for each $i > 0$, and the verification is also easy. Thus in the sequel we assume that $e_1 = 1$ and $e_2 = 0$ always hold.

Rewriting the equality (8.2) into the form of

$$S_n = -\varepsilon'_1 S_{n-1} - \varepsilon'_2 S_{n-2} - \cdots - \varepsilon'_{n-1} S_1 + \frac{1}{2}(1 - \varepsilon'_n), \tag{8.3}$$

we will prove that its right-hand side is exactly an process of counting strings, in which each word of L, which begin from 1 and of length n, contributes one, while any other strings contribute nothing, hence the result is exactly S_n.

Since $\varepsilon'_1 = \varepsilon_1 = -1$ by assumption, the first term of the right-hand side of (8.3) is S_{n-1}. It counts the number of words which are of the form $11x$ with $|11x| = n$. By $11x \in L \Leftrightarrow 1x \in L$ the contribution is exactly S_{n-1}. But the other terms are not so easy to be treated.

Introduce auxiliary sets Θ_i as follows. For $i = 1, \cdots, n-1$ define

$$\Theta_i = \{e_1 \cdots e_i x \mid x \in L, |x| = n - i \text{ and begins from } 1\}.$$

Thus $\operatorname{card}\Theta_i = S_{n-i}$ for each i, and the right-hand side of (8.3) is an algebraic sum of the cardinal number of Θ_i for i from 1 to $n-1$, and then plus the last term, which is either 1 or 0. That is to say, if $\varepsilon'_i = -1$, then we add the number of strings in Θ_i to our counting, otherwise subtract it from the counting. Since a string can belong to more than one Θ_i, we have to consider the final net effect of this counting for each string.

In order to do this counting work, we introduce notations $id(z)$ and $id_i(z)$, $i = 1, \cdots, n-1$, for each string z which is of length n:

$$id(z) = id_1(z) + \cdots + id_{n-1}(z), \text{ where}$$

$$id_i(z) = \begin{cases} +1, & \text{when } z \in \Theta_i \text{ and } \varepsilon_i' = -1, \\ -1, & \text{when } z \in \Theta_i \text{ and } \varepsilon_i' = +1, \\ 0, & \text{otherwise.} \end{cases}$$

It is obvious that the proof of this Proposition is reduced to the following statements where z is any string of length n

$$id(z) = \begin{cases} +1, & \text{when } z \in L, \text{ begins from 1, but } z \neq e_1 \cdots e_n, \\ (\varepsilon_n' + 1)/2, & \text{when } z = e_1 \cdots e_n, \\ 0, & \text{when } z \notin L. \end{cases} \tag{8.4}$$

As a matter of fact, we see that, each word $z \in L$, which is of length n and not $e_1 \cdots e_n$, contributes exactly once to the final sum of counting process, while any string $z \notin L$ contributes nothing. For $z = e_1 \cdots e_n$, if it is even, then $id(z) = +1$ and the last term of (8.3) is zero. If it is odd, then $id(z) = 0$, and this term is one. In either cases we obtain the correct contribution for it. Therefore, if (8.4) is true, then our proof is completed. For clarity we write the proof of these statements separately as follows. ∎

Proof of statements of (8.4).

Rewrite the prefix of KS by

$$e_1 \cdots e_n = 10^{n_1} 10^{n_2} \cdots 10^{n_k}$$

If $k = 1$ then each $\varepsilon_i' = -1$. In this case the equality (8.3) is

$$S_n = S_{n-1} + \cdots + S_1 + 1,$$

which can be verified directly. Thus in the sequel we assume $k > 1$. By assumption made about e_1 and e_2 and since KS is maximal we have $n_1 > 0$ and $0 \leq n_i \leq n_1$ for n_i, $i = 2, \ldots, k$. Assume $|z| = n$ and consider the following cases separately. For simplicity we write id_i for $id_i(z)$.

(1) The string $z \in L$ and 10^{n_1} is not its prefix. In this case z is of the form $z = 10^j 1x$ for some $0 \leq j < n_1$. Thus $z \in \Theta_{j+1}$ but not other Θ_i. Since $\varepsilon_{j+1}' = -1$, we have $id_{j+1} = +1$, $id_i = 0$ for $i \neq j+1$, and $id(z) = +1$.

(2) $z \in L$ and has 10^{n_1} as its prefix, but $z \neq e_1 \cdots e_n$. Then we can assume z is of the form $z = 10^{n_1} 1 \cdots 10^{n_l} y$ for some $1 \leq l < k$ while y does not have $10^{n_{l+1}}$ as prefix. Thus y has either $10^j 1$ for some $0 \leq j < n_{l+1}$ or 0 as its prefix.

Discuss the first case that y begins from 1. Since $y < 10^{n_{l+1}}$ and $z \in L$, the number l is even. From $z \in L$ and the property (D1) of Definition 2.1.1, we find that

$$z \in \Theta_{n_1+1}, z \in \Theta_{n_1+n_2+2}, \ldots, z \in \Theta_{n_1+\cdots+n_l+l}, z \in \Theta_{n_1+\cdots+n_l+l+j+1}$$

and $z \notin \Theta_i$ for other i. Thus we obtain

$$id_{n_1+1} = +1, id_{n_1+n_2+2} = -1, \ldots, id_{n_1+\cdots+n_l+l} = -1, id_{n_1+\cdots+n_l+l+j+1} = +1.$$

and all other $id_i(z) = 0$. Since l is even, we have $id(z) = +1$. The second case of y begins from 0 is similar and omitted. Note that the non-zero id_i must appear alternatingly as +1 and −1.

(3) $z = e_1 \cdots e_n$. We have $id_{n_1+1} = 1$, $id_{n_1+n_2+2} = -1, \ldots$. If z is even, then $\varepsilon'_n = 1$, k is even, and the last $id_{n_1+\cdots+n_{k-1}+k-1}(z) = -1$. Thus $id(z) = 1$. If z is odd, then we have $id(z) = 0$.

(4) $z \notin L$ and begins from 0. Then $z \notin \Theta_i$ for all i and $id(z) = 0$ evidently.

(5) $z \notin L$ and $z = 10^j y$ with $j \neq n_1$. If $j > n_1$ then the calculation is the same as in (4). If $j < n_1$ and $y = 1x$, then from $10^j 1x \notin L \Leftrightarrow y = 1x \notin L$ we obtain the same result.

(6) $z \notin L$ and is of the form $z = 10^{n_1} \cdots 10^{n_l} y$ for some $1 \le l < k$ and $y \neq 10^{n_{l+1}} \cdots$. Without loss of generality, we can assume that $y \in L$ and there exists an integer i, $1 \le i < l$, such that $10^{n_{i+1}} \cdots 10^{n_l} y \in L$ (otherwise it is the same as in (4)). Take i the smallest one for which this holds. By Proposition 3.4.2 we have

$$10^{n_i} \cdots 10^{n_l} y > e_1 \cdots e_n = 10^{n_1} 10^{n_2} \cdots 10^{n_k} \ge 10^{n_i} \cdots 10^{n_{l+1}}$$

Since $y \le 10^{n_{l+1}}$, $10^{n_i} \cdots 10^{n_l}$ must be an odd string, and hence i and l have the same parity.

For either cases of the parity of i and l, the non-zero id_i, that is, +1 and −1, appear alternatingly and their number is even, hence we obtain $id(z) = 0$. ∎

8.2.2 *Difference Equation for Periodic KS*

Although the number of recursive relations in Theorem 8.2.1 is infinite, for two important cases we can obtain one relation from them which can be used to find the growth number as in Examples of Subsection *8.1.4*.

Let $KS = (e_1 \cdots e_n)^\infty$ be a periodic kneading sequence. Here we assume (as in Section 4.3) that the string $x = e_1 \cdots e_n$ is even and minimal (in length). Note that by this assumption the case $n = 1$ corresponds $KS = 0^\infty$, the unique exception mentioned in Subsection *8.1.2*.

Since the sequence $\{\varepsilon'_i\}$ is periodic with the period n, and $\varepsilon'_n = \varepsilon_1 \cdots \varepsilon_n = 1$, for each $k \ge n$ we have

$$\begin{aligned}
&S_k + \varepsilon'_1 S_{k-1} + \cdots + \varepsilon'_{k-1} S_1 + \frac{1}{2}(\varepsilon'_k - 1) \\
&= S_k + \varepsilon'_1 S_{k-1} + \cdots + \varepsilon'_{n-1} S_{k-n+1} \\
&\quad + [S_{k-n} + \varepsilon'_1 S_{k-n-1} + \cdots + \varepsilon'_{k-n-1} S_1 + \frac{1}{2}(\varepsilon'_{k-n} - 1)] \\
&= S_k + \varepsilon'_1 S_{k-1} + \cdots + \varepsilon'_{n-1} S_{k-n+1} = 0.
\end{aligned}$$

Thus we obtain the difference equation for periodic kneading sequence.

Proposition 8.2.2 If a periodic kneading sequence $KS = (e_1 \cdots e_n)^\infty$ satisfies the conditions that $n > 1$, and $e_1 \cdots e_n$ is even and minimal, then the growth number s is the largest positive root of the difference equation

$$\lambda^{n-1} + \varepsilon_1' \lambda^{n-2} + \cdots + \varepsilon_{n-1}' = 0. \tag{8.5}$$

Proof. Define a generating function $S(t)$ by

$$S(t) = \frac{1}{2} + S_1 t + \cdots + S_k t^k + \cdots$$

From Subsection *8.1.2* and $n > 1$, $\lim_{n\to\infty} \sqrt[n]{S_n}$ exists and equals to the growth number s. Thus the convergence radius of $S(t)$ is $1/s$. Since all coefficients of $S(t)$ are positive, applying the knowledge of analytic functions (see, e.g., Titchmarsh 1939), we see that $1/s$ is a singular point of $S(t)$.

Multiplying $S(t)$ by $1, \varepsilon_1' t, \ldots, \varepsilon_{n-1}' t^{n-1}$ separately and then adding them together, we obtain an identity

$$(1 + \varepsilon_1' t + \cdots + \varepsilon_{n-1}' t^{n-1}) S(t) = \frac{1}{2}(1 + t + \cdots + t^{n-1}),$$

which holds in $|t| < 1/s$. Therefore, $1/s$ is a pole of $S(t)$, and at the same time the smallest positive root of the equation

$$1 + \varepsilon_1' t + \cdots + \varepsilon_{n-1}' t^{n-1} = 0.$$

Let $\lambda = 1/t$, the proof is completed. ∎

For Example 8.1.2, $KS = (101)^\infty$, $\varepsilon_1' = \varepsilon_2' = -1$, thus we obtain

$$\lambda^2 - \lambda - 1 = 0.$$

For Example 8.1.3, $KS = (101^{k-2})^\infty$, where $k \geq 3$ is an odd number. From $\varepsilon_1 = -1$, $\varepsilon_2 = +1$, $\varepsilon_i = -1$ for $3 \leq i \leq k$, and $\varepsilon_i' = \varepsilon_1 \cdots \varepsilon_i$, we obtain

$$\lambda^{k-1} - \lambda^{k-2} - \lambda^{k-3} + \lambda^{k-4} - \cdots - \lambda^2 + \lambda - 1 = 0.$$

8.2.3 Difference Equation for Eventually Periodic KS

Let $KS = e_1 \cdots e_m (e_{m+1} \cdots e_n)^\infty$ be an eventually periodic kneading sequence, where the string $e_{m+1} \cdots e_n$ is even and minimal in length. Thus the sequence $\{\varepsilon_i'\}$ is eventually periodic, that is, $\varepsilon_{n+i}' = \varepsilon_{m+i}'$ for $i \geq 0$.

For each $k > n$ we have

$$\begin{aligned}
&S_k + \varepsilon_1' S_{k-1} + \cdots + \varepsilon_{k-1}' S_1 + \frac{1}{2}(\varepsilon_k' - 1) \\
&= S_k + \varepsilon_1' S_{k-1} + \cdots + \varepsilon_{n-1}' S_{k-n+1} - S_{k+m-n} - \varepsilon_1' S_{k+m-n+1} - \cdots - \varepsilon_{m-1}' S_{k-n+1} \\
&\quad + [S_{k+m-n} + \varepsilon_1' S_{k+m-n-1} + \cdots + \varepsilon_{k+m-n-1}' S_1 + \frac{1}{2}(\varepsilon_{k+m-n}' - 1)] \\
&= S_k + \varepsilon_1' S_{k-1} + \cdots + \varepsilon_{n-1}' S_{k-n+1} - S_{k+m-n} - \varepsilon_1' S_{k+m-n+1} - \cdots - \varepsilon_{m-1}' S_{k-n+1},
\end{aligned}$$

and obtain the following result. Its proof is similar to that of Proposition 8.2.2 and omitted.

Proposition 8.2.3 If $KS = e_1 \cdots e_m(e_{m+1} \cdots e_n)^\infty$, where the string $e_{m+1} \cdots e_n$ is even and minimal, then the growth number s is the largest positive root of the difference equation

$$\lambda^{n-1} + \varepsilon_1' \lambda^{n-2} + \cdots + \varepsilon_{n-1}' - \lambda^{m-1} - \varepsilon_1' \lambda^{m-2} - \cdots - \varepsilon_{m-1}' = 0. \tag{8.6}$$

Remark. Since for an eventually periodic kneading sequence $KS = \rho\lambda^\infty$ the choice of ρ and λ is not unique, it seems that the best way to write the equation (8.6) is to make the length of ρ minimum, that is, $e_m \neq e_n$, and $\lambda = e_{m+1} \cdots e_n$ minimal in length. If a given KS is eventually periodic but not periodic, this is always possible.

For Example 8.1.4, $KS = 100(101)^\infty$, $m = 3$, $n = 6$, we obtain

$$\lambda^5 - \lambda^4 - \lambda^3 - 2\lambda^2 + 2\lambda + 2 = (\lambda^3 - 2)(\lambda^2 - \lambda - 1) = 0.$$

8.3 Kneading Invariant

Of course, we cannot always hope to obtain a finite recursive relationship of $\{S_n\}$ for general kneading sequences as those in the previous two subsections. As pointed out by Milnor and Thurston (1988), in the general case the growth number and topological entropy can be determined by an analytic function which is named the kneading invariant or kneading determinant.

First define two functions from $\{S_n\}$ and $\{N_n\}$ by power series

$$\begin{aligned} N(t) &= \sum_{i=0}^{\infty} N_n t^n, \text{ and} \\ S(t) &= \frac{1}{2} + \sum_{i=1}^{\infty} S_n t^n \end{aligned} \tag{8.7}$$

From the formula of growth number s, we know that both $N(t)$ and $S(t)$ have the same value $1/s$ as their convergence radius. (Except the case of $KS = 0^\infty$.) Moreover, since their coefficients are positive, $t = 1/s$ is a singular point of them.

By formula (8.1) we have an identity to connect them:

$$N(t)(1-t) = S(t) + \frac{1}{2} \tag{8.8}$$

The *kneading invariant* introduced in Milnor and Thurston (1988) is defined by

$$D(t) = 1 + \varepsilon_1' t + \varepsilon_2' t^2 + \cdots + \varepsilon_n' t^n + \cdots \tag{8.9}$$

Since each coefficient is either $+1$ or -1, the convergence radius of $D(t)$ is 1. (Since by assumption there is no c appearing in KS, thus here $D(t)$ cannot be a polynomial.)

Using the recursive relations provided by Theorem 8.2.1 we obtain the following identity

$$S(t)D(t) = \frac{1}{2}(1 + t + \cdots + t^n + \cdots).$$

Hence we can write the relation between $S(t)$, $N(t)$ and $D(t)$ as follows

$$\begin{aligned} S(t) &= \frac{1}{2(1-t)D(t)}, \\ N(t) &= \frac{1}{2(1-t)} + \frac{1}{2(1-t)^2 D(t)}. \end{aligned} \tag{8.10}$$

Therefore, the convergence radius of both $S(t)$ and $N(t)$, $1/s$, is exactly the smallest positive root of $D(t)$ if $s > 1$. We can also extend them to functions which are defined and meromorphic throughout the disk $|t| < 1$.

Remark. There is a simple relation between N_n and the lap numbers $l(f^n)$ used in Milnor and Thurston (1988). There the notation $l(f^n)$ is the lap number of f^n in the unit interval $[0, 1]$. From Proposition 3.1.1 and Definition 3.3.2 of the language $L = \mathcal{L}(KS)$ in Chapter 3 we see that N_n is exactly the lap number of f^n in the interval $[0, f(c)]$. If we define another language $\overline{L}$ by orbits starting from all points in $[0, 1]$, then it is easy to establish

$$\overline{L} = \{0\}L \cup \{1\}L.$$

Therefore, we have

$$N_n = \frac{1}{2} l_{n+1}.$$

Similarly, we can obtain

$$S_n = \frac{1}{2}\gamma_n,$$

where γ_n is the number of times that the function $f^n(x) - c$ changes sign in $(0, 1)$, since $\gamma_n = l_{n+1} - l_n$ and $S_n = N_n - N_{n-1}$ hold evidently.

Example 8.3.1 As an example, it can be verified directly that for the tent map (3.2) with parameter value $s > 1$, the following equality holds:

$$D(\frac{1}{s}) = 0.$$

Starting from critical point $x_0 = c = 1/2$, we calculate the orbit $x_0, x_1, \ldots, x_n, \ldots$. Here the iterated relation of tent map can be rewritten into the form

$$x_{n+1} = \frac{\varepsilon}{2} + \varepsilon_n s(x_n - \frac{1}{2}),$$

where $\varepsilon_n = \varepsilon(x_n - 1/2)$ and

$$\varepsilon(x) = \begin{cases} -1, & \text{when } x > 1/2, \\ 0, & \text{when } x = 1/2, \\ +1, & \text{when } x < 1/2. \end{cases}$$

Therefore, we have

$$\begin{aligned} x_1 - \frac{1}{2} &= -\frac{\varepsilon_1}{2}(1-\frac{1}{s}) + \frac{\varepsilon_1}{s}(x_2 - \frac{1}{2}) \\ &= -\frac{\varepsilon_1}{2}(1-\frac{1}{s}) - \frac{\varepsilon_1\varepsilon_2}{2s}(x_2 - \frac{1}{2}) + \frac{\varepsilon_1\varepsilon_2}{s^2}(x_3 - \frac{1}{2}) \\ &\cdots\cdots \\ &= -\frac{1}{2}(1-\frac{1}{s})(\varepsilon_1' + \frac{\varepsilon_2'}{s} + \cdots + \frac{\varepsilon_n'}{s^{n-1}}) + \frac{\varepsilon_n'}{s^n}(x_{n+1} - \frac{1}{2}). \end{aligned}$$

If $s > 1$ and none of $x_n = 1/2$, then since $x_1 = s/2$, we obtain

$$\frac{s}{2} - \frac{1}{2} = -\frac{1}{2}(1-\frac{1}{s})(\varepsilon_1' + \frac{\varepsilon_2'}{s} + \cdots + \frac{\varepsilon_n'}{s^{n-1}} + \cdots),$$

which is equivalent to $D(1/s) = 0$.

8.4 Applications

In this Section we give four results which show that in the case of unimodal maps the value of topological entropy is closely related with their dynamical behaviors.

8.4.1 A Theorem of Bowen and Frank

Here we give the proof of a theorem due to Bowen and Frank (1976) for the case of unimodal maps (see Stefan (1977) for its improvement).

Theorem 8.4.1 If a unimodal map f has a periodic orbit of primitive period $p = 2^d k$, k odd, $k \geq 3$, then the topological entropy

$$h(f) \geq \frac{1}{2^d}\log\omega > \frac{1}{p}\log 2,$$

where ω is the (unique) positive root of the equation $\lambda^k - 2\lambda^{k-2} - 1 = 0$.

First we need a Lemma as follows.

Lemma 8.4.2 If x is a maximal primitive string, and $|x| = k \geq 3$ is odd, then

$$x \geq 101^{k-2}$$

Proof. It is trivial for $k = 3$. Assume $k \geq 5$ and write $x = 101^{l_1}0\cdots$. Assume the contrary that $x < 101^{k-2}$, then $l_1 < k-2$ and l_1 is odd. Since x is maximal, then there is no substring 00 appearing in x. There are two possibilities.

(1) $x = 101^{l_1}0\cdots01^{l_n}0$,
(2) $x = 101^{l_1}0\cdots01^{l_n}$.

For the case of (1), since x is maximal, each l_i is odd (and $l_i \le l_1$). But then $|x|$ is even. Similarly, for the case of (2), each of $l_1, \ldots, l_{n-1}$ is odd, and $l_n \le l_1$. If l_n is even, then $|x|$ is even. If l_n is odd, then

$$\sigma^{|x|-l_n-2}(x) = 101^{l_n+1}0\cdots > x,$$

a contradiction. ∎

Lemma 8.4.3 If x is a maximal primitive string, and $|x| = n = 2^d k$, k odd, $k \ge 3$, then $x \ge h^d(101^{k-2})$, where $h = (1 \to 10, 0 \to 11)$.

Proof. Let $t_i = h^i(1)$ and $\bar{t}_i = h^i(0)$ (see Lemma 6.1.1). Assume the contrary that $x < h^d(101^{k-2})$, write $x = \alpha_1 \cdots \alpha_k$, where each α_i is of length $|t_d| = 2^d$. We will show that each α_i is either t_d or $\bar{t}_d$. Begin from α_1. If $\alpha_1 < t_d$, then $x^\infty < t_d^\infty$. By the proof of Lemma 6.1.6 x^∞ is either t_i^∞ for some $0 \le i < d$ or 0^∞. Either of them contradicts the primitivity of x. Thus we have $\alpha_1 = t_d$.

Using the fact of $t_{d+1} = t_d\bar{t}_d$ we obtain $\alpha_2 = \bar{t}_d$. Proceed inductively on. Assume that each of $\alpha_1, \ldots, \alpha_i$ is either t_d or $\bar{t}_d$, and consider α_{i+1}.

From $x = t_d\bar{t}_d\cdots$ there is no substring $\bar{t}_d\bar{t}_d$ appearing in x. Since $x = t_d\cdots$ is maximal, $\alpha_{i+1} \le t_d$. If $\alpha_{i+1} < \bar{t}_d$, then consider α_i. If $\alpha_i = t_d$, then since t_d is odd, we would have $\alpha_i\alpha_{i+1} > x = t_d\bar{t}_d\cdots$. If $\alpha_i = \bar{t}_d$, then α_{i-1} must be t_d, and $\alpha_{i-1}\alpha_i\alpha_{i+1} > x$ would hold too. Both cases lead to contradiction.

Therefore, we have found a string y, $|y| = k$, such that $x = h^d(y)$. The conditions about x lead to that y is maximal and primitive. Thus we know $y \ge 101^{k-2}$ by using Lemma 8.4.2. Since h is order-preserving, it follows our conclusion. ∎

Lemma 8.4.4 Let KS_1 be a kneading sequence, and h_1 the topological entropy of $L_1 = \mathcal{L}(KS_1)$. If $KS_2 = h(KS_1)$, where $h = (1 \to 10, 0 \to 11)$, then the topological entropy h_2 of $L_2 = \mathcal{L}(KS_2)$ is

$$h_2 = \frac{1}{2}h_1.$$

Proof. Using Definition (8.9) of kneading invariant, if the kneading sequence for KS_1

$$D_1(t) = 1 + \varepsilon_1' t + \cdots + \varepsilon_n' t^n + \cdots,$$

then it is easy to verify that the kneading invariant for KS_2 is

$$\begin{aligned} D_2(t) &= (1-t)(1 + \varepsilon_1' t^2 + \cdots + \varepsilon_n' t^{2n} + \cdots) \\ &= (1-t)D_1(t^2). \end{aligned}$$

Therefore, if the growth number s_1 for L_1 is greater than 1, then the growth number s_2 for L_2 is $\sqrt{s_1}$, and hence $h_2 = (1/2)h_1$. If $s_1 = 1$, then $D_1(t)$ has no zero in $[0, 1)$. From the identity (8.10) we know $s_2 = 1$ too. Thus our conclusion is also true. ∎

Proof of Theorem 8.4.1. Let KS be the kneading sequence of f and x^∞ the itinerary of periodic orbit in the condition of Theorem. Of course, we can assume x is maximal and primitive. Using Corollary of Theorem 3.5.1, $|x|$ is either p or $p/2$. Examine the claim of Theorem, it suffices to prove the first case only.

By Lemma 8.4.3 we have

$$KS \geq x^{\infty} \geq (h^d(101^{k-2}))^{\infty}$$

Using the monotone property of entropy with respect to kneading sequence (see Lemma 8.1.1), we need only to show that for $KS_1 = (h^d(101^{k-2}))^{\infty}$, its topological entropy

$$h_1 \geq \frac{1}{2^d} \log \omega > \frac{1}{n} \log 2.$$

Let $KS_2 = (101^{k-2})^{\infty}$, and use notations $L_1 = \mathcal{L}(KS_1)$ and $L_2 = \mathcal{L}(KS_2)$. If h_2 is the topological entropy of L_2, then by Lemma 8.4.4 we have

$$h_1 = \frac{1}{2^d} h_2.$$

Using Example 8.1.3 we know that $h_2 \geq \log \omega$. Since $\omega > \sqrt{2}$, we have

$$h(f) \geq h_1 = \frac{1}{2^d} h_2 \geq \frac{1}{2^d} \log \omega > \frac{1}{2^{d+1}} \log 2 > \frac{1}{2^d k} \log 2 = \frac{1}{p} \log 2$$

as desired. ∎

8.4.2 Unimodal Maps with Positive Entropy

First we show that the topological entropy of Feigenbaum attractor is zero. From Section 6.1 we have its kneading sequence

$$t_{\infty} = \lim_{n \to \infty} h^n(1).$$

Lemma 8.4.5 The topological entropy of $\mathcal{L}(t_{\infty})$, the language of Feigenbaum attractor, is zero.

Proof. From t_{∞} we can obtain the kneading invariant for Feigenbaum attractor:

$$D(t) = 1 - t - t^2 + t^3 - t^4 + t^5 + t^6 - t^7 + \cdots$$

Compare its coefficients with the Thue-Morse sequence in subsection *6.1.4*, we find that the coefficients of $D(t)$ is exactly the Thue-Morse sequence consisting of two symbols $\{-1, +1\}$. (see the transformation *6.1.4* from t_{∞} to the Thue-Morse sequence.) Recalling that the Thue-Morse sequence is generated from the homomorphism g by (6.2), it is easy to obtain $D(t)$ formally from an infinite product as follows:

$$D(t) = (1-t)(1-t^2)(1-t^4)\cdots(1-t^{2^n})\cdots$$

By the criterion of convergence in analysis we know that this infinite product is convergent and $D(t)$ has no zero when $0 \leq t < 1$. Using the relation (8.10) between $S(t)$ and $D(t)$ we know that the growth number $s = 1$ and the topological entropy $h = 0$. ∎

Theorem 8.4.6 Let KS be the kneading sequence of a unimodal map f. The following statements are equivalent:

(1) $h(f) > 0$,
(2) $KS > t_\infty$,
(3) f admits a periodic orbit whose primitive period is not a power of 2.

Proof. (1) $\Longrightarrow$ (2). If $h(f) > 0$, then by Lemmas 8.4.5 and 8.1.1 we have $KS > t_\infty$.

(2) $\Longrightarrow$ (3). Using the idea in Lemma 5.4.3 we can find a prefix x of KS such that $\bar{x}$ is a prefix of t_∞:

$$KS = x \cdots > t_\infty = \bar{x} \cdots$$

Hence x is an odd string and $\bar{x}$ is an even prefix of t_∞. Since each t_n is an odd prefix of t_∞, $t_n \neq \bar{x}$ holds, and hence $|x|$ cannot be a power of 2.

Of course, each suffix of x, say y, satisfies the condition $y \leq x$. On the other hand, if z is a proper PS of x to itself, then z is also a prefix of t_∞. At the same time, $\bar{z}$ is a suffix of $\bar{x}$, and a substring of t_∞. Therefore, z must be an odd string, and each PS of x is odd. By Corollary 1.2.12 we see that x is maximal and primitive.

Since x is odd maximal and primitive, we have

$$x^\infty \leq KS.$$

(Note that an odd prefix of KS may not be a maximal string.) Thus x^∞ is a periodic admissible sequence with respect to the given KS. Using Theorem 3.5.1, there exists a periodic orbit whose primitive period is $|x|$, which is not a power of 2.

(3) $\Longrightarrow$ (1). If a unimodal map f has a such periodic orbit, denote its period by $p = 2^d k$, where k is odd and $k \geq 3$, then we can use Theorem 8.4.1 to know that $h(f) > 0$. ∎

Using this fact and Lemma 6.1.6 we can obtain all those kneading sequences for which the topological entropy is zero.

Corollary. If x is maximal and primitive, and the topological entropy of $\mathcal{L}(x^\infty)$ is zero, then x is either 0 or $t_n = h^n(1)$ for some $n \geq 0$.

8.4.3 Periodic Orbits with Odd Primitive Period

First we need to compute the topological entropy for a special case, that is, the unimodal map which corresponds to the first inverse bifurcation point (that is to say, the 2-1 band-merging point).

Example 8.4.7 If a unimodal map f has 101^∞ as its kneading sequence, then $h(f) = \log\sqrt{2}$.

Solution 1. Using the conclusion of Proposition 8.6, we obtain the difference equation

$$\lambda^3 - \lambda^2 - 2\lambda + 2 = 0.$$

Its largest positive root $\sqrt{2}$ gives us the growth number s.

Solution 2. From the minDFA shown in Figure 4.7 we obtain directly

$$\begin{pmatrix} 1 & 1 & \\ & & 1 \\ & 2 & \end{pmatrix}$$

and obtain the same equation as in Solution 1.

Solution 3. Using the formula (8.9) we obtain the kneading invariant

$$\begin{aligned} D(t) &= 1 - t - t^2 + t^3 - t^4 + t^5 - t^6 + \cdots \\ &= 1 - t - \frac{t^2}{1+t} \\ &= \frac{1}{1+t}(1 - 2t^2), \end{aligned}$$

which has $1/\sqrt{2}$ as its smallest positive root.

Theorem 8.4.8 Let f be a unimodal map which has KS as its kneading sequence. The following statements are equivalent:

(1) $h(f) > \log\sqrt{2}$,
(2) $KS > 101^\infty$,
(3) f admits periodic orbit with odd primitive period.

Proof. (1) $\Longrightarrow$ (2). This is a consequence of Example 8.4.7 and Lemma 8.1.1.

(2) $\Longrightarrow$ (3). From (2) we have a prefix x of KS while its dual string $\overline{x}$ is the prefix of 101^∞ such that

$$KS = x \cdots > 101^\infty = \overline{x} \cdots$$

If $x = 100$, then we have $(100)^\infty \leq KS$. Otherwise, x is of the form 101^n0 where $n > 0$ is even. Since this x is maximal and primitive, $x^\infty \leq KS$ holds too. Thus x^∞ is an admissible sequence with respect to KS, and it suffices to use Theorem 3.5.1 to obtain a periodic orbit which has $|x|$ as its primitive period. Since $|x| = n+3$ is odd, (3) holds already.

(3) $\Longrightarrow$ (1). If f admits a periodic orbit with odd primitive period, then its itinerary can be written by x^∞, where x is maximal and primitive. By using Corollary of Theorem 3.5.1 we know that $|x|$ is exactly the period of the orbit of f, and hence is odd. It suffices to use Lemma 8.4.2, Example 8.1.3 and Lemma 8.1.1 to know $h(f) > \log\sqrt{2}$. ∎

8.4.4 Constant Entropy in Window

In this Subsection we will prove that in every window of bifurcation diagram (see, e.g., Figure 3.1), the entropy is constant.

First of all we need to extend the conclusion of Theorem 8.4.1 as follows.

Lemma 8.4.9 Let x be maximal primitive, and its length $|x| = p$ a power of 2. If the topological entropy h of $\mathcal{L}(x^\infty)$ is positive, then

$$h > \frac{1}{p}\log 2.$$

Proof. From Corollary of Lemma 8.4.5 we see that x is neither 0 nor $t_i = h^i(1)$ for each $i \geq 0$. Hence $|x| = p > 2$ and begins from 10.

Write $x = 10e_3e_4\cdots e_p$ and proceed in two steps.

(1) Assume that at least one of $e_{2i-1} = 0$ for some $i > 1$. Under this assumption we will show that $x > 101^\infty$.

It is trivial if $e_3 = 0$. Otherwise, let $e_3 = 1$, and there is no 00 appearing in x as it is a maximal string. Suppose i is the first integer for which $e_{2i-1} = 0$, then we must have $e_{2i-2} = 1$ and

$$x = 101^{n_1}0\cdots01^{n_l}0e_{2i}\cdots e_p.$$

Assume the contrary that $x \leq 101^\infty$. By the existence of $e_{2i-1} = 0$ for some i we can only have $x < 101^\infty$. From this inequality we find that n_1 is odd. But by the assumption about i, n_l is even, and hence $l > 1$. Combining these considerations, we have

$$101^{n_l}0\cdots > x = 101^{n_1}0\cdots,$$

which contradicts the condition of x being maximal.

From $x > 101^\infty$, the result of Example 8.4.7, and $|x| = p > 2$, we have

$$h(x^\infty) \geq h(101^\infty) = \log\sqrt{2} > \frac{1}{p}\log 2.$$

(2) Now consider general case. If in $x = 101e_4\cdots e_i\cdots e_n$ each $e_{2i-1} = 1$ holds, then we can find a new string x', such that $x = h(x')$, and if x' has the same property, then we can continue in this way.

Using the condition that $h(x^\infty) > 0$, and the entropy for each language $\mathcal{L}(h^n(1))$ is zero, we will obtain $x = h^d(y)$ for some $d > 0$ and a string y, which cannot be an image of any string by h. That is to say, if we write $y = a_1a_2\cdots$, then at least one $a_{2i-1} = 0$ for some i. Of course, y is neither 1 nor 10. Moreover, y satisfies all conditions for x in Lemma, that is, it is maximal primitive, and $|y|$ is a power of 2.

By the proof of (1) we know that $y > 101^\infty$. Thus we can use Lemma 8.4.4 to obtain

$$h(x^\infty) = \frac{1}{2^d}h(y^\infty) \geq \frac{1}{2^d}\log\sqrt{2} = \frac{1}{2^{d+1}}\log 2 > \frac{1}{p}\log 2$$

since $|x| = p = 2^d|y| > 2^{d+1}$ always holds. ∎

Theorem 8.4.10 If x is an even maximal primitive string, and when $|x|$ is a power of 2 the topological entropy of $\mathcal{L}(x^\infty)$ is positive, then

$$h(x^\infty) = h(\overline{x}x^\infty).$$

Proof. From the conditions we see that x begins from 10. Let $|x| = p > 2$. Using Formula (8.5) we have the difference equation

$$\lambda^{p-1} + \varepsilon'_1 \lambda^{p-2} + \cdots + \varepsilon'_{p-1} = 0 \tag{8.11}$$

for $KS = x^\infty$. If we denote its largest positive root by s_1 (the growth number), then $h(x^\infty) = \log s_1$.

Using Formula (8.6) it is easy to compute the difference equation for $KS = \overline{x}x^\infty$. Here $n = 2p$ and $m = p$, and we obtain

$$(\lambda^p - 2)(\lambda^{p-1} + \varepsilon'_1 \lambda^{p-2} + \cdots + \varepsilon'_{p-1}) = 0. \tag{8.12}$$

If we denote its largest positive root by s_2, then $h(\overline{x}x^\infty) = \log s_2$.

What we need to prove is $s_1 = s_2$. From these two difference equations and using the proof of Theorem 8.4.1 and Lemma 8.4.9, we see that

$$s_1 > \sqrt[p]{2}$$

holds for each maximal primitive string x, and it follows that $s_1 = s_2$ is true. ∎

From this result we see that the topological entropy of kneading sequences in each window is constant. Using the $*$-composition law (see Definition 4.5.2), in any given window there are infinitely many windows. A *primary window* is a window which is not contained in other windows. It has been proved that the converse of Theorem 8.4.10 is also true that if $KS_1 < KS_2$ and both of them have the same positive topological entropy, then they must be in the same (primary) window (Jonker and Rand 1980).

On the other hand, if the values of topological entropy are beyond all primary windows, that is to say, it cannot be taken by any kneading sequence in windows, then the corresponding kneading sequences are uniquely determined (cf. the original proof in Milnor and Thurston 1988).

Proposition 8.4.11 Let s be the growth number of a language $\mathcal{L}(KS)$. If s is not an algebraic unit satisfying a polynomial equation of the form $s^n \pm s^{n-1} \pm \cdots \pm 1 = 0$, then the kneading sequence KS is uniquely determined by s.

Proof. Assume the contrary that there exist two distinct kneading sequences, say KS_1 and KS_2, having the same value s as their growth number. Suppose that $KS_1 < KS_2$ holds. Using Lemma 5.4.3, that is, comparing their prefixes, we find that $KS_1 = x\cdots$ and $KS_2 = \overline{x}\cdots$. Here $\overline{x}$ is odd maximal primitive, and we obtain the order relations

$$KS_1 \leq x^\infty < \overline{x}^\infty \leq KS_2.$$

By Lemma 8.1.1 and $h(KS_1) = h(KS_2) = s$, the new kneading sequence x^∞ (and $\overline{x}^\infty$) also has s as its growth number. But then s satisfies a difference equation of the form (8.5), a contradiction. ∎

Part III

GRAMMATICAL COMPLEXITY OF CIRCLE HOMEOMORPHISMS

CHAPTER 9

LANGUAGES OF CIRCLE HOMEOMORPHISMS

Beginning from this chapter a systematic presentation of the grammatical complexity of circle homeomorphisms is given in Part III (Chapter 9–11).

The study of circle homeomorphisms in the spirit of coarse-grained description can be found in references such as Morse and Hedlund (1940), Gambaudo, Lanford III, and Tresser (1984), Procaccia, Thomae, and Tresser (1987), Friedmann (1991), Zheng (1989c, 1991), Crutchfield (1994), Siegel, Tresser, and Zettler (1992), and books like Guckenheimer and Holmes (1983), Hao (1989), de Melo and van Strien (1993). The material in the sequel is based on Xie (1995c).

After transforming the real orbits of a circle homeomorphism into the symbolic sequences over $\{0, 1\}$ and defining the language of circle homeomorphism, we study four aspects of this language:

1. The structure and number of words of the same length.
2. The structure of infinite sequences (admissible sequences) and the cardinal number of the set of all such sequences.
3. The grammatical complexity of the language, that is, the set of all finite strings, in the Chomsky hierarchy.
4. The structure of automaton accepting this language.

This chapter is organized as follows.

In Section 9.1 we define the concepts used below, including admissible sequences, languages, and words. Some basic properties about them are given.

The next section is devoted to discussion about the rotation number and the topological entropy. For the completeness and the consistency of method we present such results by proofs that are based on purely symbolic arguments. We hope this discussion will contribute some new insight to these concepts. The main result of this section is a formula about the number of words of the same length, which becomes essential for discussion later.

In Section 9.3 the nested block structure, which is peculiar to circle homeomorphisms, is explored. This structure appeared in references, e.g., Procaccia, Thomae, and Tresser 1987. The presentation about circle homeomorphisms in Part III is to take this structure as the starting point of the combinatorial study of them. Based on this structure a method of computation for periodic admissible sequences is developed, its relation with the Farey address and the continued fraction is discussed thoroughly in Subsection *9.3.3*.

The operators and notations introduced in Section 1.1, including cyclic shift, shift, primitive and maximal string, and operations in Subsection *1.1.5*, are used in the sequel.

9.1 Definition of Languages

The circle considered here is seen as a unit torus $S = R/Z$, that is, the set of real numbers mod (1). Thus we can identify the circle S with the unit interval $[0, 1]$ with the condition that its boundary points coincides. Therefore, every orientation-preserving homeomorphism f (if not an identity map) of the circle S can be equivalently seen as a bijective map f of the interval $J = [0, 1)$ onto itself as shown in Figure 9.1.

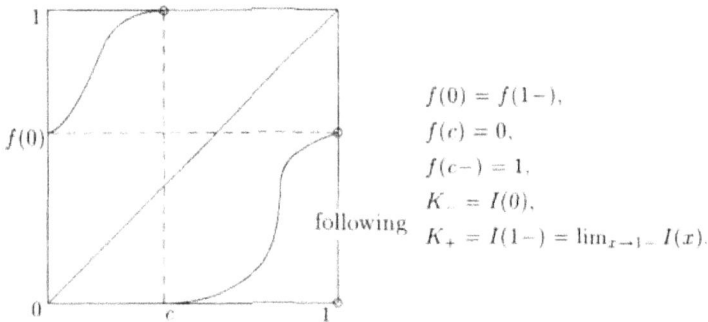

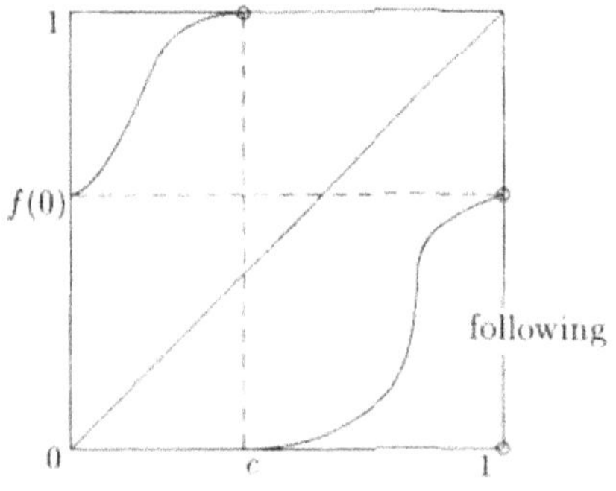

$f(0) = f(1-)$,
$f(c) = 0$,
$f(c-) = 1$,
$K_- = I(0)$,
$K_+ = I(1-) = \lim_{x \to 1-} I(x)$.

Figure 9.1: A circle homeomorphism f.

This bijective f satisfies the following properties:

1. f has a unique point of discontinuity $c(f)$, which is an interior point of J and $f(c-) = 1$ and $f(c) = 0$.
2. f is strictly monotone increasing in each component of $J\backslash\{c\}$ and $f(1-) = f(0)$.

Using the coarse-grained description method, every orbit of f starting from a point $a \in J$, $\{f^n(a)\}_{n \geq 0}$, can be transformed into an infinite sequence over the alphabet $\Sigma = \{0, 1\}$, in which the symbol 0 represents the left component $[0, c)$, and the symbol 1 represents the right component $[c, 1)$. We call this sequence the *itinerary* of point a and denote it by $I(a)$.

Two special infinite sequences play the dominant role in this study. One is the itinerary of the left boundary point 0 of J, denoted by $K_- = I(0)$, and the other one is the limit $I(1-0)$, denoted by K_+. The properties of f guarantees the existence of this limit. We call K_- and K_+ the *lower kneading sequence* and the *upper kneading sequence* respectively.

It is evident that these two sequences are closely related. Using the properties (listed above) of f we obtain the following basic result.

Proposition 9.1.1 1. If the point c is not a periodic point of f, then

$$K_- = 0u \text{ and } K_+ = 1u,$$

where u is an infinite suffix of them.

2. If c is a periodic point of f but not a fixed point, then

$$K_- = (0v1)^\infty \text{ and } K_+ = (1v0)^\infty,$$

where v is a string in Σ^* and the length of $0v1$, $|0v1|$, is the primitive period of c.

Proof. Denote K_+ and K_- by

$$K_+ = 1e_1 \cdots e_n \cdots \text{ and } K_- = 0e'_1 \cdots e'_n \cdots$$

If c is not a periodic point, then $f^n(0) \neq c$ for all $n > 0$. Since $f(c) = 0$ and f is bijective, $f^n(0) \neq 0$ is also true for all $n > 0$. Fix n. Since $f(0) = f(1-)$, there exists a neighborhood $O(0) = [0, \varepsilon) \cup (1 - \varepsilon, 1)$ with $\varepsilon > 0$ such that all f, f^2, f^{n-1} are injective on $O(0)$, and every image $f^i(O(0))$ for $i = 1, \ldots, n-1$ is included either in $(0, c)$ or in $(c, 1)$. Thus we obtain $e_n = e'_n$. Since this holds for every n, we obtain $\sigma(K_+) = \sigma(K_-) = u$ as desired.

If c is a periodic point but not fixed, then using the condition $f(c) = 0$, and assuming that its primitive period is $p > 1$, we have $K_- = I(0) = (0e'_1 \cdots e'_{p-1})^\infty$. From $f^p(0) = 0$ and $f(c) = 0$, it follows that $f^{p-1}(0) = c$, thus $e'_{p-1} = 1$, and K_- is of the form $(0v1)^\infty$ as desired.

Using the foregoing argument for the case of c being not periodic, it is easy to obtain $e_i = e'_i$ for all i, $1 \le i \le p-2$. Since $f^{p-1}(0) = c$, there exists a number $\delta > 0$ such that all $f, f^2, \ldots, f^{p-1}$ are injective on the subinterval $(1-\delta, 1)$, and $f^{p-1}(1-\delta, 1) \subset (0, c)$. Thus we obtain $e_{p-1} = 0$. Since $f(c-) = 1$, the discussion of e_i for all $i \ge p$ is similar and leads to $K_+ = (1v0)^\infty$ ∎

In order to discuss further we have to introduce some order relation between itineraries. Using the notion of free monoid in Chapter 1, in symbolic study of circle homeomorphisms we meet again the free monoid Σ^*, but with a different ordering structure from that in Section 1.2, which is generated from unimodal maps.

The lexicographical order is introduced between strings (and sequences) from Σ. First we define that $0 < 1$ is the order between the symbols of $\Sigma = \{0, 1\}$. If $s = s_1 s_2 \cdots s_n s_{n+1} \cdots$ and $t = t_1 t_2 \cdots t_n t_{n+1} \cdots$ have the following properties that $s_i = t_i$ for $i = 1, 2, \ldots, n$ and $s_{n+1} \neq t_{n+1}$, then we define $s < (>) t$ by the fact $s_{n+1} < (>) t_{n+1}$. If there is no such n can be found for two strings s and t, then we have $s = t$. This order relation have effect for infinite sequences as well as for finite strings over Σ.

Remark. An important point is that if two strings s and t are not of the same length, then the relation $s = t$ in the lexicographical order means simply that the shorter one is a prefix of the longer one. This fact will be used frequently in the sequel.

It is easy to see that the corresponding relation between the points and their itineraries is order-preserving in the following sense:

$$\alpha > \beta \Rightarrow I(\alpha) \geq I(\beta).$$

One consequence of this result is that $K_- \leq I(\alpha) \leq K_+$ holds for every point $\alpha \in J$.

Introducing a shift operator σ on sequences by $\sigma(s_1 s_2 \cdots s_n \cdots) = s_2 \cdots s_n \cdots$, then we have also that

$$K_- \leq \sigma^i(I(\alpha)) \leq K_+ \text{ for } n \geq 0$$

(see Subsection *1.1.4*). Since in the beginning we do not know if an infinite sequence (e.g. K_+), which satisfying these relations, must be an itinerary for some point in J, we take the following extension as our starting point.

Definition 9.1.2 An infinite sequence s over Σ is said to be an *admissible sequence* of f if

$$K_- \leq \sigma^i(s) \leq K_+ \text{ for } i \geq 0.$$

Remark. It is evident that
(1) If s is admissible then each $\sigma^i(s)$ $(i \geq 0)$ is admissible,
(2) Every itinerary is admissible,
(3) Both K_- and K_+ are admissible.

Now we define the formal language of a circle homeomorphism f.

Definition 9.1.3 The language of f, denoted by $L(f)$, or simply L, is a subset of Σ^* that

$$L = \{x \mid x \text{ is a substring of an admissible sequence } s\}.$$

If $x \in L$, x is called a *word* of L.

It is obvious that the language L defined thus far is a dynamical language (see Chapter 2 for Definition of dynamical language).

The language L can be determined by K_+ and K_- directly as follows.

Proposition 9.1.4 A string $x \in \Sigma^*$ belongs to L if and only if

$$K_- \leq \sigma^i(x) \leq K_+ \text{ for } i = 0, 1, \ldots, |x| - 1.$$

Proof. The "only if" part is trivial from Definition 9.1.2. If x satisfies the conditions above, then we can decompose it into $x = x'x''$ such that x'' is the longest suffix of x, which is a prefix of either K_+ or K_-, that is, x'' is the LPS with respect to K_+ and K_-. It means that we have

$$K_- < \sigma^i(x) < K_+ \text{ for } 0 \leq i < |x'|.$$

If x'' is a prefix of K_+, then $s = x'K_+$ is an admissible sequence and has x as its prefix. This proves that $x \in L$. The case of x'' being a prefix of K_- can be discussed similarly. ∎

One of the most important properties of languages for circle homeomorphisms is as follows. Sometimes it is called the *Morse-Hedlund property* (Siegel, Tresser, and Zettler 1992, Morse and Hedlund 1940).

Proposition 9.1.5 If $x, y \in L$ and $x = 0x'$ and $y = 1y'$ in which neither of x', y' is empty, then $x' \geq y'$.

Proof. Using Proposition 9.1.1 we discuss the following two cases respectively.

1. For the case of $K_- = 0u$ and $K_+ = 1u$, we have $x \geq K_- \Rightarrow x' \geq u$ and $y \leq K_+ \Rightarrow y' \leq u \leq x'$.

2. Otherwisely, for the case of $K_- = 0v1\cdots$ and $K_+ = 1v0\cdots$, we have $x' \geq v1\cdots$ and $y' \leq v0\cdots$ and thus finish our proof. ∎

We will also list two facts below, their proofs are lengthy but similar to the corresponding proof of Theorem 3.3.3 and so omitted.

Proposition 9.1.6 If x is a word of L, then there exists a point $\alpha \in J$ such that x is a prefix of the itinerary $I(\alpha)$.

Proposition 9.1.7 If an admissible sequence s does not have K_+ as its suffix, then s is an itinerary of some point of J.

9.2 Rotation Number and Entropy

In this Section first we give a symbolic treatment to the notion of rotation number, the most important concept for circle homeomorphisms. But the main result here is Theorem 9.2.5, which gives the answer to the first question listed in the beginning of this chapter about the number of words of the same length in the language L. It implies also that the topological entropy of each homeomorphism of circle must be zero. The result of Theorem 9.2.5 is also essential for many results later.

9.2.1 Rotation Number of Circle Homeomorphisms

Following Guckenheimer and Holmes (1983), for each infinite sequence s we use the notation

$$N(s, k) = \operatorname{card}\{s_i : s_i = 1 \text{ for } 1 \leq i \leq k\}$$

and define the rotation number of s by the limit of $N(s,k)/k$ when k tends to infinity. For each admissible sequence the existence of this limit is guaranteed as follows.

Proposition 9.2.1 The limit

$$\lim_{k\to\infty} \frac{N(s,k)}{k}$$

exists for each admissible sequence s and independent of s.

From this Proposition we obtain the most important concept for circle homeomorphisms.

Definition 9.2.2 For a circle homeomorphism f, the limit in Proposition 9.2.1 is called the *rotation number* of f.

The proof of Proposition 9.2.1 depends on the following fact, which also appeared in Guckenheimer and Holmes (1983). Since the subject here are admissible sequences (not itineraries), our proof is different.

Proposition 9.2.3 If s and t are two admissible sequences of f, then

$$|N(s,k) - N(t,k)| \leq 1$$

holds for each integer $k \geq 0$.

Proof. Let $s = s_1 s_2 \cdots$ and $t = t_1 t_2 \cdots$. If $k = 1$ the claim is trivial. Proceed inductively on k. Suppose the claim holds for $k \leq n$ and consider the case of $k = n+1$. If $s_1 = t_1$, then we can use the inductive hypothesis for $s_2 \cdots s_{n+1}$ and $t_2 \cdots t_{n+1}$. If $s_1 \neq t_1$, then assume $s_1 = 0$ and $t_1 = 1$ without loss of generality. Using proposition 9.1.5 we have

$$s_2 s_3 \cdots s_{n+1} \geq t_2 t_3 \cdots t_{n+1}.$$

If the equality holds in it, then $N(s, n+1) = N(t, n+1) - 1$. Otherwise there exists an integer i, $2 \leq i \leq n$, that

$$s_2 \cdots s_i = t_2 \cdots t_i \text{ and } s_{i+1} = 1 > t_{i+1} = 0.$$

Now we have $N(s, i+1) = N(t, i+1)$. If $i = n$ then the proof is finished. If $i < n$ we can use the inductive hypothesis to $s_{i+2} \cdots s_{n+1}$ and $t_{i+2} \cdots t_{n+1}$ and complete the proof. ∎

Remark. In Siegel, Tresser, and Zettler (1992) an infinite sequence s is called *Sturmian* if

$$|N(\sigma^i(s), k) - N(\sigma^j(s), k)| \leq 1$$

holds for any i, j (cf. Morse and Hedlund 1940). it is proved there that a sequence is Sturmian if and only if it satisfies the Morse-Hedlund property (see Proposition 9.1.5). In this sense the conclusion of Proposition 9.2.3 is a stronger result.

Proof of Proposition 9.2.1.

From the definition of the notation $N(s,k)$ it is evident that

$$N(s, k+l) = N(s,k) + N(\sigma^k(s), l)$$

holds for each pair of positive integers k and l.

Now taking a fixed positive integer k, using the above equality and Proposition 9.2.3 we can write every integer $m = nk + l$ for $0 \leq l < k$ and obtain

$$nN(s,k) + N(s,l) - n \leq N(s, nk+l) \leq nN(s,k) + N(s,l) + n.$$

Dividing them by $m = nk + l$ and let m tend to infinity, we find that

$$\frac{N(s,k)}{k} - \frac{1}{k} \leq \limsup_{m\to\infty} \frac{N(s,m)}{m} - \liminf_{m\to\infty} \frac{N(s,m)}{m} \leq \frac{N(s,k)}{k} + \frac{1}{k}.$$

It implies that

$$0 \leq \limsup_{m\to\infty} \frac{N(s,m)}{m} - \liminf_{m\to\infty} \frac{N(s,m)}{m} \leq \frac{2}{k}.$$

Since k is arbitrary, we see that the claim of Proposition is true for s. Using the result of Proposition 9.2.3 and dividing it by k, it is evident that the limits for s and t are the same. ∎

In the sequel we will denote the rotation number of f by $\rho(f)$ or, simply, ρ.

Remark. In the proof of Proposition 9.2.1 the main tool is Proposition 9.1.5. It implies that if every substrings taken from an infinite sequence or a set of infinite sequences satisfy the conclusion of this Proposition, then the rotation number exists for every such infinite sequence and has the same value.

9.2.2 Topological Entropy of Circle Homeomorphisms

As in Section 2.5 we have the following definition.

Definition 9.2.4 Let f be a homeomorphism of circle S and L its language. The entropy of f is the limit

$$\lim_{n\to\infty} \frac{\log N_n}{n},$$

where $N_n = \text{card}\{x : x \in L \text{ and } |x| = n\}$.

A well-known fact about circle homeomorphism is that its topological entropy must be zero (see, e.g., Walters 1982). In the following we will give a symbolic proof for it. As a matter of fact, it is a consequence of a much stronger result in the following Theorem which is essential in the rest of Part III.

Theorem 9.2.5 let f be a homeomorphism of circle S. If the point c is not a periodic point of f, then

$$N_n = n + 1 \text{ for } n \geq 0.$$

If c is a periodic point of primitive period q, then

$$N_n = n + 1 \text{ for } n < q \text{ and } N_n = q \text{ for } n \geq q - 1.$$

Remark. The formula $N_n = n + 1$ appeared in Crutchfield (1994, p. 38) without proof. A similar formula also appeared in Morse and Hedlund (1940, Theorem 3.5-3.7) for Sturmian series and beam studied there.

The proof of this Theorem depends on a basic Lemma, which means that under certain condition, a word of L is uniquely left-prolongable (see the property D2 in Definition 2.1.1).

Lemma 9.2.6 Let K_+ be the upper kneading sequence of f and $x \in L$. If x is not a prefix of $\sigma(K_+)$, then there exists a unique $a \in \Sigma = \{0, 1\}$ such that $ax \in L$.

Proof. According Proposition 9.1.1, we discuss two cases.

1. Write $K_+ = 1u$ and $K_- = 0u$, where $u = \sigma(K_+)$ is also an admissible sequence. If neither of $1x$, $0x$ belongs to L, then we have $1x > K_+$ and $0x < K_-$, which leads to $u < x < u$, a contradiction. On the other hand, if both $1x$ and $0x$ belong to L, then we have $u \leq x \leq u$. This leads to $x = u$ and implies that x is a prefix of u, which contradicts our condition.

2. Write $K_+ = 1v0\cdots$ and $K_- = 0v1\cdots$. If neither of $1x$, $0x$ belongs to L, then we have $v0 < x$ and $x < v1$, which cannot be true under the condition that x is not a prefix of u. Otherwise we have $v1 \leq x$ and $x \leq v0$, which hold only when x is a prefix of v, a contradiction again. ∎

Proof of Theorem 9.2.5. Since L is a dynamical language, if $x \in L$ and $|x| > 0$, then we can write $x = ax'$ for some $a \in \Sigma$ and $x' \in L$. So we can use Lemma 9.2.6 to calculate N_n as follows.

Discuss two cases respectively.

1. Write $K_+ = 1u$ and $K_- = 0u$ as before. Proceed inductively on n. It is trivial for $n = 0$ and $n = 1$. Assume that the claim is true for n and discuss the case of $n + 1$. Consider all words of length n, $x_1, x_2, \ldots, x_{n+1}$. It is evident that a prefix u' of u, which is of length n, belongs to L, and both $1u'$ and $0u'$ belong to L. Denoting this u' by x_1, and using Lemma 9.2.6 to $x_2, \ldots, x_{n+1}$, we obtain $N_{n+1} = n + 2$ and complete our induction.

2. Here $K_+ = 1v0\cdots$, $K_- = 0v1\cdots$ and $|0v1| = q$, the period of point c (and point 0). If $n < q - 1$, the argument is the same as in the case 1. Consider $n \geq q - 1$ and the prefix x of $\sigma(K_+)$, which is of length n. Since x has $v0$ as its prefix, the string $0x = 0v0\cdots$ is less than K_- and does not belong to L. So we can use Lemma 9.2.6 to other strings of length n and prove $N_n = q$ for each $n \geq q - 1$ inductively. ∎

9.3 Nested Block Structure

From the viewpoint of symbolic dynamics, the most important symbolic feature of circle homeomorphisms is the nested block structure in their admissible sequences. It appeared more or less explicitly in Gambaudo, Lanford III and Tresser (1984), Procaccia, Thomae and Tresser (1987), Zheng (1989c, 1991), Hao (1989), and was proved in Siegel, Tresser, and Zettler (1992).

In this section we will give a rigorous discussion about this structure and take this fact as the starting point to obtain our main results in analysis of grammatical complexity later.

9.3.1 Existence of Nested Block Structure

In this Subsection we will prove that for a homeomorphism f of circle S all admissible sequences have a natural nested block structure, which decides all combinatorial properties of L. The first Lemma appeared in, e.g., Gambaudo, Lanford III and Tresser (1984).

Lemma 9.3.1 If either $10^k1, 10^{k'}1 \in L$ or $01^k0, 01^{k'}0 \in L$, then $|k - k'| \leq 1$.

Proof. If $k \neq k'$, then we can assume $k' < k$. Using the proposition 9.1.5, for the first case we have $0^{k-1}1 \geq 0^{k'}1 \Rightarrow k - 1 \leq k'$, and for the second case $1^{k'}0 \geq 1^{k-1}0$, thus obtain the same conclusion, that is, $|k - k'| = 1$. ∎

Using this Proposition and Proposition 9.1.5, we can easily obtain all admissible sequences for some simple cases. In the following Proposition we list some of them, since the calculation involved there is quite elementary their proof is omitted.

Proposition 9.3.2 1. For $\rho = 0$, there are only two kinds of admissible sequences: (a) 0^∞, (b) 0^i10^∞ for $i \geq 0$.

2. For $\rho = 1$, there are only two kinds of admissible sequences: (a) 1^∞, (b) 1^i01^∞ for $i \geq 0$.

3. For $\rho = 1/k$ and $k \geq 2$, there are only three kinds of admissible sequences: (a) $(10^{k-1})^\infty$ and their shifts, (b) $(0^{k-1}1)^i0^k(10^{k-1})^\infty$ for $i \geq 0$ and their shifts containing 0^k, (c) $(10^{k-1})^i10^{k-2}1(0^{k-1})^\infty$ for $i \geq 0$ and their shifts containing $10^{k-2}1$.

4. For $\rho = 1 - 1/k$ and $k > 2$, there are only three kinds of admissible sequences: (a) $(01^{k-1})^\infty$ and their shifts, (b) $(1^{k-1}0)^i1^k(01^{k-1})^\infty$ for $i \geq 0$ and their shifts containing 1^k, (c) $(01^{k-1})^i01^{k-2}0(1^{k-1})^\infty$ for $i \geq 0$ and their shifts containing $01^{k-2}0$.

For a given f with the rotation number of the cases 3 and 4, the latter two kinds, (b) and (c), of admissible sequences cannot exist simultaneously.

In order to deal with more general cases, we have to be able to use such block structure repeatedly. Here the following fact that these structures are nested one by one plays the key role.

Lemma 9.3.3 Suppose f has its rotation number ρ.

1. If an admissible sequence s contains blocks $10^{k-1}1$ and 10^k1 for some $k \geq 2$, then after a possible prefix of 0^i is discarded and a block renaming of $10^{k-1} \to 1, 10^k \to 0$ is performed, a new sequence s' over $\{0,1\}$ is obtained and has its rotation number ρ'. The relation of ρ, k, and ρ' is as follows:

$$\text{if } \frac{1}{k+1} < \rho < \frac{1}{k}, \text{ then } \rho' = 1 - \left(\frac{1}{\rho}\right) \text{ and } k = \left[\frac{1}{\rho}\right]$$

2. If s contains blocks $01^{k-1}0$ and 01^k0 for some $k \geq 2$, then after a possible prefix of 1^i is discarded and a block renaming of $01^{k-1} \to 0, 01^k \to 1$ is performed, a new sequence s' is obtained and has its rotation number ρ'. The relation of ρ, k, and ρ' is as follows:

$$\text{If } 1 - \frac{1}{k} < \rho < 1 - \frac{1}{k+1}, \text{ then } \rho' = \left(\frac{1}{1-\rho}\right) \text{ and } k = \left[\frac{1}{1-\rho}\right].$$

Proof. Because the discussion is similar for both cases, we only give the proof for case 2. From Proposition 9.1.5 we know that every admissible sequence consists of blocks 01^{k-1}, 01^k, and a possible prefix of 1^i. From the Remark before Definition 9.2.4 we see that in order to prove the rotation number exists for the new sequence s', it suffices to prove that Proposition 9.1.5 still holds for all such s'. Assuming that $x = 0x'$ and $y = 1y'$ are two substrings of some s'. Let a homomorphism h be

$$h = \{1 \to 01^k, 0 \to 01^{k-1}\}.$$

From the procedure of renaming rule we see that $01^{k-1}h(x')$ and $01^k h(y')$ belong to L. Using Proposition 9.1.5 we have $h(x') \geq h(y')$, which is equivalent to $x' \geq y'$. The relations between ρ, k, and ρ' for two cases is an easy exercise of counting and omitted. ∎

Now we define two homomorphisms on the free monoid $\{0, 1\}^*$ as follows.

$$h_{r,k} = \{1 \to 10^{k-1}, 0 \to 10^k\},$$

$$h_{l,k} = \{1 \to 1^k 0, 0 \to 1^{k-1}0\},$$

for some $k \geq 0$. As a fact, they are the reverse transformations of the blocks renamings in the above proposition with slight modification to the case 2, which is convenient for finding maximal sequence as follows.

Theorem 9.3.4 1. If ρ is rational and $0 < \rho < 1$, then each admissible sequence s has a suffix, which is a periodic admissible sequence w^∞ and independent of s. This string w is a maximal word of L and has a expression of

$$w = h_1 \cdots h_n(1),$$

where each $h_i, i = 1, \ldots, n$, is either h_{r,k_i} or h_{l,k_i}.

2. If ρ is irrational, then the upper kneading sequence K_+ has a expression of

$$K_+ = \lim_{n \to \infty} h_1 \cdots h_n(1),$$

where each h_i is either h_{r,k_i} or h_{l,k_i} for each $i \geq 1$.

Proof. 1. It holds already for the cases listed in Proposition 9.3.2. For general case of rational ρ we can use Lemma 9.3.3 to obtain new admissible sequences and new rotation number. Because ρ is rational, after finite steps we will arrive to one of two case: either $\rho = 1/k$ or $\rho = 1 - 1/(k+1)$ for some $k > 2$. From Proposition 9.3.2 we know that for the former case there is a periodic suffix of $(10^{k-1})^\infty$, the latter case a periodic suffix of $(1^k 0)^\infty$. This decides that the previous homomorphism is $h_{r,k}$ or $h_{l,k}$. Reversing the above procedure, which is equivalent to apply homomorphisms either $h_{r,k}$ or $h_{l,k}$, we obtain the required w^∞, which is an infinite suffix of the original sequence s. Its independence of s is implied by Lemma 9.3.1.

2. In the case of irrational ρ the procedure of block renaming as described in Lemma 9.3.3 will be going on indefinitely without end. Since we know that K_+ is a

maximal sequence, using the procedure of block renaming to K_+ and let h_i be the reverse transformation of block renaming at the i-th step, $i = 1, 2, \ldots$. Define

$$w^{(n)} = h_1 \cdots h_n(1),$$

then $w^{(n)}$ is a prefix of K_+. Since $w^{(n)}$ is a proper prefix of $w^{(n+1)}$ for each $n > 0$ so K_+ is the limit of $\{w^{(n)}\}$ when n tends to infinity. ∎

Corollary 9.3.5 If ρ is rational, then each admissible sequence is eventually periodic, and its periodic suffix is w^∞ in the proof of Theorem 9.3.4 or its shift. (On the other hand, it is already known that if there is one admissible sequence being eventually periodic, then the rotation number is rational.)

9.3.2 Farey Tree and Continued Fraction

It seems that the notions of Farey tree and Farey address appeared first in Ostlund and Kim (1985) (and some unpublished papers), and have been widely used since then in discussions about circle homeomorphisms (Hao 1989, Zheng 1989c, 1991, Siegel, Tresser, and Zettler 1992).

On the other hand, the notion of *Farey composition* appeared often in books of number theory (see, e.g., Hardy and Wright 1981). Recall that a Farey composition of two rational fractions p/q and p'/q' is defined by

$$\frac{p}{q} \oplus \frac{p'}{q'} = \frac{p+p'}{q+q'}.$$

Applying this composition rule to $0/1$ and $1/1$, the two extremes of the interval $[0, 1]$, one gets $1/2$, which is considered as the top of the Farey tree, and is the zeroth level. Reapplying this rule to $0/1$, $1/2$, and $1/1$, we obtain $1/3$ and $2/3$, which consist of the first level of the Farey tree. Proceed on in this way, we obtain the *Farey tree* as shown in Figure 9.2 through level 5. In this Figure the two "ancestors" $0/1$ and $1/1$ are omitted.

There are 2^k members on the k-th level. It can be proved that each rational number p/q generated in this procedure is irreducible, which means that p and q are prime to each other. Each member in the tree is the Farey composition of two numerically closest members in the tree above and two ancestors $0/1$ and $1/1$. It is easy to find that this tree is symmetrical to the central vertical, that is to say, the values of two members which are symmetrically located to the central vertical yields the sum 1.

Denoting the members of the n-th level by a_i^n, i from 0 to $2^n - 1$, we put a_{2i}^n below and to the left and a_{2i+1}^n below and to the right of a_i^{n-1}, and use bonds to connect the three members as shown in Figure 9.2, in which the left bond is labeled "-1" and the right "$+1$".

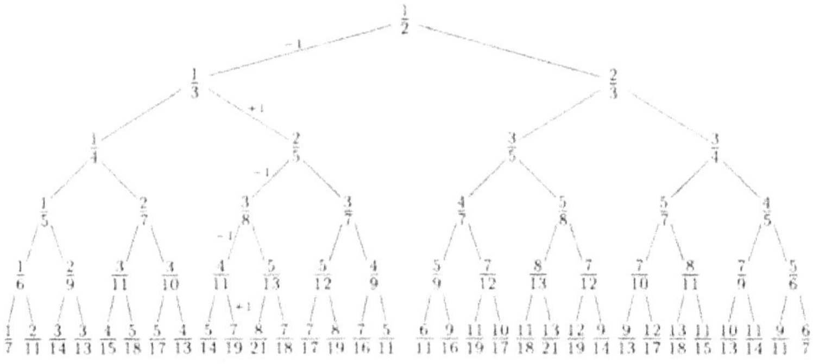

Figure 9.2 The Farey tree through level 5.

The *Farey address* of a rational number is defined to be the string of ± 1 obtained by reading the path from the top of the tree to the number. We use the notation $Ad(\rho) = \langle b_1, b_2, \ldots \rangle$ to denote the Farey address of number ρ. An example is shown in Figure 9.2 that the Farey address of 7/19 is $\langle -1 + 1 - 1 - 1 + 1 \rangle$.

Although there are only rational numbers appearing in Farey tree, but if infinite paths are considered then we can use Farey address to obtain irrational number between 0 and 1.

The Farey address is closely related with the continued fraction representation of real numbers. The details about this classical subject can be found, for instance, in Hao (1989), Hardy and Wright (1981).

Each real number $\rho \in (0, 1)$ can be represented by a (simple) continued fraction as follows:

$$\rho = \cfrac{1}{a_1 + \cfrac{1}{a_2 + \cfrac{1}{a_3 +}}} = [a_1, a_2, a_3, \ldots],$$

where the integers $a_0, a_1, \cdots$ are called the first, second, ... partial quotients of the continued fraction. If ρ is a rational number, then its continued fraction is finite. In this case $\rho = p/q$ has exactly two continued fraction representations as

$$p/q = [a_1, a_2, \ldots, a_n] = [a_1, a_2, \ldots, a_n - 1, 1],$$

which are called the short and the long continued fraction representations of p/q.

Some basic facts about Farey tree, Farey address and their relations with continued fraction are listed below. For simplicity we omitted the proofs for them.

1. The sum of all 2^n members of the n-th level in Farey tree is 2^{n-1}.
2. If $h/k < l/m$ are two numerically consecutive numbers in Farey tree through level n, then $kl - hm = 1$.

3. If $h/k < r/s < l/m$ are three numerically consecutive numbers in Farey tree through level n, then

$$\frac{r}{s} = \frac{h+l}{k+m},$$

but here $h + l$ and $k + m$ need not be prime to each other.

4. If ρ and ρ' have the Farey addresses $\langle b_1, b_2, \ldots\rangle$ and $\langle b'_1, b'_2, \ldots\rangle$, and $b_k + b'_k = 0$ holds for each k, then $\rho + \rho' = 1$.
5. Every rational number $p/q \in (0, 1)$ appears in the Farey tree once and only once.
6. Every real number $\rho \in (0, 1)$ is uniquely determined by its Farey address in Farey tree.
7. A rational number $p/q \in (0, 1)$ is a member of the n-th level in Farey tree if and only if the sum of its quotients in the continued fraction is $n + 2$.
8. If $p/q \in (0, 1)$ has the Farey address $\langle b_1, \ldots b_k\rangle$, and its long continued fraction representation is written in the form of $[1 + a_1, a_2, \ldots, a_n, 1]$, then

$$b_1 b_2 \cdots b_k = \begin{cases} (-1)^{a_1}(+1)^{a_2}\cdots(+1)^{a_n}, & \text{when } n \text{ is even,} \\ (-1)^{a_1}(+1)^{a_2}\cdots(-1)^{a_n}, & \text{otherwise.} \end{cases}$$

An example of the last item listed above is shown below.

$$Ad(\frac{7}{19}) = \langle -1 + 1 - 1 - 1 + 1\rangle, \text{ and}$$

$$\rho = \cfrac{1}{2 + \cfrac{1}{1 + \cfrac{1}{2 + \cfrac{1}{1 + \cfrac{1}{1}}}}} = [2, 1, 2, 1, 1].$$

Remark. In contrast with the Farey tree and Farey address, the material of *Farey series* can be found in many textbooks of number theory. The nth Farey series $\mathcal{F}_n$ is defined as the ascending sequence of all irreducible rational numbers in $[0, 1]$ with denominator at most equal to $n \geq 1$. For example,

$$\mathcal{F}_5 = \left(\frac{0}{1}, \frac{1}{5}, \frac{1}{4}, \frac{1}{3}, \frac{2}{5}, \frac{1}{2}, \frac{3}{5}, \frac{2}{3}, \frac{3}{4}, \frac{4}{5}, \frac{1}{1}\right)$$

Two rational numbers are *Farey neighbors* if they are consecutive in some Farey series.

The basic results about Farey series are as follows.

1. If $h/k < l/m$ are Farey neighbors, then $kl - hm = 1$
2. If $h/k < r/s < l/m$ are three consecutive numbers in some Farey series, then
$$\frac{r}{s} = \frac{h+l}{k+m}.$$
(Note that $h+l$ and $k+m$ need not be prime to each other.)
3. If $h/k < r/s < l/m$, $h/k, l/m \in \mathcal{F}_n$, and $r/s \in \mathcal{F}_{n+1}$ but not $\mathcal{F}_n$, then $r = h + l$ and $s = k + m$.

It should be pointed out that the facts about Farey tree and Farey address cannot be deduced directly from these well-known results about Farey series and have to be proved independently. As an example, it is not difficult to prove that each pair of two consecutive members on the same level of Farey tree are not Farey neighbor in some Farey series.

9.3.3 Relation of Nested Block Structure with Farey Address

Here we explore the relation between the nested block structure and the Farey address. This discussion also clarify its relation with continued fraction representation of rotation number. In fact, Theorem 9.3.7 offers a new method to compute the Farey address.

A method to compute the word w in Theorem 9.3.4 was proposed in Hao (1989) as follows. Define two *Farey transformations* by

$$\tau_{+1} = \{1 \to 1, 0 \to 10\},$$

$$\tau_{-1} = \{1 \to 10, 0 \to 0\}.$$

The word w can then be obtained as follows: If $Ad(\rho)$ is $(b_1 \cdots b_n)$, then (see pp. 220-221 of Hao 1989)

$$w = \tau_{b_1} \cdots \tau_{b_n}(10).$$

The relation between the homomorphisms $h_{r,k}$, $h_{l,k}$ and Farey transformation $\tau_{\pm 1}$ is simple:

$$h_{r,k} = \tau_{-1}^{k-1}\tau_{+1},$$

$$h_{l,k} = \tau_{+1}^{k-1}\tau_{-1}.$$

Example 9.3.6 For $\rho = 4/17$. Using the method of Subsection *9.3.1* we can obtain $h_1 = h_{r,4}$ and $\rho' = 3/4$, then $h_2 = h_{l,3}$. So we have

$$\begin{aligned} w &= h_1 h_2(1), \\ &= h_1(1110), \\ &= 10001000100010000 \end{aligned}$$

On the other hand, its continued fraction representation is $[4, 3, 1]$ (or $[4, 4]$). Its Farey address is $\langle -1, -1, -1, +1, +1, +1\rangle$. The formula of computation of w by Hao (1989) is

$$w = \tau_{-1}\tau_{-1}\tau_{-1}\tau_{+1}\tau_{+1}\tau_{+1}(10)$$

which gives the same result.

Next we give a rigorous proof for the relation between these two methods. Define a transformation from homomorphisms $h_{r,k}, h_{l,k}$ to strings of -1 and $+1$ as follows.

Assume ρ is a given rational rotation number and $0 < \rho < 1$. If $w = h_1 \cdots h_n(1)$ is the word specified in Theorem 9.3.4, then define

$$T(h_i) = \begin{cases} (-1)^{k-1}(+1) & \text{if } h_i = h_{r,k} \text{ and } i < n, \\ (-1)^{k-2} & \text{if } h_i = h_{r,k} \text{ and } i = n, \end{cases}$$

and

$$T(h_i) = \begin{cases} (+1)^{k-1}(-1) & \text{if } h_i = h_{l,k} \text{ and } i < n, \\ (+1)^{k-1} & \text{if } h_i = h_{l,k} \text{ and } i = n. \end{cases}$$

Theorem 9.3.7 If ρ and w are given and transformation T is defined as above, then the Farey address of ρ is obtained as follows

$$Ad(\rho) = T(h_1 \cdots h_n) = T(h_1) \cdots T(h_n).$$

Two simple Propositions are helpful (their proofs are simple and omitted).

Lemma 9.3.8 If $\rho = [1, a_2, a_3, \dots]$, then

$$[\frac{1}{1-\rho}] = a_2 + 1 \text{ and } (\frac{1}{1-\rho}) = [a_3, \dots].$$

Lemma 9.3.9 If $\rho = [a_1, a_2, a_3, \dots]$ where $a_1 > 1$, then

$$[\frac{1}{\rho}] = a_1 \text{ and}$$

$$1 - (\frac{1}{\rho}) = \begin{cases} [a_3 + 1, \dots] & \text{if } a_2 = 1 \\ [1, a_2 - 1, a_3, \dots] & \text{if } a_2 > 1. \end{cases}$$

Proof of Theorem 9.3.7. Proceed inductively on n. If $n = 1$, there are two cases. If $\rho = 1/k$ for $k \geq 2$, then $h = h_{r,k}$ and $h(1) = 10^{k-1}$. By the rule we have $T(h) = (-1)^{k-2}$. Since $\rho = [k-1, 1]$ the claim is true. If $\rho = 1 - (1/(k+1))$ for $k \geq 2$, then $h = h_{l,k}$ and $h(1) = 1^k0$. By the rule we have $T(h) = (+1)^{k-1}$. Since $\rho = [1, k-1, 1]$ the claim is also true.

Now assume that the claim holds for n and consider the case of $n + 1$.

Case 1. If $1 - (1/k) < \rho < 1 - (1/(k+1))$ for $k \geq 2$, then $\rho = [1, a_2, \dots]$. Using Lemma 9.3.8 we see that $k = a_2 + 1$ and $\rho' = (1/(1-\rho)) = [a_3, \dots]$. Using the inductive hypothesis to $h_2, \ldots, h_{n+1}$, we have $T(h_2 \cdots h_{n+1}) = (-1)^{a_3-1} \ldots$. Since $T(h_1) = T(h_{l,a_2+1}) = (+1)^{a_2}(-1)$ the claim is true.

Case 2. If $1/(k+1) < \rho < 1/k$ for $k \geq 2$, then $\rho = [a_1, a_2, \ldots]$ and $a_1 > 1$. Using Lemma 9.3.9 we have $k = a_1$ and $h_1 = h_{r,a_1}$. There are two possibilities.

If $a_2 = 1$, then $\rho' = 1 - (1/\rho) = [a_3 + 1, a_4, \ldots]$. So we have

$$T(h_2 \cdots h_{n+1}) = (-1)^{a_3}(+1)^{a_4} \ldots$$

Since $T(h_1) = (-1)^{a_1-1}(+1)$ the claim is true.

If $a_2 > 1$, then $\rho' = [1, a_2 - 1, a_3, \ldots]$. So we have

$$T(h_2 \cdots h_{n+1}) = (+1)^{a_2-1}(-1)^{a_3} \ldots$$

and finish our proof. ∎

CHAPTER 10

COMPLEXITY LEVELS OF CIRCLE HOMEOMORPHISMS

It is well-known that the language of circle homeomorphism is regular if and only if its rotation number is rational (see, e.g., Friedmann 1991). In Part III this fact will be proved by two methods. The first proof is given in Section 10.1 and depends on discussion about the structure of all words and admissible sequences. The second proof will be given by Theorem 11.1.2 in the next chapter, which depends on the Myhill-Nerode Theorem.

In Section 10.2 the non-regular languages of circle homeomorphisms are discussed. It is proved that for these languages their lowest level in Chomsky hierarchy is context-sensitive. The structure of all words and admissible sequences is analyzed as well.

Section 10.3 is completely devoted to an important example, the case of golden mean ratio, that is, the rotation number

$$\rho = (\sqrt{5} - 1)/2.$$

The set of all DEB's, the admissible sequences and the infinite automaton accepting its language are calculated.

10.1 Regular Languages for Rational Rotation Numbers

For the case of rational rotation numbers the maximal string w appearing in Theorem 9.3.4 plays a central role in the sequel. This maximal string w will be referred to as the *basic word* of the language L when its rotation number ρ is rational. For simplicity it is also said that w is the basic word for the rotation number ρ, if ρ is rational.

The main result of this section is the following Theorem.

Theorem 10.1.1 Let f be a homeomorphism of circle and ρ be its rotation number. If ρ is rational and $0 < \rho < 1$, then the basic word w can be written in the form of $1v0$ and $|w| = q \geq 2$.

1. The formal language L is regular if and only if ρ is rational.

2. If $\rho = 0$ or 1 then L is uniquely determined, but if ρ is rational and $0 < \rho < 1$, then L can be one of three possibilities, i.e., its kneading sequences is one of the following cases:

(a) $K_+ = (1v0)^\infty, K_- = (0v1)^\infty$,

(b) $K_+ = (1v0)^\infty, K_- = 0v0(1v0)^\infty$,
(c) $K_+ = 1v1(0v1)^\infty, K_- = (0v1)^\infty$

3. There are only q different admissible sequences for case (a), and countable for cases (b) and (c). Each admissible sequence is an itinerary of some point in the interval $J = [0, 1)$.

Remark. The claim 1 of this theorem was obtained first by Friedmann (1991) for some special homeomorphisms of circle.

10.1.1 Properties of Basic Word

Before we give the proof for the main Theorem, it is necessary to know more about the basic word w.

Lemma 10.1.2 If $w = h_1 \cdots h_n(1)$, where $h_i = h_{r,k_i}$ or h_{l,k_i}, $i = 1, \ldots, n$, then w is a primitive string, that is, there is no string x such that $w = x^m$ for some $m > 1$.

Proof. We will prove a stronger statement that p, the number of 1 in w, and q, the length of w, are coprime. Proceed inductively on n. It is trivial for $n = 1$. It suffices to show that if it is true for w, then it is also true for $h(w)$, where h is either $h_{r,k}$ or $h_{l,k}$ for some $k > 0$.

Let p and q be coprime. If $h = h_{r,k} = \{1 \to 10^{k-1}, 0 \to 10^k\}$, then it is easy to calculate p', the number of 1 in $h(w)$, and q', the length of $h(w)$, as follows,

$$p' = q, \text{ and } q' = kq + (q - p) = (k + 1)q - p,$$

so p' and q' are coprime still.

If $h = h_{l,k} = \{1 \to 1^k 0, 0 \to 1^{k-1}0\}$, then we have

$$p' = (k - 1)q + p, \text{ and } q' = kq + p,$$

and obtain the same conclusion. ∎

From the fact that w^∞ is an admissible sequence and Proposition 9.2.1 we can obtain ρ from w directly.

Corollary 10.1.3 If w is the basic word for the rational rotation number ρ, then $\rho = p/q$, where $q = |w|$ and p is the number of symbol 1 appearing in w.

Lemma 10.1.4 Let $w = 1v0$ be the basic word for a rational rotation number ρ, where $0 < \rho < 1$, then the string

$$0v1 = \min_{0 \le i < |w|} \sigma^i(w),$$

where σ is the cyclic shift operator.

Proof. Proceed inductively on n in the expression of $w = h_1 \cdots h_n(1)$.

It is trivial for $n = 1$. Now it suffices to argue that if the claim is true for w, then it is also true for $h(w)$, where h is either $h_{r,k}(w)$ or $h_{l,k}(w)$. We only give the proof for $h_{r,k}$, as the discussion of $h_{l,k}$ is similar.

Denote $w = w_1 w_2 \cdots w_q$, where each $w_i \in \{0, 1\}$ and $q = |w|$. For simplicity write $h_{r,k}$ by h. Using the obvious fact of $w_1 = 1, w_q = 0$, we have

$$h(w) = 10^{k-1}h(w_2)\cdots h(w_{q-1})10^k$$

From the inductive hypothesis that $0w_2 \cdots w_{q-1}1$ is a cyclic shift of w, so the string

$$10^k h(w_2)\cdots h(w_{q-1})10^{k-1}$$

is also a cyclic shift of $h(w)$. This implies directly that $0^k h(w_2)\cdots h(w_{q-1})10^{k-1}1$ is a cyclic shift of $h(w)$.

The remaining part of proof is to show that

$$0^k h(w_2)\cdots h(w_{q-1})10^{k-1}1 \le \sigma^i(h(w)) \text{ for all } 0 \le i < |h(w)|.$$

It is equivalent to

$$10^k h(w_2)\cdots h(w_{q-1})10^{k-1} \le h(w_i)\cdots h(w_q)h(w_1)\cdots h(w_{i-1}) \text{ for all } 0 \le i < q.$$

Since h is order-preserving, this is true from the inductive hypothesis that

$$0w_2 \cdots w_{q-1}1 \le \sigma^i(w)$$

holds for every $0 \le i < q$. ∎

10.1.2 *Proof of Theorem 10.1.1*

The claim 2 of Theorem 10.1.1 is proved by the following Proposition.

Proposition 10.1.5 If ρ is rational, $0 < \rho < 1$ and $w = 1v0$, then there are only three possibilities for the upper and lower kneading sequences:

(a) $K_+ = (1v0)^\infty, K_- = (0v1)^\infty$,

(b) $K_+ = (1v0)^\infty, K_- = 0v0(1v0)^\infty$,

(c) $K_+ = 1v1(0v1)^\infty, K_- = (0v1)^\infty$

Proof. From the proof of Theorem 9.3.4, w^∞ is an admissible sequence. Using the result of Lemma 10.1.4, $(0v1)^\infty$ is also admissible. Now we can prove that each word x in L satisfies conditions that

$$0v0 \le x \le 1v1$$

As a matter of fact, if $x < 0v0$ then writing $x = 0x'$ we have $x' < v0$. Since $1v0 \in L$, using Proposition 9.1.5 leads to a contradiction. Similarly it is impossible that $x > 1v1$. It is also clear from Proposition 9.1.5 that $0v0$ and $1v1$ cannot belong to L simultaneously.

Now there are three possibilities.

(a) Neither of $0v0$, $1v1$ belongs to L. Then $(1v0)^\infty$ and $(0v1)^\infty$ are maximal and minimal admissible sequence respectively, and must be K_+ and K_-.

(b) The string $0v0$ belongs to L. In this case $(1v0)^\infty$ is the maximal admissible sequence and must be K_+ itself. On the other hand, K_- must begin from $0v0$. Let $K_- = 0v0s$ with an infinite suffix s. Using Proposition 9.1.5 again and $K_+ = (1v0)^\infty$, we have $s \geq (1v0)^\infty$, which implies that $s = (1v0)^\infty$.

(c) The string $1v1$ belongs to L. This discussion is similar to case (b). ∎

Lemma 10.1.6 If $w = 1v0$ and $x \in L$ satisfy the conditions that

$$0v1 \leq x \leq 1v0 \text{ and } |x| \leq |w|,$$

then x is a prefix of $\sigma^i(w)$ for some i.

Proof. It suffices to discuss the case of $|x| = |w| = q$. From Theorem 9.3.4 we know that w^∞ is admissible. It implies that each $\sigma^i(w) \in L$ for $0 \leq i < |w|$. Using Lemma 10.1.2 they provide q different words of L. Combining the results of Theorem 9.2.5 and Proposition 10.1.5, we see that there are exactly q words of length $|w|$ between $w = 1v0$ and $0v1$ and complete the proof. ∎

Corollary 10.1.7 If $w = 1v0$ and s is an admissible sequence which satisfies the conditions that

$$(0v1)^\infty \leq \sigma^i(s) \leq (1v0)^\infty \text{ for all } i \geq 0,$$

then $s = \sigma^i(w^\infty)$ for some $0 \leq i < |w|$.

The claim 3 of Theorem 10.1.1 is proved in the following Proposition.

Proposition 10.1.8 For the case (a) there are only q different admissible sequences, but for the cases (b) and (c) the set of all admissible sequences is countable.

Proof. The conclusion of case (a) is a consequence of Corollary 10.1.7 and Proposition 10.1.5. For the case (b), it is easy to see that, except those admissible sequences of case (a), there are another kind of admissible sequences which contain the word $0v0$. But if s is such an admissible sequence, then its suffix after $0v0$ can only be $(1v0)^\infty$ as shown in the proof of Proposition 10.1.5.

Denoting $w' = 0v1$ and using Lemmas 10.1.8 and 9.2.6 we obtain the general forms of them:

$$w'^m(0v0)w^\infty$$

and their shift containing $0v0$.

Using Lemma 10.1.6 we know that there exists no other admissible sequence any more. the discussion about case (c) is similar and omitted.

From the discussion and Lemma 9.2.6 we see that if the rotation number is rational, then every admissible sequence must be an itinerary of some point in the interval $J = [0, 1)$. ∎

Proof of the claim 1 of Theorem 10.1.1. The "if" part. It is very easy to write out the regular expression of L in each case (see Subsection A.2.2).

If $\rho = 0$, then from Proposition 9.3.2 we can obtain the regular expression of language L:

$$L = 0^* + 0^*10^*$$

Similarly, for $\rho = 1$, we have

$$L = 1^* + 1^*01^*$$

For other value of ρ there are three different cases.

(a) $K_+ = (1v0)^\infty, K_- = (0v1)^\infty$. We have obtained all admissible sequences in Corollary of Lemma 10.1.6 and so

$$L = S(w)w^*P(w),$$

where the notations $S(w)$ and $P(w)$ mean to take all suffixes and prefixes of w.

(b) $K_+ = (1v0)^\infty, K_- = 0v0(1v0)^\infty$. Using Lemma 10.1.8 we have

$$L = S(w)w^*P(w) + S(w')w'^*(0v0)w^*P(w).$$

(c) $K_+ = 1v1(0v1)^\infty, K_- = (0v1)^\infty$. Similarly we can obtain

$$L = S(w)w^*P(w) + S(w)w^*(1v1)w'^*P(w').$$

The "only if" part. If L is a regular language, then using Lemma A.4.1 (the so-called pumping Lemma of regular language), there exists a constant n such that for every word $x \in L$ with $|x| \geq n$, there exists a decomposition $x = uvw$ in such a way that $|uv| \leq n$, $|v| \geq 1$, and $uv^iw \in L$ for all $i \geq 0$. Since L is dynamical, we see that $uv^i \in L$ for all $i \geq 0$. This implies that uv^∞ is an admissible sequence of f. Since this sequence is eventually periodic, the rotation number of f is rational. ∎

10.2 Non-Regular Languages for Irrational Rotation Numbers

In this Section the rotation number ρ is always supposed to be an irrational number. From Theorem 10.1.1 we know that under this condition the language is non-regular. The main result is the following Theorem.

Theorem 10.2.1 Assume f is a homeomorphism of circle and its rotation number ρ is irrational.

1. The language $L(f)$ is uniquely determined by ρ. It is not a context-free language, and is recursive if and only if the rotation number ρ is computable.
2. Every word of language L is a substring of every admissible sequence.
3. The set of all admissible sequences is uncountable, and except at most a countable subset of it, each admissible sequence is an itinerary of some point of the interval J.

In the following discussion we use the notations in Theorem 9.3.4 that $K_+ = \lim_{n \to \infty} w^{(n)}$, where $w^{(n)} = h_1 \cdots h_n(1)$ and $K_+ = 1u$ and $K_- = 0u$ (from Proposition 9.1.1).

First we establish a lemma which asserts that there exists a special kind of prefixes of K_+. As a fact, this result will also play an important role in next Chapter.

Lemma 10.2.2 If ρ is irrational, then for each number N, there exists a positive integer k such that

$$K_+ = yy \ldots, \text{ where } y = w^{(k)} \text{ and } |y| > N.$$

Proof. At first we choose n sufficient large such that $|w^{(n)}| > N$. Since $w^{(n)}$ is a proper prefix of $w^{(n+1)}$, if h_{n+1} is an $h_{l,k} = \{1 \to 1^k0, 0 \to 1^{k-1}0\}$ for some $k > 1$, then $h_{n+1}(1)$ has 11 as its prefix and K_+ has $h_1 \cdots h_n(11) = yy$ with $y = w^{(n)}$ as prefix. In this case take $k = n$. Otherwise if h_{n+2} is an $h_{l,k}$ then we can take $y = w^{(n+1)}$ and $k = n + 1$. Finally, if

$$h_{n+1} = \{1 \to 10^{k_1 - 1}, 0 \to 10^{k_1}\}$$

and

$$h_{n+2} = \{1 \to 10^{k_2 - 1}, 0 \to 10^{k_2}\},$$

then

$$\begin{aligned} w^{(n+2)} &= h_1 \cdots h_{n+2}(1) \\ &= h_1 \cdots h_{n+1}(10^{k_2 - 1}) \\ &= h_1 \cdots h_n(10^{k_1 - 1}10^{k_1} \ldots) \\ &= h_1 \cdots h_n h_{n+1}(11) \ldots, \end{aligned}$$

so we can still take $y = w^{(n+1)}$ and $k = n + 1$. ∎

The claim 2 of Theorem 10.2.1 is proved separately by following Proposition.

Proposition 10.2.3 Every word of language L is a substring of every admissible sequence.

Proof. The proof is divided in two steps.

Step 1. For any given word x of L, using Lemma 10.2.2 we can find $K_+ = yy \ldots$, where $y = w^{(k)} = h_1 \cdots h_k(1)$ for some k and $|y| > |x|$. Writing $y = 1v0$, we have $K_+ = 1v01v0 \ldots$ and $K_- = 0v01v0 \ldots$ by Proposition 9.1.1. From the fact that $x \in L$ we have $0v0 \le x \le 1v0$. If x is a prefix of $0v$, then it is, by Lemma 10.1.4, a substring of $v01v0$. Otherwise we have $0v1 < x \le 1v0$. Since $yy = 1v01v0$ is a prefix of K_+, every $\sigma^i(1v0)$ belong to L, from Lemma 10.1.6 we see that x is a proper prefix of $\sigma^i(1v0)$ for some $0 \le i < |y|$ and a substring of $\sigma(K_+)$. As a matter of fact, we have proved that x is a substring of string $v01v$.

Step 2. Using Lemma 10.2.2 once more we can find $k' > k$ such that $K_+ = y'y' \ldots$, where $y' = w^{(k')} = h_1 \cdots h_{k'}(1)$ and $|y'| > |yy|$. Writing $y' = 1v'0$ we see that v' has $v01v$ as its prefix. Now consider any admissible sequence s. We have the inequalities:

$$K_- = 0v'0 \ldots \le \sigma^i(s) \le K_+ = 1v'0 \ldots \quad \text{for } i \ge 0.$$

If s contains $0v'0$ as its substring, then it contains, by step 1, x as well. Otherwise each substring of s, which is of length $|y'|$, is between $0v'1$ and $1v'0$. Since K_+ has $y'y'$ as its prefix, we can use Lemma 10.1.6 again and conclude that any such substring of s is a cyclic shift of y'. It means clearly that $s = \sigma^i(y'^\infty)$ and contains y' as its substring too. So s contains also v' as its substring and x as well. ∎

Remark. This result is not generally true for the case of rational rotation number. As a fact, among the three kinds of languages in Theorem 10.1.1 the claim of the Proposition 10.2.3 holds only for the first kind.

Now consider the proof of claim 1 of Theorem 10.2.1. The uniqueness of language L for a given irrational rotation number is already implied in Theorem 9.3.4, since K_+ is uniquely determined there. Using the same method in the proof of "only if" part of claim 1 of Theorem 10.1.1, it is very easy to prove that L cannot be a context-free language. The only modification needed here is to use the pumping lemma of context-free languages instead of Lemma A.4.1 (see Lemma B.2.7 and Examples B.2.8 and B.2.9).

The proof of the second part of claim 1 is performed in following Proposition. A rigorous definition of computable real number can be found, e.g., in Ko (1991). Here we only need to know that, a computable real number is one which can be effectively approximated to any desired degree of precision by a computer program given in advance.

Proposition 10.2.4 The language of f, L, is recursive if and only if its rotation number ρ is a computable real number.

Proof. If L is a recursive language, then there exists a Turing machine that can generate all words of L in increasing size and no other strings. So we can obtain the prefix of K_+ with length n in increasing order by taking

$$\max\{x : x \in L \text{ and } |x| = k\}.$$

From Proposition 9.2.1 we have

$$|\rho - \frac{N(K_+, k)}{k}| \le \frac{1}{k},$$

which means that the number ρ is computable.

If ρ_0 is a computable real number, then we can find the prefix of K_+ with increasing length, and generate all words through Proposition 10.2.3 in the order of increasing size. As a matter of fact, from Lemma 9.3.3 we can define a function G as follows:

$$G(\rho) = \begin{cases} 1 - (\frac{1}{\rho}) & \text{for } \frac{1}{k+1} < \rho < \frac{1}{k}, k \ge 2 \\ (\frac{1}{1-\rho}) & \text{for } 1 - \frac{1}{k} < \rho < 1 - \frac{1}{k+1}, k \ge 2 \end{cases}$$

This function G is defined in $(0, 1)$ except at all points $1/k$ and $1 - 1/(k+1)$ for $k \ge 2$. For a given irrational number ρ_0 in $(0, 1)$ and for a given natural number n there is a

$\varepsilon > 0$ such that for each number ρ in $(\rho_0 - \varepsilon, \rho_0 + \varepsilon)$, the sequences of homomorphisms $h_1, h_2, \cdots, h_n$ are the same. Since ρ_0 is computable, we can compute its approximate value to any precision, so we can find the prefix of K_+ with any required length. This makes it possible to generate all words of L in increasing size and establishes that L is a recursive language. ∎

The claim 3 of Theorem 10.2.1 consists of three parts as follows.

3.1 The set of all admissible sequences is uncountable.

3.2 It is possible that some admissible sequences are not itineraries of any point of the interval $J = [0, 1)$.

3.3 The set of those admissible sequences which are not itineraries is at most countable.

The claim 3.1 will be proved in Proposition 11.2.3 of next Chapter, since the main idea is taken from the possible structure of infinite machine which is discussed there.

The claim 3.3 is proved in the following Lemma.

Lemma 10.2.5 If the rotation number ρ is irrational, then the set of those admissible sequences which are not itineraries is at most countable.

Proof. From Proposition 9.1.7 we see that if an admissible sequence is not an itinerary, then it must have K_+ as its suffix. Since ρ is irrational, K_+ cannot be eventually periodic. Combining this observation with Lemma 9.2.6, we can prove that K_+ and each admissible sequence which has K_+ as its suffix are uniquely left-prolongable, and hence the conclusion required.

Assume

$$s = v1u, \text{ where } K_+ = 1u,$$

is admissible, then from Lemma 9.2.6 we know that, if there exists any prefix of s which is not a prefix of $u = \sigma(K_+)$, then there is a unique $a \in \Sigma = \{0, 1\}$ such that as is also admissible. Assume the contrary that each prefix of s is also a prefix of u, then we have $u = (v1)^\infty$ and $K_+ = 1u = 1(v1)^\infty$, which contradicts the condition of ρ being irrational. Hence our claim is true. ∎

Remark. Moreover, we can prove that if s is an admissible sequence which has K_+ as its suffix, then s can be explicitly written as $s = s'K_+$, where string s' is obtained from a prefix of K_- written backward.

The truth of claim 3.2, which depends on f, is implied in the following discussion.

Proposition 10.2.6 If the upper kneading sequence K_+ is an itinerary of some point of the interval $J = [0, 1)$, then the homeomorphism f is not ergodic, that is, there are wandering intervals for f.

Proof. If $K_+ = I(\alpha)$ for some point $\alpha \in J$, then $\alpha \in [c, 1)$. By Proposition 9.1.1 we have $K_+ = 1u$ and $K_- = 0u$ for some admissible sequence u. But then

$$u = I(f(0)) = I(f(\alpha)).$$

Since $f(0) = f(1-0)$ we have $f(0) > f(\alpha)$. Consider the interval $[f(\alpha), f(0)]$. Every point of this interval has u as its itinerary. Since K_+ cannot be eventually periodic, the forward iterates of this interval, $f^i([f(\alpha), f(0)])$, are all disjoint. It implies that there exist wandering intervals of f and finishes our proof. ∎

Finally we may say that in almost each concrete examples of circle homeomorphisms the language considered is context-sensitive. As a fact, the only known proofs that certain recursive languages being not context-sensitive are ultimately based on diagonalization (cf. Hopcroft and Ullman 1979).

We can give a positive claim about the case of periodic continued fractions below.

Proposition 10.2.7 If ρ has a periodic continued fraction representation, then its language L is a context-sensitive language.

Proof. From the discussion about the relation between Theorem 9.3.4 and Farey address in Subsection *9.3.3*, we see that if ρ has a periodic continued fraction representation, then $\{h_n\}$, the sequence of homomorphisms, will has an eventually periodic structure, that is, $h_l = h_{l+k}$ for a fixed positive k and all $l > N$. So we can construct a Turing machine to recognize the language L as follows. Using finite control of it this machine can generate prefixes of K_+ in increasing size. Now for a given string x this machine can compute the prefix of K_+ which is of the same length with x. After that it can check the conditions

$$K_- \leq \sigma^i(x) \leq K_+ \text{ for } i = 0, 1, \ldots, |x| - 1$$

to accept x or not. It is clear that this Turing machine can be realized by a linear bounded automaton of Subsection *B.1.4*, and hence L is a context-sensitive language as desired. ∎

10.3 The Case of Golden Mean Ratio

There are many discussions about the circle homeomorphism with rotation number $\rho = (\sqrt{5} - 1)/2$, the golden mean ratio (see, e.g., Procaccia, Thomae and Tresser (1987), Crutchfield (1994)).

From Proposition 10.2.7 we know that the language for the golden mean ratio is context-sensitive, but not context-free. Here we show some new results.

Using the formula in Lemma 9.3.3, we find that $G(\rho)$ $(= 1/(1-\rho)$ for $\frac{1}{2} < \rho < 1)$ has $(\sqrt{5} - 1)/2$ as one of its infinite fixed points. So we obtain $h_n = \{1 \to 110, 0 \to 10\}$ for every $n \geq 1$ and denote it by h hereafter. Using Theorem 9.3.4 we have $K_+ = \lim_{n\to\infty} h^n(1) = h^\infty(1)$. Some prefixes of this K_+ are

$$\begin{aligned} h(1) &= 110, \\ h^2(1) &= 11011010, \\ h^3(1) &= 110110101101101011010. \end{aligned}$$

Recalling the concept of distinct excluded block (DEB) introduced in Section 2.2 for all dynamical languages, a string x is a distinct excluded block for a language if x does not belong to this language, but every proper string of x does.

Here we need all operators on non-empty string x which are introduced in Subsection *1.1.5*: π (in two ways), $\overline{x}$ and $\underline{x}$.

Lemma 10.3.1 The set of distinct excluded blocks of the language L for $\rho = (\sqrt{5}-1)/2$ consists of two families:

$$\{\overline{h^n(1)}\}_{n>0} \text{ and } \{\underline{h^n(1)}h^n(0)\}_{n\geq 0}.$$

We can also obtain the explicit expression for its uncountable admissible sequences.

Lemma 10.3.2 A sequence s is admissible if and only if it can be written in one of the following forms:

$$1.\ s = 0^{n_0}[h(0)]^{n_1}\cdots[h^k(0)]^{n_k}K_+,$$

$$2.\ s = 0^{n_0}[h(0)]^{n_1}\cdots[h^i(0)]^{n_i}\cdots,$$

where the sequence $\{n_i\}_{0\leq i\leq k}$ and $\{n_i\}_{i\geq 0}$ satisfies the following conditions:

(1) each n_i is taken from $\{0, 1, 2\}$,

(2) there is no subsequence of the form 21^l2 for each $l \geq 0$,

(3) there is no prefix of the form $1^{l'}2$ for each $l' > 0$. Furthermore, an admissible sequence is a shift of K_+ (and K_-) if and only if the sequence $\{n_i\}_{i>0}$ has 01^∞ as its suffix.

The proofs of these two Lemmas involve many inductive details and so omitted.

Remark. There is some similarity between this example and the language of Feigenbaum attractor in Chapter 6.

Now we will determine explicitly the structure of the infinite automaton for this language, which have been introduced in Section 5.2 for unimodal maps. It amounts to find all such states which have two arcs to arrive at accepting states.

Lemma 10.3.3 Let $v_n = \pi h^n(1)\pi$ for $n \geq 1$. If x is a proper prefix and suffix of v_n and $|x| > |v_{n-1}|$, then $x = \pi h^n(0)\pi$.

Proof. Using an easy verified identity $h^n(1)h^n(0) = \overline{h^n(0)}\underline{h^n(1)}$, and writing

$$h^n(1) = h^{n-1}(1)h^{n-1}(1)h^{n-1}(0) = h^{n-1}(1)\overline{h^{n-1}(0)}\underline{h^{n-1}(1)},$$

we have $v_n = v_{n-1}0\overline{h^{n-1}(0)}0v_{n-1}$. It implies that x must have v_{n-1} as its prefix and suffix. From Lemmas 10.1.4 and 10.1.2 we know that $0v_{n-1}1$ appears in $v_{n-1}01v_{n-1}$ and only once. In our case we can obtain an explicit expression:

$$1v_{n-1}01v_{n-1}0 = \overline{h^{n-1}(0)}0v_{n-1}1h^{n-2}(1).$$

Combining these expressions, it proves that $\pi h^{n-1}(0)0v_{n-1}$ is a prefix and suffix of v_n and longer than $|v_{n-1}|$, so we obtain

$$x = \overline{\pi h^{n-1}(0)}0v_{n-1} = v_{n-1}0h^{n-1}(0)\pi = \pi h^{n-1}(0)\pi.$$

It is simple to verify that there is no other substring of v_n which satisfies all conditions in Lemma and so omitted. ∎

The structure of this machine can be described as follows. Among all prefixes of $\sigma(K_+)$ the following prefixes corresponding those states which have two arcs to arrive at accepting states:

$$v_n = \pi h^n(1)\pi \text{ and } x_n = \pi h^n(0)\pi.$$

The states they arrive at are determined through $v_n 1 R_L x_{n+1} 1$ and $x_n 0 R_L v_n 0$.

Let $v_n = x_n a_n$ and $x_{n+1} = v_n b_n$ then it is easy to obtain relations $a_n b_n = a_{n+1}$ and $b_n a_{n+1} = b_{n+1}$ inductively. Notice that each a_n begins from 1 and each b_n begins from 0.

From the above discussion we see that each admissible sequences can also be written as:

$$(a_1)^{\delta_1}(b_1)^{\delta_1'} \cdots (a_n)^{\delta_n}(b_n)^{\delta_n'} \cdots,$$

where each δ_i, δ_i' is 0 or 1, but if any one is 0, then its successor must not be 0.

Using the results obtained thus far it is easy to construct the infinite automaton for the golden mean ratio. The general structure of infinite automata for circle homeomorphisms is the topic of next Chapter.

Example 10.3.4 In Figure 10.1 we draw the initial part of this infinite automaton, which is essentially the same machine obtained by Crutchfield (1994, Fig. 10(a)).

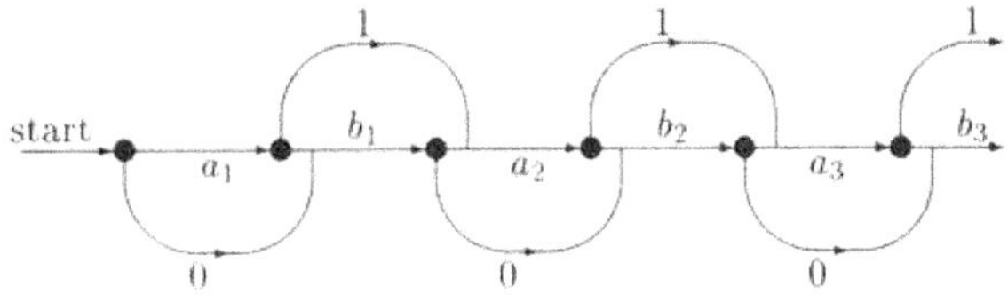

Figure 10.1: The initial part of infinite automaton of golden mean ratio.

The strings in Figure 10.1 are:

$$\begin{aligned} a_1 &= 1, & a_2 &= 101, & a_3 &= 10101101, \\ b_1 &= 01, & b_2 &= 01101, & b_3 &= 0110110101101, \end{aligned}$$

and

$$K_+ = 1u = 1a_1b_1a_2b_2a_3b_3\cdots$$

The general formula for a_n and b_n are as follows.

$$a_n = 10(\pi h^{n-1}(1)\pi) \text{ and } b_n = 0h^n(0)\pi,$$

where the homomorphism $h = (1 \to 110, 0 \to 10)$.

CHAPTER 11
AUTOMATA OF CIRCLE HOMEOMORPHISMS

In Section 11.1 the structure of minDFA's which accept regular languages generated by circle homeomorphisms is obtained. Here the main tools are the same as those used in Chapter 4 for regular languages generated from unimodal maps.

For non-regular languages generated from circle homeomorphisms we have to consider some kind of infinite automata. In Chapter 5 it is found that the infinite automata accepting non-regular languages generated by unimodal maps are determined by the kneading maps (see Section 5.2). For circle homeomorphisms, however, we find that there exists a universal structure of infinite automata which is completely determined by the continued fraction representation of rotation numbers. Sections 11.2 and 11.3 are devoted to this theory.

11.1 Finite Automata of Circle Homeomorphisms

In this section we discuss the finite automata which accept regular languages of circle homeomorphisms. The minimal deterministic finite automaton (min DFA) $M = (Q, \Sigma, \delta, q_0, F)$ is obtained, and the regular complexity, $N = \operatorname{card} Q$, is calculated.

Since the cases of $\rho = 0$ or 1 are too simple, we suppose $0 < \rho < 1$ holds in the following discussion.

The main tool in this section is the natural equivalence relation and the Myhill-Nerode Theorem (see Section A.3).

Given a language L we can introduce a natural equivalence relation R_L into S^*: xR_Ly if and only if for each $z \in S^*$, xz is in L exactly when yz is in L.

At first we discuss the basic case (a) in Proposition 10.1.5. Here we have an analogy with Lemma 4.2.6 for the languages generated from unimodal maps.

Lemma 11.1.1 Suppose $0 < \rho < 1$. If $K_+ = (1v0)^\infty$ and $K_- = (0v1)^\infty$, and $x \in L$, then there exists a prefix y of $\sigma(K_+)$ such that xR_Ly.

Proof. It is trivial if x itself is a prefix of $\sigma(K_+)$. Otherwise we can use Lemma 9.2.6 to find a unique $a \in S = \{1, 0\}$ such that ax is a word of L. It is easy to prove that xR_Lax holds. In fact, if $axz \in L$ for some $z \in S^*$, then $xz \in L$ is true. Conversely, if $xz \in L$, then since x is not a prefix of $\sigma(K_+)$, so xz is also not a prefix of $\sigma(K_+)$. Again using Lemma 9.2.6 there exist an unique $b \in S$ such that bxz belong to L. It implies that $bx \in L$ too. Using the uniqueness of symbol a, we have $b = a$ and $axz \in L$. From the proof of Lemma 10.1.6 we see that x is a substring of $\sigma(K_+)$, so after finite steps we will find a prefix y of $\sigma(K_+)$ such that $yR_L x$, where x is a suffix of y. ∎

This result implies that in order to find all equivalence classes with respect to the language L, we can restrict our attention to all prefixes of $\sigma(K_+)$. Thus we have a situation similar to the corresponding problem for unimodal map (see Lemma 4.2.6). It is clear that the unique remaining equivalence class is the complement of L, denoted by L', by Lemma 2.2.11.

Theorem 11.1.2 Let the rotation number $\rho \in (0,1)$ of a circle homeomorphism f be rational. For the three possibilities shown in Theorem 10.1.1 the index of R_L is $2n$ for the first case, and $3n$ for the latter two cases, where n is the length of basic word $w = 1v0$.

Proof. Case (a). Here $K_+ = (1v0)^\infty$ and $K_- = (0v1)^\infty$. Because both of $1v1$ and $0v0$ do not belong to L, by Lemma 10.1.6 every $x \in L$ is a substring of K_+.

Consider all prefixes of $\sigma(K_+) = v0(1v0)^\infty$. If a prefix y of $\sigma(K_+)$ is of length $|y| \geq 2n-1$, then we can write it as $v0(1v0)^i P(w)$, where $P(w)$ is a proper prefix of w and $i > 0$. Using the primitivity of w by Lemma 10.1.2, it is evident that $v0(1v0)^i P(w) R_L v0P(w)$ for each $i > 0$. In particular we have obtained

$$v01v0R_L v0. \tag{11.1}$$

Now consider all $2n-1$ prefixes of $v01v$, including ε, the empty word. We will show that any two of them are not equivalent with respect to L.

First consider $v01v$ and v. Taking $z = 1$, since $v01v1 \notin L$ but $v1 \in L$, so

$$v01vR_L v \text{ does not hold.} \tag{11.2}$$

Furthermore, suppose $v01v = y_1y_2y_3$ with $|y_2| \neq 0$ and $y_1 R_L y_1y_2$. Using the property of right invariant of R_L (see Subsection 4.3.1), we have $y_1y_2y_3R_L y_1y_2^2y_3$ and inductively $y_1y_2^ky_3 \in L$ for each $k > 0$. So $y_1y_2^\infty$ is admissible and $|y_2| = |w|$ by Proposition 9.2.1. It implies that there exists a decomposition of $v = v_1v_2$ such that $v_1 = y_1$ and $v_201v_1 = y_2$. But it leads to $vR_L v01v$, a contradiction to (11.2). Combining this discussion with (11.1), we obtain $2n$ equivalence classes that consist of all prefixes of $v01v$, including ε, and L', and there is no other class remained.

Case (b). $K_+ = (1v0)^\infty$ and $K_- = 0v0(1v0)^\infty$. In this case there are additional n equivalence classes that are represented by words $0v0P(w)$ with all proper prefixes $P(w)$ of w. Its proof is easy and hence omitted.

Case (c). $K_+ = 1v1(0v1)^\infty$ and $K_- = (0v1)^\infty$. Here there are additional n equivalence classes that are represented by words $1v1P(0v1)$, in which $P(0v1)$ taking all proper prefixes of $0v1$. ∎

Using this Theorem we can construct min DFA for L. As shown in Appendix A, a finite automaton consists of five elements: a finite set of states Q, an alphabet Σ of input symbols, an initial state $q_0 \in Q$, a transition function δ, and an accepting set $F \subset Q$.

First consider Case (a). There are $2n-1$ accepting states and one non-accepting state. Each accepting state is an equivalence class, and has a prefix of $v01v$ as its representative element. Write $v01v = x_1x_2\cdots x_{2n-2}$ and $x^{(i)} = x_1\cdots x_i$ for $i = 1,\ldots,2n-2$ and $x_{(0)} = \varepsilon$. Denote $q_i = [x^{(i)}]$, $i = 0, 1, \ldots, 2n-2$, the accepting states and $Q = \{q_0, q_1, \cdots, q_{2n-1}\}$, where q_{2n-1} is the equivalence class L', the non-accepting state.

The remaining problem is how to decide the transition function δ mapping $Q \times S$ to Q. That is, $\delta(q, a)$ is a state for each state q and input symbol a. It is a consequence of L being dynamical that $\delta(q_{2n-1}, a) = q_{2n-1}$ for both 0 and $a = 1$. That is to say, q_{2n-1} is a dead state (see Lemma 2.1.3).

Thus it suffices to decide the transition from the accepting set $F = Q\backslash\{q_{2n-1}\}$ to itself.

It is easy to see that

$$\delta(q_i, x_{i+1}) = q_{i+1} \text{ for } i = 0, 1, \ldots, 2n-3.$$

Using the result of (11.1), that is, $v01v0R_Lv0$, we have $\delta(q_{2n-2}, 0) = q_{n-1}$. From $v01v1 \notin L$ we have $\delta(q_{2n-2}, 1) = q_{2n-1}$.

In order to decide other transition from F to itself, we need the following result. (Compare it with Proposition 5.1.3 which leads to the theory of kneading map.)

Lemma 11.1.3 For Case (a) the state $\delta(q_i, \overline{x_{i+1}})$, $0 \le i \le 2n-3$, is accepting if and only if $x^{(i)}$ is both a prefix and a suffix of v, that is, a PS of v.

Proof. First consider a word $z \in L$ such that both $z1$ and $z0$ belong to L. Using Theorem 9.2.5 we know that z is unique among all words of length $|z|$. Since both $v0$ and $v1$ belong to L, we see that if $|z| \le |v|$, then v must have z as its suffix. Otherwise, if $|z| > |v|$, then z must have v as its proper suffix. In this case, one of strings $1v1$, $0v0$ must appear in either $z1$ or $z0$. This contradicts the fact that in Case (a) neither of $1v1$, $0v0$ belongs to L. This proves the "only-if" part.

Now suppose $x^{(i)}$ is a suffix of v. Because both $v0$ and $v1$ belong to L, so the claim is trivially true. ∎

Using this result we can complete the construction of minDFA for Case (a). The state $x^{(i)}$ satisfying the condition of Lemma 11.1.3 has two arcs to arrive at accepting states.

For Case (b), $K_+ = (1v0)^\infty$ and $K_- = 0v0(1v0)^\infty$. Some modifications are in need. First add n new accepting states to Q. Let $q_1', \cdots, q_n'$ be the equivalence classes with the representative words $0v0P(1v0)$, where $P(w)$ take all proper prefixes of w. Since in this Case an admissible sequence beginning from $0v0$ must be K_-, each q_i' has only one way to q_{i+1}' for $i = 0, \ldots, n-1$ and q_n' has only one way to q_1'.

The second change is to find the state q_i, whose representative word is both a prefix of $v01v$ and a suffix of $v10v$. From Lemmas 10.1.2 and 10.1.4 we know that $1v0$ is primitive, and $0v1$ is a cyclic shift of $1v0$, hence this state is uniquely determined. Since now $0v0$ belongs to L, we have $\delta(q_i, \overline{x_{i+1}}) = q_1'$.

For Case (c), $K_+ = 1v1(0v1)^\infty$ and $K_- = (0v1)^\infty$. Similarly, let $q'_1, \cdots, q'_n$ be the new states with the representative words $1v1P(0v1)$ and decide their transition naturally. The second change in this case is to the state q_{2n-2}, whose representative word is $v01v$. Here we have $\delta(q_{2n-2}, 1) = q'_1$.

Example 11.1.4 Let $\rho = 3/10$. Construct the minDFA for the Case (a).

Using the method of Lemma 9.3.3, we obtain the basic word

$$\begin{aligned} w &= h_1h_2(1) \\ &= 1001001000, \end{aligned}$$

where two homomorphisms are

$$\begin{aligned} h_1 &= \{1 \to 100, 0 \to 1000\}, \\ h_2 &= \{1 \to 110, 0 \to 10\}. \end{aligned}$$

For Case (a) we have

$$\begin{aligned} K_+ &= (1001001000)^\infty, \text{ and} \\ K_- &= (0001001001)^\infty \end{aligned}$$

From $v = 00100100$ and Lemma 11.1.3 we can find five strings, each of them is a PS of v:

$$\varepsilon, 0, 00, 00100, 00100100 = v.$$

Now we can construct the minDFA for this example as shown in Figure 11.1. There

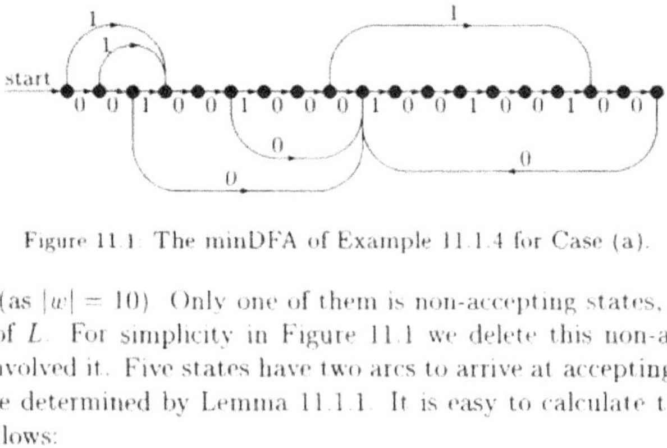

Figure 11.1 The minDFA of Example 11.1.4 for Case (a).

are 20 states (as $|w| = 10$). Only one of them is non-accepting states, that is L', the complement of L. For simplicity in Figure 11.1 we delete this non-accepting state and all arcs involved it. Five states have two arcs to arrive at accepting states, whose transitions are determined by Lemma 11.1.1. It is easy to calculate the equivalence relation as follows:

$$\begin{gathered} \varepsilon 1 R_L 001, \\ 01 R_L 001, \\ 000 R_L 001001000, \\ 001000 R_L 001001000, \\ 001001001 = v1 R_L 0010010001001001. \end{gathered}$$

Other transitions are determined by the rules given above and $v01v0 R_L v0$.

Example 11.1.5 In Figure 11.2 (a), (b), (c) we give minDFA for all cases of $\rho = 1/3$. Here the basic word is $w = 100$, and $v = 0$. By the discussion above the basic transition relation (11.1) is $00100R_L00$. For simplicity the unique non-accepting state is not shown there.

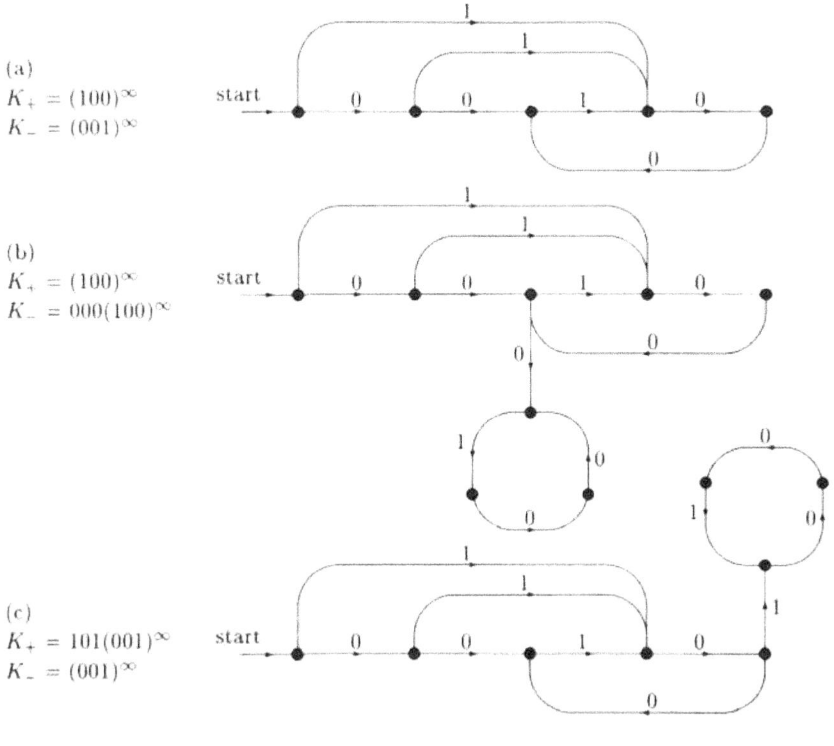

Figure 11.2 minDFA of Example 11.1.5 for $\rho = 1/3$.

11.2 Infinite Automata of Circle Homeomorphisms

An Example of infinite automaton has been given at the end of Chapter 10 for the circle homeomorphism with the golden mean ratio as its rotation number.

From this section we will begin to establish a general theory about such infinite automata. First we give some basic facts about equivalence classes of R_L. Then as an application we prove that for irrational rotation number there exist uncountably infinite admissible sequences.

Let

$$K_+ = \lim_{n \to \infty} h_1 \cdots h_n(1) = 1u \text{ and } K_- = 0u.$$

A direct consequence of Lemma 9.2.6 is the following result.

Lemma 11.2.1 If the rotation number ρ is irrational, and $x \in L$, then there is a prefix p of $u = \sigma(K_+)$ such that $xR_L p$.

Proof. It is trivial if x itself is a prefix of u. Otherwise, by the claim 2 of Theorem 10.2.1, there is a prefix p of u such that x is a proper suffix of p. Suppose that p is the shortest prefix of u that has x as its suffix. Then by applying Lemma 9.2.6 finite times as in proof of Lemma 11.1.1 we obtain $xR_L p$. ∎

Another basic fact here is the following observation.

Lemma 11.2.2 Any two distinct prefixes of $u = \sigma(K_+)$ do not belong to the same equivalence class of R_L.

Proof. Assume the contrary that we have $u = pyu'$ with $y \neq \varepsilon$ such that $pR_L py$ holds. Using the property of right invariant of R_L, we see that $pR_L py^2$, and by induction $pR_L py^n$ for all $n \geq 0$. It implies that $py^n \in L$ for each $n \geq 0$ and py^∞ is an admissible sequence, which is eventually periodic. Using Proposition 9.2.1 leads to a contradiction to the condition that the rotation number ρ is irrational. ∎

These results implies that in order to construct infinite automata for the case of irrational rotation numbers, we can simply take all prefixes of u and L' to be states of this machine. In order to decide the transition function we can proceed as follows. Recalling the notation $w^{(n)} = h_1 \cdots h_n(1)$ and the content of Lemma 10.2.2, we can find a subsequence $\{w^{(n_k)}\}_{k\geq 1}$ of $\{w^{(n)}\}_{n\geq 1}$ such that

$$w^{(n_k)} = h_1 \cdots h_{n_k}(1) = 1v_k01v_k0\ldots.$$

Using the idea of Lemma 11.1.3, it can be shown that each v_k is a prefix and suffix of the next v_{k+1}, so we can decide the transition function inductively, if ρ is computable.

As a theoretical application we will use this infinite automaton to prove that the cardinal number of the set of all admissible sequences is uncountable. (This completes our proof of Theorem 10.2.1).

Proposition 11.2.3 If ρ is irrational, then the set of all admissible sequences is uncountable.

Proof. It suffices to find a subset of admissible sequences, which has the cardinal of continuum.

Denote by v_1 the string in $K_+ = 1v_101v_10\ldots$ in Lemma 10.2.2. By Lemma 10.1.4 we can decompose $v_1 = v_{11}v_{12}$ such that $v_101v_1 = v_{11}0v_11v_{12}$. From Lemmas 11.2.1 and 11.2.2 we have $v_11R_L v_{11}0v_11$. Let $v_{11}0v_11 = v_1b_1$ and take a string v_2 such that $K_+ = 1v_201v_20\ldots$ and $|v_2| > |v_1b_1|$. Let $v_2 = v_1b_1c_1$. Arguing as above to the string v_2 and so on, inductively we have $\{v_n\}_{n>0}$, $\{b_n\}_{n>0}$ and $\{c_n\}_{n>0}$ such that

$$v_n1R_L v_nb_n \text{ and } v_{n+1} = v_nb_nc_n,$$

in which b_n begins from 0. Now consider infinite sequences of the form

$$s = v_1(b_{n_1}c_1)(b_{n_2}c_2)\cdots(b_{n_i}c_i)\cdots,$$

where $\{n_i\}_{i>0}$ is an infinite sequence over $\{0,1\}$, and

$$b_{n_i} = \begin{cases} 1 & \text{if } n_i = 0, \\ b_i & \text{if } n_i = 1. \end{cases}$$

From the procedure of finding v_n, b_n and c_n it is easy to verify that

$$\begin{aligned} v_1(b_{n_1}c_1)(b_{n_2}c_2)\cdots(b_{n_i}c_i)R_L\, & v_1(b_1c_1)(b_{n_2}c_2)\cdots(b_{n_i}c_i) \\ R_L\, & v_2(b_{n_2}c_2)\cdots(b_{n_i}c_i) \\ & \cdots\cdots \\ R_L\, & v_{i+1} \in L, \end{aligned}$$

and hence every s is admissible. Since the set of sequences $\{n_i\}$ over $\{0,1\}$ is uncountable, the proof is finished. ∎

11.3 Universal Structure of Automata of Circle Homeomorphisms

In this Section we will prove that for languages generated by circle homeomorphisms the automata accepting them have a universal structure which is determined completely by the continued fraction representation of the rotation number. Since the case of rational rotation numbers is simple, we focus our attention to the case of irrational rotation numbers.

Assume f is a homeomorphism and its rotation number ρ is irrational. In this case the language $L(f)$ is completely determined by ρ (by Theorem 10.2.1). The upper kneading sequence and lower kneading sequence of f are denoted by $K_+ = \lim_{n\to\infty} h_1\cdots h_n(1) = 1u$, and $K_- = 0u$, where the sequence u is written as

$$u = \sigma(K_+) = e_1\cdots e_n\cdots, \text{ and } a^{(n)} = e_1\cdots e_n \text{ for } n \geq 1,$$

and $a^{(0)} = \varepsilon$. From Myhill-Nerode Theorem we need only to compute all equivalence classes with respect to R_L. The notation $[x]$ is the equivalence class containing x, and x is said to be a representative of the class $[x]$. From Lemmas 11.2.1 and 11.2.2 we can take $\{a^{(n)}\}$ $(n \geq 0)$ as the (representative of) accepting states of automaton. If the transition function of automaton is denoted by δ, then it is trivial that

$$\delta([a^{(n)}], e_{n+1}) = [a^{(n+1)}].$$

For simplicity we use $x \sim y$ to denote xR_Ly in the sequel. For a nonempty string s, say $s = 10110$, the following notations are introduced in Subsection *1.1.5*

$$\pi 10110 = 0110,\ 10110\pi = 1011,\ \pi 10110\pi = 011,$$
$$\overline{10110} = (1011\pi)\bar{0} = 10111,\ \underline{10110} = \bar{1}(\pi 0110) = 00110$$

Definition 11.3.1 If $p = a^{(n)}$ is a prefix of u and has the property that

$$\overline{p\bar{a}}_{n+1} = a^{(n)}\overline{\bar{e}}_{n+1} \in L,$$

then p is called a prefix of the first kind (of u) (Otherwise, p is of the second kind.)

It is always true that the first (or shortest) prefix of the first kind of u is ε.

Two homomorphisms $h_{r,m}$ and $h_{l,m}$ defined in Section 9.3 are used in the sequel. For simplicity, we often use the notations h_r and h_l instead to denote them, that is, $h_r = (1 \to 10^{m-1}, 0 \to 10^m)$, and $h_l = (1 \to 1^m 0, 0 \to 1^{m-1}0)$ for some $m \geq 2$. We also say some h is of type h_r (h_l), if $h = h_{r,m}$ ($h_{l,m}$) for some $m \geq 2$.

11.3.1 Main Results

The universal structure of infinite automata is described by two Theorems below.

Theorem 11.3.2 There exist two sequences of prefixes of $u = \sigma(K_+)$,

$$\{x_n\}_{n \geq 1} \text{ and } \{y_n\}_{n \geq 1},$$

which satisfy the following conditions:

1. Each x_n has 1 as its last symbol, and each y_n has 0 as its last symbol. The lengths of $\{x_n\}$ and $\{y_n\}$ satisfy the inequalities: $|x_n| < |y_n| < |x_{n+1}|$ for $n \geq 1$.

2. Each $x_n\pi$ is of the first kind and $\bar{x}_n \sim y_n$. Moreover, if p is also a prefix of the first kind, and satisfies the condition $|x_n\pi| < |p| < |y_n\pi|$, then $p1$ is a prefix of u, and

$$p0 \sim y_n.$$

3. Each $y_n\pi$ is of the first kind and $\bar{y}_n \sim x_{n+1}$. Moreover, if p' is also a prefix of the first kind, and satisfies the condition $|y_n\pi| < |p'| < |x_{n+1}\pi|$, then $p'0$ is a prefix of u, and $p'1 \sim x_{n+1}$.

4. If $x_1\pi \neq \varepsilon$, then there exists an extra $y_0 = 0$ which satisfies the conditions above.

The meaning of this Theorem is explained in Figure 11.3.

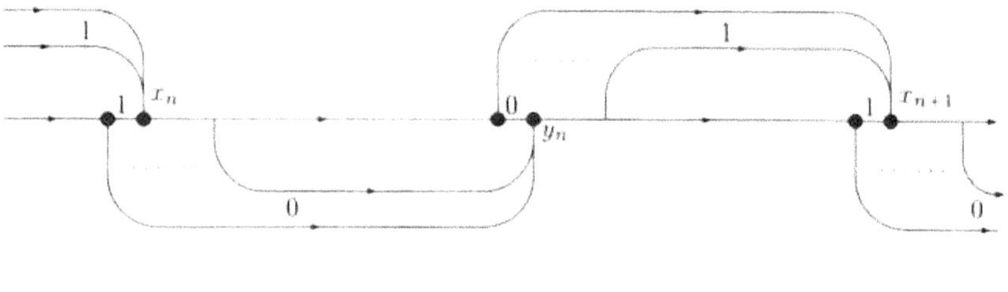

Figure 11.3 An interpretation of Theorem 11.3.2

Introduce a *counting system* about the prefixes of the first kind of $u = \sigma(K_+)$ in the order of size.

Firstly, let $c(\cdot)$ be a function from prefixes of u to an alphabet set $\{-1, +1\}$ defined by

$$c(a^{(n)}) = \begin{cases} \varepsilon, & \text{when } a^{(n)}\bar{e}_{n+1} \notin L, \\ +1, & \text{when } e_{n+1} = 1 \text{ and } a^{(n)}0 \in L, \\ -1, & \text{when } e_{n+1} = 0 \text{ and } a^{(n)}1 \in L. \end{cases}$$

Secondly, let $C(\cdot)$ be a function from prefixes of u to $\{-1, +1\}^*$ defined by:

$$C(a^{(n)}) = \prod_{n=0}^{n} c(a^{(n)}) = c(a^{(0)})c(a^{(1)}) \cdots c(a^{(n)}).$$

Finally, define an infinite sequence $C(u)$ by

$$C(u) = \lim_{n\to\infty} C(a^{(n)}) = \prod_{n=0}^{\infty} c(a^{(n)}).$$

Theorem 11.3.3 If the rotation number ρ has its continued fraction representation $[a_1, a_2, \ldots, a_n, \ldots]$, then

$$C(u) = (-1)^{a_1 - 1}(+1)^{a_2} \cdots (-1)^{a_{2k-1}}(+1)^{a_{2k}} \cdots$$

Remark. From the formula of $C(u)$ we see that the simplest infinite automaton for languages of circle homeomorphisms is:

1. $\rho = (\sqrt{5} - 1)/2 = [1, 1, \ldots]$ (cf. Figure 10.1), and
2. $\rho = (3 - \sqrt{5})/2 = [2, 1, \ldots]$.

11.3.2 Strings $h_1 \cdots h_k(1)$ and $h_1 \cdots h_k(0)$

Let two strings a and b be defined by

$$a = h_1 \cdots h_k(1) \text{ and } b = h_1 \cdots h_k(0)$$

for some $k \geq 0$ and homomorphisms $h_1, \cdots, h_k$ (as defined in Section 9.3). An obvious fact about them is that both a and b has 1 as their first symbol, and 0 as their last symbol.

Lemma 11.3.4 If $a = h_1 \cdots h_k(1)$ and $b = h_1 \cdots h_k(0)$, then both strings a and b are primitive.

Proof. The conclusion about a is already obtained by Lemma 10.1.2. Since each b can be seen as $h_1 \cdots h_{k-1}h'_k(1)$ with a modified h'_k, the conclusion of b holds too. ∎

Lemma 11.3.5 If $a = h_1 \cdots h_k(1)$ and $b = h_1 \cdots h_k(0)$, then

$$ab = \bar{b}\bar{a}.$$

Proof. For the case of $k = 1$ it is simple to verify the identity directly. For $h = (1 \to 1^m0, 0 \to 1^{m-1}0)$, then

$$ab = 1^m01^{m-1}0, \text{ and } ba = 1^{m-1}01^m0,$$

and $ab = \bar{b}\underline{a}$ holds. Similarly, it holds for $h = (1 \to 10^{m-1}, 0 \to 10^m)$.

Assume the identity holds for k already, and discuss the case of $k + 1$. Calculate the string ab as follows:

$$\begin{aligned} ab &= h_1 \cdots h_{k+1}(10) \\ &= h_1(h_2 \cdots h_{k+1}(10)) \\ &= h_1(h_2 \cdots h_{k+1}(1)h_2 \cdots h_{k+1}(0)) \\ &= h_1(\overline{h_2 \cdots h_{k+1}(0)}\underline{h_2 \cdots h_{k+1}(1)}) \end{aligned}$$

For $h_1 = (1 \to 1^m0, 0 \to 1^{m-1}0)$, we have

$$\begin{aligned} h_1(\overline{h_2 \cdots h_{k+1}(0)}) &= h_1((h_2 \cdots h_{k+1}(0)\pi)1) \\ &= \overline{h_1h_2 \cdots h_{k+1}(0)}0. \\ h_1(\underline{h_2 \cdots h_{k+1}(1)}) &= h_1(0(\pi h_2 \cdots h_{k+1}(1)) \\ &= \pi h_1 \cdots h_{k+1}(1), \text{ and} \\ 0(\pi h_1 \cdots h_{k+1}(1)) &= \underline{h_1 \cdots h_{k+1}(1)}; \end{aligned}$$

For $h_1 = (1 \to 10^{m-1}, 0 \to 10^m)$, we have

$$\begin{aligned} h_1(\overline{h_2 \cdots h_{k+1}(0)}) &= h_1((h_2 \cdots h_{k+1}(0)\pi)1) \\ &= h_1h_2 \cdots h_{k+1}(0)\pi, \\ h_1(\underline{h_2 \cdots h_{k+1}(1)}) &= h_1(0(\pi h_2 \cdots h_{k+1}(1))) \\ &= 1\underline{h_1 \cdots h_{k+1}(1)}, \text{ and} \\ (h_1h_2 \cdots h_{k+1}(0)\pi)1 &= \overline{h_1 \cdots h_{k+1}(0)}. \end{aligned}$$

Thus our induction is completed. ∎

Corollary 11.3.6 If $a = h_1 \cdots h_k(1)$ and $b = h_1 \cdots h_k(0)$, then

$$ab \neq ba, \text{ and } |a| \neq |b|.$$

11.3.3 Square Prefixes of K_+

Recalling Lemma 10.2.2 in the previous Chapter that the upper kneading sequence K_+ has $y^2 = yy$ as its prefix for some string y, and the length of y can be greater than any given number. We will call yy a *square prefix* of K_+. It turns out that these special prefixes will play a very important role in proof of Theorem 11.3.2. In order to use them, however, we have to study them more deeply in this Subsection.

Lemma 11.3.7 The upper kneading sequence $K_+ = \lim_{n\to\infty} h_1 \cdots h_n(1)$ has A^2 as its square prefix for some string $A = h_1 \cdots h_k(1)$ if and only if either h_{k+1} is of type h_l or both h_k and h_{k+1} are of type h_r.

Proof. The "if" part is easy. As a matter of fact, if h_{k+1} is of some h_l, then $h_{k+1}(1) = 11\cdots$, and $K_+ = h_1 \cdots h_k(11)\cdots$, we have $K_+ = AA\cdots$ with $A = h_1 \cdots h_k(1)$. Otherwise, if

$$h_k = (1 \to 10^{m-1}, 0 \to 10^m), \text{ and } h_{k+1} = (1 \to 10^{m'-1}, 0 \to 10^{m'})$$

for some m and m', then

$$h_k h_{k+1}(1) = h_k(10^{m'-1}) = 10^{m-1}(10^m)^{m'-1} = (10^{m-1})^2 \cdots,$$

and we have $K_+ = AA$ with $A = h_1 \cdots h_{k-1}(10^{m-1}) = h_1 \cdots h_k(1)$.

The "only if" part. Assume the contrary that the claim is wrong, then $K_+ = AA\cdots$ for $A = h_1 \cdots h_k(1)$, but h_{k+1} is of type h_r, and h_k is of type h_l. Suppose

$$h_k = (1 \to 1^m 0, 0 \to 1^{m-1}0), \text{ and } h_{k+1} = (1 \to 10^{m'-1}, 0 \to 10^{m'})$$

for some m and m'. Let

$$a = h_1 \cdots h_{k-1}(1), \text{ and } b = h_1 \cdots h_{k-1}(0),$$

then we have

$$\begin{aligned} K_+ = h_1 \cdots h_{k+2}(1) \cdots &= h_1 \cdots h_{k+1}(1)a \cdots \\ &= a^m b (a^{m-1} b)^{m'-1} a. \end{aligned}$$

Compare this expression with $K_+ = AA\cdots$ where $A = h_1 \cdots h_k(1) = a^m b$, we would have $ab = ba$, which contradicts Lemma 11.3.4. ∎

A direct application of this Lemma solves the problem of finding the next square prefix of K_+ for a given square prefix.

Proposition 11.3.8 Let $K_+ = AA\cdots$ where $A = h_1 \cdots h_k(1)$. An integer j is the first one after k for which $K_+ = A'A'\cdots$ and $A' = h_1 \cdots h_j(1)$ if and only if one of the following possibilities happens:

1. both h_k and h_{k+1} are of type h_l and $j = k+1$,
2. both h_{k+1} and h_{k+2} are of type h_r and $j = k+1$,
3. h_{k+1} is of type h_l, h_{k+2} is of type h_r, and $j = k+2$.

11.3.4 Calculation of $\{x_n\}$

Now we define the sequences $\{x_n\}$ mentioned in Theorem 11.3.2.

For a given K_+, by Lemmas 10.2.2, 11.3.7 and Proposition 11.3.8, there exists a sequence of $\{A_n\}$ such that $K_+ = A_n A_n \cdots$ for each $n \geq 1$, $A_n = h_1 \cdots h_{k_n}(1)$, $|A_{n+1}| > |A_n|$, and each A_{n+1}^2 is the next square prefix of K_+ after A_n^2.

If $K_+ = 11\cdots$, then $A_1 = 1$. In this case let $x_1 = 1$. In general case, The definition of x_n is given as follows. First we write

$$A_n = 1v_n0 \text{ where } v_n = \pi A_n \pi.$$

Since $A_n^2 \in L$, using Lemma 10.1.4 we know that the string $v_n1 \in L$ too. Thus each v_n is a prefix of the first kind of

$$u = \sigma(K_+) = v_n01v_n0\cdots$$

From Lemmas 11.2.1 and 11.2.2 there exists a unique prefix of u, denoted by x_n, such that

$$v_n1 \sim x_n.$$

We can calculate x_n explicitly as shown by the next lemma, and prove that each $x_n\pi$ is a prefix of the first kind of u. (Evidently, if $x_1 = 1$, then $x\pi = \varepsilon$ is of the first kind.) For simplicity, we often omit the subscripts "n" of A_n and x_n below if no confusion happens.

Lemma 11.3.9 Let $K_+ = AA\cdots$, in which $A = h_1\cdots h_k(1)$ for some $h_1, \ldots, h_k$. If $A = 1v0$, $v1 \sim x$, where x is a prefix of $u = \sigma(K_+)$, then x can be computed from A and $B = h_1\cdots h_k(0)$ as follows:

$$x = \begin{cases} \pi A\overline{B}, & \text{when } |A| > |B|, \\ \pi\overline{B}, & \text{when } |A| < |B|, \end{cases}$$

and $x\pi$ is a prefix of the first kind.

Proof. By Corollary 11.3.6 $|A| = |B|$ is impossible. Let $a = h_1\cdots h_{k-1}(1)$ and $b = h_1\cdots h_{k-1}(0)$ and discuss other cases separately.

1. If h_k is of type h_l, denoting $h_k = (1 \to 1^m0, 0 \to 1^{m-1}0)$, we have $A = a^mb$, $B = a^{m-1}b$, and hence $|A| > |B|$. From Lemma 11.3.5 we have

$$1v01v0 = a^mba^mb = a^{m-1}(ab)a^{m-1}(ab) = a^{m-1}\bar{b}(\underline{a}a^{m-1}\bar{b})\underline{a} = a^{m-1}\bar{b}(0v1)\underline{a}.$$

From $v1 \sim x$ and x being a prefix of u, by the proof of Lemma 11.1.1 we see that x has $v1$ as its suffix, and is the shortest prefix of u having this property. Since $0v1$ is primitive (cf. Lemma 1.1.11), it appears in $1v01v0$ only once. Therefore we obtain

$$x = \pi a^{m-1}\bar{b}(0v1) = \pi a^mba^{m-1}\bar{b} = \pi A\overline{B}.$$

2. If h_k is of type h_r, denoting $h_k = (1 \to 10^m, 0 \to 10^{m+1})$, we have $A = ab^m$, $B = ab^{m+1}$, and hence $|A| < |B|$. Similarly we have

$$1v01v0 = ab^mab^m = (ab)b^{m-1}(ab)b^{m-1} = \bar{b}(\underline{a}b^{m-1}\bar{b})\underline{a}b^{m-1} = \bar{b}(0v1)\underline{a}b^{m-1},$$

and thus obtain

$$x = \pi\bar{b}\underline{a}b^{m-1}\bar{b} = \pi ab^m\bar{b} = \pi\overline{B}.$$

From these explicit expression of x, it is easy to verify that $x\pi$ is of the first kind. As a fact, if $x = \pi A\overline{B}$, then $x\pi = \pi AB\pi$. From $K_+ = h_1\cdots h_kh_(k+1)(1)\cdots$ it is obvious that $AB = h_1\cdots h_k(10) \in L$, and hence $\pi AB = \overline{x} \in L$. The discussion for the case of $\pi\overline{B}$ is quite the same. ∎

11.3.5 Calculation of Prefixes of the First Kind

Here the basic tool is the following lemma.

Lemma 11.3.10 If both strings p and q are prefixes of $u = \sigma(K_+)$, $|p| < |q|$, and q is of the first kind, then p is also of the first kind if and only if p is a suffix of q (and hence p is a PS of q).

Proof. The "only if" part. If both $p0$ and $p1$ belong to L, take a suffix p' of q such that $|p'| = |p|$. Since both $q0$ and $q1$ belong to L, we have also that both $p'0$ and $p'1$ belong to L. From Theorem 9.2.5 and its proof, it can only be $p = p'$.

The "if" part is obviously true since L is a dynamical language (see Definition 9.1.3). ∎

In the previous Subsection We have seen that each v_n is of the first kind. Since $|v_n|$ is increasing, we need only to compute all prefixes of the first kind whose lengths are between those of v_n and v_{n+1}.

Lemma 11.3.11 Let A, B and x be those strings defined in Lemma 11.3.9.

1. If $|A| > |B|$, $\pi AB\pi$ can only have its substring $\pi A\pi$ as either prefix or suffix.
2. If $|A| < |B|$, $\pi B\pi$ can only have its substring $\pi A\pi$ as either prefix or suffix.

Proof. Discuss the two situations stated in Lemma 11.3.9 separately.

1. If $|A| > |B|$, using the notations there we have $A = a^m b$, $B = a^{m-1}b$ and

$$x\pi = \pi AB\pi = \pi a^m b a^{m-1} b\pi.$$

From Lemma 11.3.5 it is easy to verify that $\pi A\pi$ is a PS of $x\pi$. Suppose $\pi AB\pi = z_1(\pi A\pi)z_2$ and write it into the form of

$$\pi a^m b a^{m-1} b\pi = z_1(\pi a)(a^{m-1}b\pi)z_2.$$

Since $B = a^{m-1}b$ is primitive (by Lemma 11.3.4), the substring $a^{m-1}b\pi$ can appear in $a^{m-1}ba^{m-1}b\pi$ only twice, namely, as its prefix and suffix. By the same reason the string a is also primitive. Therefore, we can only have either $z_1 = \varepsilon$ or $z_2 = \varepsilon$.

2. If $|A| < |B|$, then $A = ab^m$ and $B = ab^{m+1}$. Using Lemma 11.3.5 the string $\pi A\pi$ is a PS of $x\pi = \pi B\pi$. Assume $x\pi = z_1(\pi A\pi)z_2$, and write it as

$$x\pi = \pi ab^{m+1}\pi = (\pi a)b^2(b^{m-1}\pi = (z_1(\pi a))b(b^{m-1}\pi)z_2$$

Using the fact that the string b is primitive, we have either $z_1 = \varepsilon$ or $z_2 = \varepsilon$. ∎

Lemma 11.3.12 If A and B are defined as above, and $v = \pi A\pi$, $v' = \pi A^m B\pi$, then all proper PS of v' whose length is greater than $|v|$ are

(1) $\pi A^{m-1}B\pi, \ldots, \pi AB\pi$ when $|A| > |B|$,

(2) One more string $\pi B\pi$ added to the strings above when $|A| < |B|$

Proof. Assume z is a proper PS of v', and $|z| > |v|$. Since v is a prefix of v', z has $v0$ as its prefix.

(1) Using Lemma 11.3.5 it is easy to verify that those strings listed above are proper PS of v'. On the other hand, using the fact that A being primitive and Lemma 11.3.11, there is no other PS of v' which can have $v0$ as its prefix.

(2) When $|A| < |B|$, there exists another possibility. From Lemma 11.3.5 we have

$$AB = AAb = \overline{B}\underline{A},$$

which implies that both AB and A^2 have $B\pi$ as their prefix, thus one more proper PS of v' is $\pi B\pi$. ∎

Two results which are dual to Lemma 11.3.11 and 11.3.12 are listed below without proof.

Lemma 11.3.13 Let A, B and x be those defined in Lemma 11.3.9.

1. If $|A| > |B|$, $\pi A\pi$ can only have its substring $\pi B\pi$ as either prefix or suffix.
2. If $|A| < |B|$, $\pi AB\pi$ can only have its substring $\pi B\pi$ as either prefix or suffix.

Lemma 11.3.14 If A and B are defined as above, and $v = \pi A\pi$, $v' = \pi AB^m\pi$, then all proper PS of v' whose length is greater than $|v|$ are

(1) $\pi AB^{m-1}\pi, \ldots, \pi AB\pi$ when $|A| > |B|$,

(2) One more string $\pi B\pi$ to the strings above when $|A| < |B|$.

Remark. From the conclusions of Lemmas 11.3.12 and 11.3.14 we know that the next prefix of the first kind after v is $x\pi$ in each case.

11.3.6 Proof of Theorem 11.3.2

First use Proposition 11.3.8 to obtain all square prefixes of K_+, namely, all A_n, and all $v_n = \pi A_n\pi$. From $v_n 1 \sim x_n$ determine all x_n. Since each $x_n\pi$ is of the first kind, we can use $\overline{x}_n \sim y_n$ to determine all y_n. We will give the explicit expression of y_n and verify each claims for the three cases listed in Proposition 11.3.8.

Case 1. Both h_k and h_{k+1} are of type h_l and $j = k + 1$.

Let $A = h_1 \cdots h_k(1) = 1v0$ and $A' = h_1 \cdots h_{k+1}(1) = 1v'0$. (For convenience we denote v_n, v_{n+1} by v, v', and x_n by x, etc.) Denoting $h_{k+1} = (1 \to 1^m0, 0 \to 1^{m-1}0)$ for some $m > 1$, then $A' = A^mB = 1v'0$.

Since $x \sim v1$, $\overline{x}$ has $A = 0v0$ as its suffix. From $\overline{x} \sim y$ we see that y also has A as its suffix. By Lemma 11.3.11 and $v'0 = \pi A^m B$, we have

$$y = v'0 = \pi A^m B.$$

By Lemma 11.3.12 we have all prefixes of the first kind between v and v':

$$\pi A^{m-1}B\pi, \cdots, \pi AB\pi \text{ (and } \pi B\pi \text{ also when } |A| < |B|).$$

Each of them is a prefix of $u = \pi A^m B \cdots = v'0 \cdots$ and the next symbol followed these prefix in u is 1 (by $AB = \overline{B}\underline{A}$). If p is any one of them, then $p0$ is a suffix of $v'0$. Using Lemma 11.1.1 we obtain

$$p0 \sim v'0.$$

For the case 1 the claim of Theorem 11.3.2 is proved and shown in Figure 11.4.

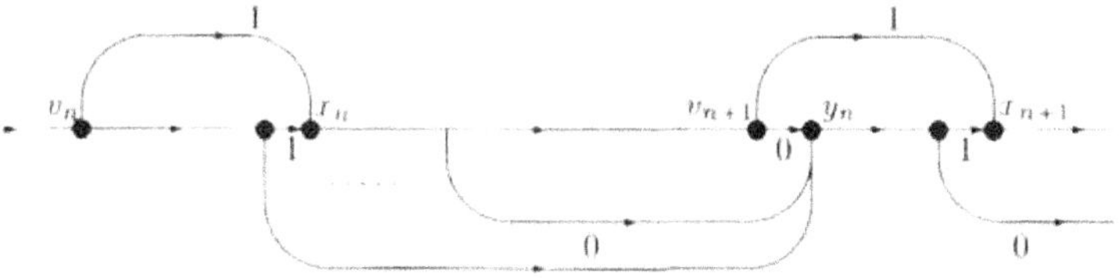

Figure 11.4 The situation of automata for Case 1.

Case 2. Both h_k and h_{k+1} are of type h_l and $j = k + 1$.

Take $A = h_1 \cdots h_k(1)$ as above. Since h_k is of type h_r, we have $|B| > |A| = |1v0|$ and $A' = AB^m = 1v'0$. By Lemma 11.3.9 $x = \pi B$.

Determine y from $\overline{x} = \pi B \sim y$ which is a prefix of $u = \sigma(K_+) = v'0 \cdots = \pi AB^m$. Using Lemma 11.3.13 leads to $y = \pi AB$.

Using Lemma 11.3.14, all PS of $v' = \pi AB^m \pi$ whose length is longer than $v = \pi A\pi$ can be listed as follows.

$$\pi B\pi, \pi AB\pi, \ldots, \pi AB^{m-1}\pi, \pi AB^m\pi = v',$$

where the first one is $x\pi$, and the second one is $y\pi$. From Lemma 11.3.9 we have

$$v'1 = \pi A\overline{B}^m \sim x' = \pi\overline{B}' = \pi AB^m\overline{B}.$$

By Lemma 11.3.13 we find that

$$\pi A\overline{B}^i \sim x^i \text{ for } i = 1, \ldots, m.$$

Thus for the case 2 the claim of Theorem 11.3.2 is proved and shown in Figure 11.5.

Case 3. h_{k+1} is of type h_l, h_{k+2} is of type h_r, and $j = k + 2$.

In this case we have

$$A' = h_1 \cdots h_{k+2}(1) = A^m B(A^{m-1}B)^{m'-1} \text{ for } m > 1, m' > 1,$$

where A and B are defined as above.

Proceed as in the case 1, we obtain $y = \pi A^m B$. The discussion from $v = \pi A\pi$ to $y\pi = \pi A^m B\pi$ is also similar to that of case 1.

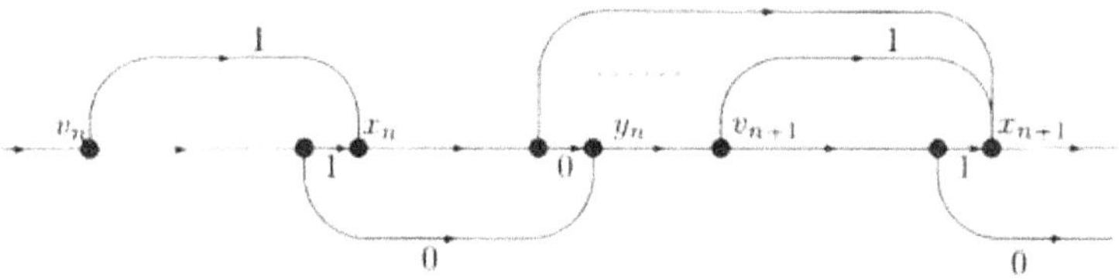

Figure 11.5: The situation of automata for Case 2.

Consider the prefixes of the first kind of u between $y\pi$ and $v' = \pi A'\pi$. Let $A'' = A^m, B'' = A^{m-1}B$. We have

$$y\pi = \pi A''\pi,\ v' = \pi A''(B'')^{m'-1}\pi.$$

Using Lemma 11.3.14, all PS of v' which is longer than $y\pi$ can be listed as follows:

$$\pi A''B''\pi, \ldots, \pi A''(B'')^{m'-2}, v'$$

Using the same argument as in Case 2, we know that for each of them, say p, $p1 \sim x'$ holds. The situation of Case 3 is shown in Figure 11.6.

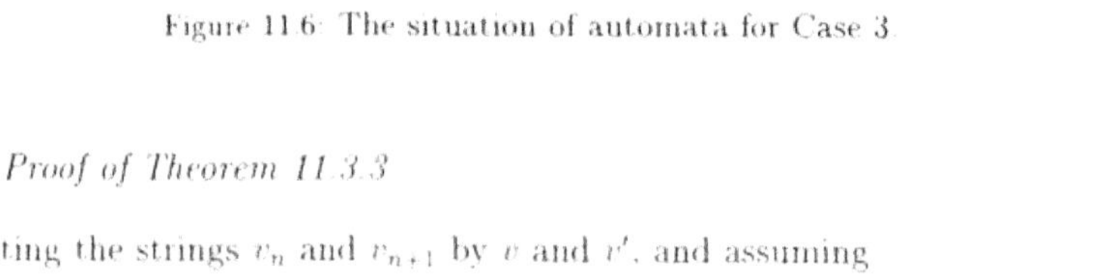

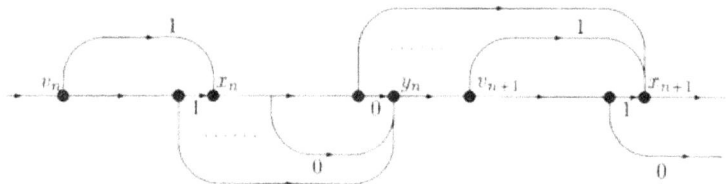

Figure 11.6: The situation of automata for Case 3.

11.3.7 *Proof of Theorem 11.3.3*

Denoting the strings v_n and v_{n+1} by v and v', and assuming

$$A_n = 1v0 = h_1 \cdots h_k(1), \text{ and } A_{n+1} = 1v'0 = h_1 \cdots h_j(1),$$

where j is either $k+1$ or $k+2$ as determined by Proposition 11.3.8, we can use the results for Cases 1–3 in Theorem 11.3.2 to find the relation between $C(v)$ and $C(v')$ as follows.

For Case 1, we have $j = k+1$, both h_{k+1} and h_{k+2} are of type h_l. If $h_{k+1} = (1 \to 1^m0, 0 \to 1^{m-1}0)$ for some m, then

$$C(v') = \begin{cases} C(v)(+1)^{m-1}(-1), & \text{when } h_k \text{ is of type } h_l, \\ C(v)(+1)^m(-1), & \text{otherwise.} \end{cases} \tag{11.3}$$

For Case 2, we have $j = k + 1$, and both h_k and h_{k+1} are of type h_r. If $h_{k+1} = (1 \to 10^{m-1}, 0 \to 10^m)$ for some m, then

$$C(v') = C(v)(+1)(-1)^{m-1} \tag{11.4}$$

For Case 3, we have $j = k + 2$, h_{k+1} is of type h_l, and h_{k+2} is of type h_r. If $h_{k+1} = (1 \to 1^m 0, 0 \to 1^{m-1}0)$, and $h_{k+2} = (1 \to 10^{m'-1}, 0 \to 10^{m'})$ for some m, m', then

$$C(v') = \begin{cases} C(v)(+1)^{m-1}(-1)^{m'}, & \text{when } h_k \text{ is of type } h_l, \\ C(v)(+1)^m(-1)^{m'}, & \text{otherwise.} \end{cases} \tag{11.5}$$

As in Subsection *9.3.3* we introduce a mapping T from h to strings of $(-1, +1)^*$:

$$T(h) = \begin{cases} (-1)^{m-1}(+1), & \text{when } h \text{ is of type } h_r \text{ and } h = (1 \to 10^{m-1}, 0 \to 10^m), \\ (+1)^{m-1}(-1), & \text{when } h \text{ is of type } h_l \text{ and } h = (1 \to 1^m 0, 0 \to 1^{m-1}0). \end{cases}$$

Now we can calculate $C(u)$ for a given K_+.

Proposition 11.3.15 If $K_+ = \lim_{n\to\infty} \prod_{k=1}^n h_k(1)$, then

$$C(u) = T(\prod_{k=1}^{\infty} h_k) = \prod_{k=1}^{\infty} T(h_k).$$

Proof. We claim that if $A_n = 1v_n0 = h_1 \cdots h_k(1)$, then

$$C(v_n) = \begin{cases} \prod_{i=1}^k T(h_i), & \text{when } h_k \text{ is of type } h_l, \\ (\prod_{i=1}^k T(h_i))\pi, & \text{otherwise.} \end{cases} \tag{11.6}$$

Proceed inductively on n.

(1) If h_1 is of type h_r, say $h_1 = (1 \to 10^{m-1}, 0 \to 10^m)$, then, no matter what h_2 may be, $A_1 = h_1(1) = 10^{m-1}$ and $v_1 = 0^{m-2}$. It is easy to verify that $C(v_1) = T(h_1)\pi$.

If h_1 is of type h_l, say $h_1 = (1 \to 1^m 0, 0 \to 1^{m-1}0)$, then the first square prefix of K_+ is 11, and $A_1 = 1$. We should verify the claim from $n = 2$. If h_2 is also of type h_l, then $A_2 = h_1(1) = 1^m 0$, and we can use the result of Case 1 in Theorem 11.3.2 to obtain $C(v_2) = (+1)^{m-1}(-1) = T(h_1)$.

If h_1 is of type h_l as above, but h_2 is of type h_r, say $h_2 = (1 \to 10^{m'-1}, 0 \to 10^{m'})$, then, no matter what h_3 may be, we have $A_2 = h_1h_2(1)$. Using the result of Case 3 in Theorem 11.3.2 we can obtain $C(v_2) = T(h_1)T(h_2)\pi$ as desired.

(2) Assume our claim is true for n, and consider the case of $n + 1$.

Case 1. Here $j = k + 1$ and both h_{k+1}, h_{k+2} are of type h_l. Using (11.3) and the inductive hypothesis, for h_k being h_l, we have

$$C(v') = C(v)(+1)^{m-1}(-1) = (\prod_{i=1}^{k} T(h_i))T(h_{k+1}),$$

and for h_k being h_r, we have

$$\begin{aligned} C(v') &= C(v)(+1)^m(-1) = C(v)(+1)T(h_{k+1}) \\ &= ((\prod_{i=1}^{k} T(h_i))\pi)(+1)T(h_{k+1}) \\ &= \prod_{i=1}^{k+1} T(h_i). \end{aligned}$$

Case 2. Here $j = k + 1$ and both h_k, h_{k+1} are of type h_r. Using (11.4) and the inductive hypothesis we have

$$\begin{aligned} C(v') &= C(v)(+1)(-1)^{m-1} \\ &= ((\prod_{i=1}^{k} T(h_i))\pi)(+1)(-1)^{m-1} \\ &= (\prod_{i=1}^{k+1} T(h_i))\pi. \end{aligned}$$

Case 3. Here $j = k + 2$, h_{k+1} is of type h_l, and h_{k+2} is of type h_r. Using (11.5), if h_k is of type h_l, then

$$\begin{aligned} C(v') &= C(v)(+1)^{m-1}(-1)^{m'} \\ &= (\prod_{i=1}^{k} T(h_i))T(h_{k+1})(T(h_{k+2})\pi) \\ &= (\prod_{i=1}^{k+2} T(h_i))\pi, \end{aligned}$$

otherwise, we have

$$\begin{aligned} C(v') &= C(v)(+1)^m(-1)^{m'} \\ &= ((\prod_{i=1}^{k} T(h_i))\pi)(+1)(+1)^{m-1}(-1)(-1)^{m'-1} \\ &= (\prod_{i=1}^{k+2} T(h_i))\pi. \end{aligned}$$

Thus the induction is completed. Let $k \to \infty$ in (11.6) we obtain the conclusion required. ∎

Using the same discussion as those in Theorem 9.3.7 we can establish the following fact, by which and Proposition 11.3.15 our proof of Theorem 11.3.3 is finished.

Proposition 11.3.16 If ρ is an irrational rotation number which has the continued fraction representation $[a_1, a_2, \ldots, a_n, \ldots]$, and $K_+ = \lim_{n\to\infty} h_1 \cdots h_n(1)$, then

$$\prod_{k=1}^{\infty} T(h_k) = (-1)^{a_1-1}(+1)^{a_2} \cdots (-1)^{a_{2k-1}}(+1)^{a_{2k}} \cdots$$

APPENDIX A
FINITE AUTOMATA AND REGULAR LANGUAGES

Languages and automata are the main tools throughout the whole book.

If we consider strings over an alphabet set Σ, then each subset of (the free monoid) Σ^* is referred to as a (formal) language over Σ. This simple definition of languages reveals that a study of languages is just a study of sets of symbolic strings, and the knowledge of languages in the theoretical computer science may be useful to the study of strings in other fields.

The automata, including the famous Turing machines (see Appendix B), in this book are considered as computation devices used to accept or recognize languages, and in this context the degree of complexity of automata is a measure for the complexity of languages accepted.

In this appendix we present the basic material about regular languages and finite automata. Of course, this is not a complete introduction for these topics. The choice of material is determined by the need of the presentation in the book.

The standard references about languages and automata are Hopcroft and Ullman (1979), Salomaa (1973, 1981, 1985), Davis, Sigal and Weyuker (1994).

A.1 Finite Automata

A.1.1 Deterministic Finite Automata

There are several ways to describe regular languages. The first one presented here is by finite automata.

A *finite automaton* (FA) may be considered as an abstract machine shown in Figure A.1, which consists of an *input tape* and a *finite state control device*. This finite control may be in any one of a finite number of internal states, and it has a read head to read symbolic strings written on the tape. This input tape is divided into squares, each holding a single symbol from an alphabet.

The working of finite automata (and other automata introduced in the sequel) is at discrete times. Thus we may speak of the time "$t = 0$", "$t = 1$" or the "next" time instant. In one move a finite automaton reads a symbol from the tape, enters a new state, and moves its read head one square to the right.

It seems that a finite automaton is a memoryless device, where no memory device is present. But since its output, namely its state, depends not only on the current

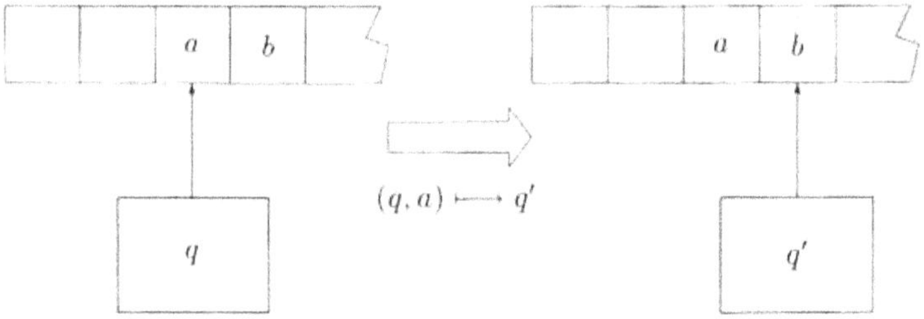

Figure A.1 A finite automaton

input symbol but also its previous state, so a finite automaton can remember its past input to a certain extent.

Now suppose a finite automaton is initially in state q_0. Its read head scans the first symbol in the first square of the input tape. Generally, as shown in Figure A.1 the finite automaton is in state q and its read head scans a square which contains a symbol a. Depending on the state q and the input symbol a the finite automaton enters a new state q' and moves its read head to the next right square. In Figure A.1 we denote this transition by

$$(q, a) \longmapsto q',$$

and write $q' = \delta(q, a)$ hereafter. Let Q be the set of finite internal states of a finite automaton, and Σ the alphabet set. If the state $q' = \delta(q, a)$ is uniquely determined by q and a, that is to say, for each input symbol in Σ there is only one possible transition out of each state, then we say this finite automaton is a *deterministic finite automaton* (DFA). In this case the transition rules δ is called a *transition function* which mapping $Q \times \Sigma$ to Q.

We can use finite automata as acceptors or recognizers for languages, that is, sets of symbolic strings. For this purpose we have to divide the state set Q into two subsets, F, a subset of accepting states, and its complement in Q, $Q - F$, a subset of non-accepting states. We call F the *accepting set*. If after reading a string from the input tape the finite automaton enters an accepting state, then we say this string is *accepted*, otherwise it is not accepted. In this way a given finite automaton accepts a subset of Σ^*, and excludes each string beyond this subset. If we can construct a finite automaton for a given language, then we obtain an acceptor or recognizer of this language.

In summary, a finite automaton consists of five elements: a finite set of states Q, an alphabet Σ of input symbols, an initial state $q_0 \in Q$, a transition function δ, and an accepting set $F \subseteq Q$. Usually we may denote a finite automaton by a 5-tuple $(Q, \Sigma, \delta, q_0, F)$. If a finite automaton $M = (Q, \Sigma, \delta, q_0, F)$ is given, then the language it accepts may be denoted by $L(M)$.

Now we can define the regular languages (RGL).

Definition A.1.1 A language is *regular* if it is accepted by a finite automaton.

If a language L cannot be accepted by any finite automaton, then L is referred to as a non-regular language in the sequel.

Two simplest regular languages are as follows.

Example A.1.2 Take $M = (\{q_0\}, \Sigma, \delta, q_0, \{q_0\})$, where the transition function δ is defined by $\delta(q_0, a) = q_0$ for each symbol $a \in \Sigma$, then it is obvious that this finite automaton will accept each strings over Σ. The language accepted by this finite automaton M is exactly the free monoid introduced in Chapter 1), namely, $L = L(M) = \Sigma^*$ This is the largest language over a given alphabet Σ in the sense of inclusion.

Remark. It seems that this language is too trivial to have any use. This is not true. In Chapter 4 we will show that the language of surjective unimodal map is the Σ^* itself (for $\Sigma = \{0, 1\}$).

We can also obtain the same language as follows. For each given finite automaton $M = (Q, \Sigma, \delta, q_0, F)$, if we modify its accepting set by letting $F = Q$, then any string over Σ will be accepted and the language $L(M)$ is Σ^*.

Example A.1.3 Conversely, if we modify a given finite automaton M by letting $F = \emptyset$, then M cannot accept any strings, and the language accepted by this finite automaton is the smallest language $L = L(M) = \emptyset$ for each Σ^*

A more complex language, which has dynamical meaning, is as follows.

Example A.1.4 Let L be the set $\{0^m 1^n \mid m \geq 0, n \geq 0\}$. We show that L is a regular language. Let $M = (Q, \Sigma, \delta, q_0, F)$, where $Q = \{q_0, q_1, q_2\}$, $F = \{q_0, q_1\}$, $\Sigma = \{0, 1\}$, and δ is defined as follows:

$$\delta(q_0, 0) = q_0, \quad \delta(q_0, 1) = q_1, \quad \delta(q_1, 1) = q_1,$$
$$\delta(q_1, 0) = q_2, \quad \delta(q_2, 0) = q_2, \quad \delta(q_2, 1) = q_2.$$

It is easy to see that $L = L(M)$.

In the rest of the book we will use the abbreviation FA and DFA for finite automaton and deterministic finite automaton.

A.1.2 Extensions of DFA

There are several extensions of DFA, which are useful in applications.

If the transition $(q, a) \mapsto q'$ in Figure A.1 is not deterministic as in DFA of the previous subsection, that is, the transition function δ may not be defined for some pairs of $(q, a) \in Q \times \Sigma$, or it may not be single valued for other pairs of (q, a). It means that for some internal state of FA and input symbol the FA may stop and not move again, while for other state and input symbol there may exist more than one possible states which the finite automaton may enter.

Using the notation 2^Q to denote the set of all subsets of Q, we may say that now the transition function δ is a mapping from a subset of $Q \times \Sigma$ to 2^Q. We call this new automaton a *nondeterministic finite automaton* (NFA).

NFA may also be used as acceptors or recognizers of languages. But here there is a difference between DFA and NFA. For a given DFA, an input string will uniquely determine the transition path, that is, the sequence of transition through states which begins from q_0. For a NFA, however, an input string may correspond several transition paths, and some of them may break off before the scanning of string has been finished. We say a string is accepted by this NFA if at least one of the paths, which corresponds to the whole string, begins from q_0 and leads to an accepting state. Thus we may define $L(M)$ as before.

A further extension of DFA is called the NFA *"with ε-moves"*, in which the transition of state of FA may happen even no input symbols are read from the input tape. We may call such move by *"transition on the empty input ε"*. It means that in Figure A.1 the state q may be changed to a new state q' without reading the current input symbol a. At the same time the read head does not move to right. In this case the transition function δ is a mapping from $Q \times (\Sigma \cup \{\varepsilon\})$ or its subsets to 2^Q. The languages accepted by such FA are defined similarly.

It seems no doubt that the two extensions of DFA above would increase the capability of machines to do their jobs. But a main theorem in the theory of FA shows that this observation is not true.

Theorem A.1.5 A language L is accepted by an NFA with or without ε-move if and only if L is accepted by a DFA.

The proof of this theorem can be found, for instance, in Hopcroft and Ullman (1979). It does not mean that the concept of NFA is useless, as a matter of fact, NFA is a very convenient tool in many applications, but the capability of accepting languages is not increased by the extensions above.

A.1.3 Transition Diagrams of FA

A *transition diagram* of FA is a directed graph which gives a vivid representation of FA. If we can construct a transition diagram of a finite automaton which accept a given language L, then it is very easy to decide from this diagram if a given string s belongs to L. Therefore, in this book a transition diagram of FA is often referred to as the FA itself.

Figure A.2 is the transition diagrams for Examples A.1.2 ($\Sigma = \{0, 1\}$) and A.1.4.

A new example of FA and its transition diagram is shown in Figure A.3.

Example A.1.6 In Figure A.3(a) the finite set of state is $Q = \{q_0, q_1, q_2, q_3\}$, in which q_0 is the initial state and indicated by the arrow labeled "start", the input alphabet set is $\Sigma = \{0, 1\}$, the accepting set is $F = \{q_0, q_1, q_2\}$, and the transition function δ is shown in Figure A.3(b).

Now we can use Figure A.3 to prove the following proposition.

Proposition A.1.7 If a language L is accepted by the FA of Figure A.3, then a string $s \in (0+1)^*$ belongs to L if and only if s does not contain 100 as its substring.

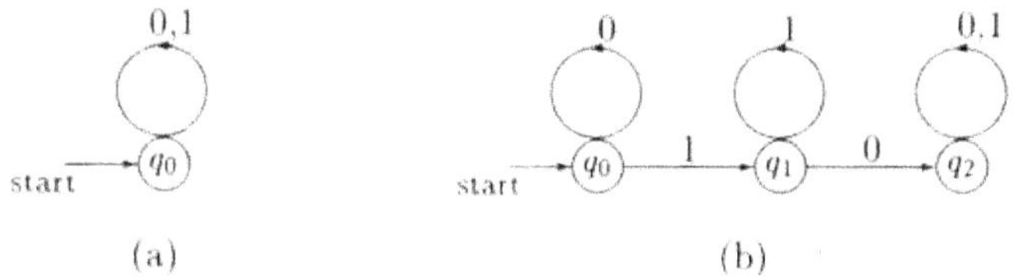

Figure A.2: Transition diagrams of (a) Example A.1.2 and (b) Example A.1.4

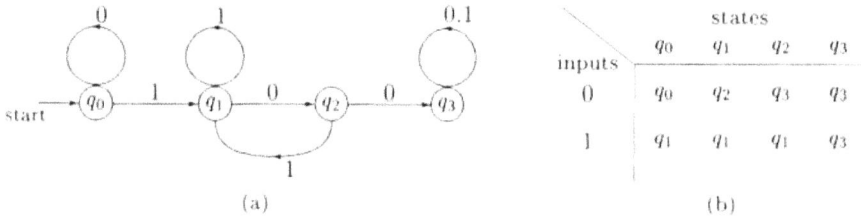

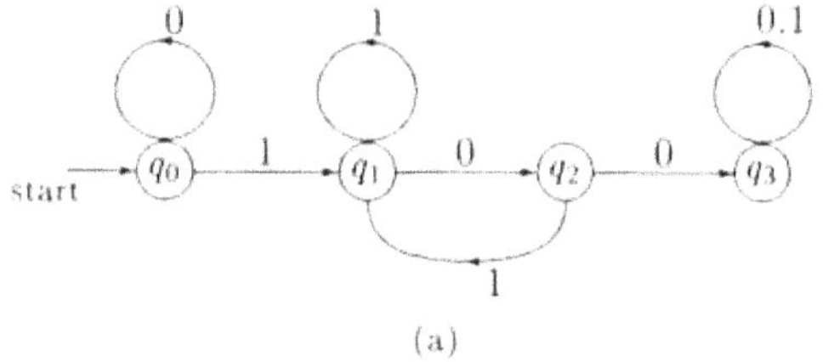

inputs \ states	q_0	q_1	q_2	q_3
0	q_0	q_2	q_3	q_3
1	q_1	q_1	q_1	q_3

(b)

Figure A.3: (a) An example of DFA, (b) its transition function.

Proof. The "only if" part is easy. It depends on the observation that, using Figure A.3 we may directly verify that starting from each state q_i and reading 100 will lead to the unique non-accepting state q_3. It simply means that if s contains 100 as its substring, then the transition path corresponding to s will enter q_3 and remains there, this contradicts to $s \in L$.

Consider the "if" part. Let $s = s_1 s_2 \cdots s_n$ and proceed inductively on $n = |s|$. It may be directly verified that $s \in L$ for $n \leq 3$. Assume the claim is true for $n \leq k$ and consider the case $n = k + 1$. Consider the state q_i of FA after it has read $s_1 s_2 \cdots s_k$, a prefix of s. If this state is not q_2, then it is easy to see from Figure A.3 that our FA cannot enter the non-accepting state q_3 whatever the last symbol s_{k+1} is.

If the state after reading $s_1 \cdots s_k$ is q_2, the unique chance that the next state being q_3 is $s_{k+1} = 0$. But the only chance to enter q_2 is from state q_1 and reading $s_k = 0$, and the only chance to enter q_1 is reading symbol 1, whatever the previous state is, we conclude that if the claim were wrong then the string s would have 100 as its suffix, a contradiction. ∎

A.2 Grammars and Expressions

Although finite automata is the main method in this book to describe regular languages, we still need some other methods and introduce them in this Section.

A.2.1 Right-Linear Grammars

Here we introduce the second method to describe regular languages, that is, right-linear grammars.

Generally speaking, a *grammar* is denoted by $G = (V, \Sigma, P, S)$, where both V and Σ are finite sets of symbols, P is a finite set of grammar rules, $S \in V$ is a special start symbol. Assume that $V \cap \Sigma = \emptyset$. The symbols of V are called *variables* or *non-terminals*, while the symbols of Σ are called *terminals*.

If beginning from the starting variable S, and applying the grammar rules finite times we obtain a string which consists entirely of terminals of Σ, then this string is said to be a *word* generated by the grammar G. A language generated from the grammar G is a set consists of all words generated from G and denoted $L = L(G)$.

If each grammar rule of P is one of the two forms:

$$A \to \beta B, \quad A \to \beta,$$

where $A, B \in V$, $\beta \in \Sigma^*$, then the grammar G is called a *right-linear grammar*.

The basic result in this aspect is as follows (see, e.g., Hopcroft and Ullman 1979)

Theorem A.2.1 A language is regular if and only if it can be generated by a right-linear grammar.

In Table A.1 we list the right-linear grammars for four Examples presented in the previous Section. (Here the convention | is that, for instance, $S \to A \mid B$ means that $S \to A$ and $S \to B$.)

Table A.1 Grammatical Rules of Examples A.1.2 – A.1.6

Example	Grammatical Rules
A.1.2	$S \to \varepsilon \mid 0S \mid 1S$
A.1.3	$S \to S$
A.1.4	$S \to \varepsilon \mid 0S \mid 1A, \quad A \to \varepsilon \mid 1A$
A.1.6	$S \to \varepsilon \mid 0S \mid 1A, \quad A \to \varepsilon \mid 1A \mid 0B, \quad B \to \varepsilon \mid 1A$

A.2.2 Regular Expressions

Besides finite automata and right-linear grammars, the third equivalent ways to describe regular languages is regular expressions.

Here we need some operations on languages. Let L_1 and L_2 be formal languages over an alphabet Σ. The *union* or *sum* of L_1 and L_2, denoted $L_1 + L_2$, is the set $\{x \mid x \in L_1 \text{ or } x \in L_2\}$. The *product* or *concatenation* of L_1 and L_2, denoted $L_1 \cdot L_2$ or L_1L_2, is the set $\{xy \mid x \in L_1 \text{ and } y \in L_2\}$. For a language L over Σ we define its *powers*

$$L^0 = \varepsilon, \quad L^i = LL^{i-1} \text{ for } i \geq 1,$$

and its *closure*

$$L^* = L^0 \cup L^1 \cup \cdots \cup L^n \cup \cdots = \{s_1 \cdots s_k \mid k \geq 0, s_1, \ldots, s_k \in \Sigma\}.$$

Some notes are: 1. $\{\varepsilon\}$ is a language which contains one element, the empty string ε, and $\emptyset$ is a language which contains no strings. 2. The operation of "*" is quite similar to the corresponding operation in the monoid (see 1).

If a language L contains only finite strings then it is called a *finite language*. Now we have the basic result in this aspect.

Theorem A.2.2 A language is regular if and only if it can be obtained from finite languages by applying the three operations $+, \cdot, *$ a finite number of times.

Now we can give the definition of *regular expressions*.

1. $\emptyset$ is a regular expressions and denotes the empty set.
2. ε is a regular expressions and denotes the set $\{\varepsilon\}$.
3. Each $a \in \Sigma$ is a regular expressions and denotes the set $\{s\}$.
4. If r and s are regular expressions denotes the languages R and S, then $(r + s)$, (rs), and (r^*) are regular expressions that denote the sets $R + S$, RS, and R^* respectively.
5. No expression is regular unless it may be generated using a finite number of applications of 1–4.

In Table A.2 we list the regular expressions for Examples A.1.2, A.1.3, A.1.4, and A.1.6.

Table A.2 Regular Expressions of Examples A.1.2 – A.1.6

Example	Regular Expression
A.1.2	$(0+1)^*$
A.1.3	$\emptyset$
A.1.4	0^*1^*
A.1.6	$1^* + 0^*1(1+01)^* + 0^*1(1+01)^*0$

There are methods to compute the regular expression of a regular language $L = L(M)$ from a given FA M, or to construct a finite automaton for a given regular expression (Hopcroft and Ullman 1979).

A.3 Minimum States DFA

In the previous two Sections we see that for a given regular language there exist several ways to describe it. If a language L, however, is given in any other way, for example, L is generated from a dynamical system, how can we know that L is a regular language?

Moreover, It is easy to see that for a regular language, there may exist many FA's which accept the same language, and similar situations happen for the other two representation — regular expressions and right-linear grammars.

It is important in many applications to be able to determine if a given language is regular, and if it is, how to obtain a minimal description for it. Here there exists a beautiful theory centered around an equivalence relation R_L and the Myhill-Nerode Theorem, which gives the (unique) minimal DFA for a given regular language.

A.3.1 A Natural Equivalence Relation R_L

Let Σ be an alphabet, and L be a language over Σ. Note that here the language L need not be regular.

Definition A.3.1 Let a language $L \subset \Sigma^*$ be given. A relation R_L is introduced on Σ^* by L such that xR_Ly if and only if for each z, either both or neither of xz and yz is in L.

It is easy to see that R_L is an *equivalence relation*, that is to say, R_L is reflexive, symmetric, and transitive as follows:

(1) xR_Lx for all $x \in \Sigma^*$.
(2) xR_Ly implies yR_Lx.
(3) xR_Ly and yR_Lz imply xR_Lz.

The meaning of R_L is clear that it divides the free monoid Σ^*, the set of all finite strings over the alphabet Σ, into disjoint nonempty equivalence classes. That is a partition

$$\Sigma^* = S_1 \cup S_2 \cup \cdots,$$

where S_i, $i = 1, 2, \ldots$ are equivalence classes which satisfy the following conditions

(1) $S_i \cap S_j = \emptyset$ for each $i \neq j$.
(2) xR_Ly holds for each x and y in the same S_i.
(3) xR_Ly is false for each $x \in S_i$ and $y \in S_j$ if $i \neq j$.

We call this relation R_L the *natural equivalence relation generated by L on* Σ^*.

Some simple but important properties of R_L are:

(R1) $xR_Ly \to xzR_Lyz$ for each $z \in \Sigma^*$, that is to say, *right invariant* with respect to concatenation.
(R2) xR_Ly and $x \in L$ implies $y \in L$, i.e., the language L (and its complement) is exactly the union of several classes.
(R3) Both $x, y \in L$ (or $x, y \notin L$) and xaR_Lya for each symbol $a \in \Sigma$ imply xR_Ly.

A.3.2 The Role of R_L

At first sight it seems that this R_L is just an abstract concept in mathematics. But it is not so. As a matter of fact, this equivalence relation R_L is closely related to a practical question: how to construct a (model of) machine to decide, that is, to accept or recognize, if a string belonging to a given language or not. In some sense it amounts to find what structure a given language has.

Imagine that we have a string $s = s_1 \cdots s_n$ over an alphabet Σ as the input string to our machine. If we have received its prefix $s_1 \cdots s_k$ for some $k < n$, then what should we store into our memory (whatever it may be) at this point to make the final decision in the future if $s \in L$ when the input of this string has finished. A simple method is to put the prefix $s_1 \cdots s_k$ itself into our memory. But it is easy to realize that this is not a welcome idea.

Of course, if L is a finite language and contains only a few words, then we may prepare a table to list all of them together, and to look at it if a given string is on the table. Even in this case we still don't wish to store the whole string first and to make our decision only after its input is finished. Instead we will look at the prefix which has been received step by step to see if we can make the negative decision as early as possible. Usually a language contains too many words (often infinitely many words) to make this look-out method meaningful.

The notion of R_L reveals a new idea to solve this question. If we have received a prefix of an input string s, then what we need to store is not necessarily the prefix itself, but the equivalence class which the prefix belongs. (Here we don't consider what the partition of $\Sigma^* = S_1 \cup S_2 \cup \cdots$ is and how we can fix the special class S_i to which the prefix $s_1 \cdots s_k$ belongs. These problems involve many technical details and are discussed in Parts II and III of this book for several kinds of languages.) The important point we make here is that, if we write $s = xz$ and have received x, then the information we should put into memory to decide in the future if $s \in L$ is simply the class S_i containing x (if we can). This is obviously true from the definition of R_L that, in order to decide if $xz \in L$ or not, the decisive element here is the equivalence class S_i which contains x but not the prefix x itself.

Furthermore, if the partition of Σ^* induced through relation R_L is finite, that is to say, there are only finite different equivalence classes, then we may calculate the quantity of information which we need to store. For example, if for a certain language L we have the partition of Σ^* as

$$\Sigma^* = S_1 \cup S_2 \cup \cdots \cup S_N,$$

then the quantity of information we need to store is only $\log_2 N$ bit (Shannon 1949).

In the next Subsection we will see that this discussion can be transformed into a rigorous theory, which is the content of the Myhill-Nerode theorem.

A.3.3 The Myhill-Nerode Theorem

In this Subsection we present the Theorem in the title and give its proof. Since the ideas in the proof are indispensable in constructing minDFA (cf. Chapters 4 and 9), it is appropriate to include it completely.

First we need some notions: the index of an equivalence relation, and minDFA.

The *index* of an equivalence relation is the number of its equivalence classes. It may be infinite.

For a given language L, we use minDFA to denote the deterministic finite automaton which accepts L and has minimum states.

Theorem A.3.2 (Myhill-Nerode) Let a language L be given, then

(1) L is regular if and only if R_L is of finite index.

(2) If L is regular, then the minimum states DFA accepting L is unique up to an isomorphism, that is, a renaming of the states, and the number of its states is the index of R_L.

Proof. We divide this proof in several steps.

(a) First a given DFA M may be used to introduce a new equivalence relation R_M into Σ^*, the set of all finite strings over Σ: xR_My if after reading either x or y the DFA M is in the same state. It is obvious that the partition induced by R_M on Σ^* is determined completely by M, and the index of R_M is just the number of states of M.

Now we make a basic observation, if L is accepted by a DFA M, x and y are two input strings such that, after reading either x or y, the DFA M is in the same state, then xR_Ly. As a matter of fact, since M is deterministic, reading either xz or yz will make M enters the same state for each string z, and then either both or neither of xz and yz will be accepted.

This observation simply means that

$$xR_My \Longrightarrow xR_Ly,$$

and each equivalence class of R_M is a subset of some equivalence class of R_L. This implies in turn that the index of R_L will not be greater than the index of R_M. Thus, the existence of a DFA M for a given L guarantees that the equivalence relation R_L is of finite index. This finishes the proof of the "only if" part of claim (1).

(b) Let the language L be given. Assume that the index of R_L is finite and denote it N. Let the equivalence classes be denoted $[x_1], [x_2], \cdots, [x_N]$, where each x_i is an (arbitrary) representative string of the equivalence class $[x_i]$. Without loss of generality, assume that $\varepsilon \in [x_1]$, i.e., $[x_1] = [\varepsilon]$. Now we may construct a DFA $M = (Q, \Sigma, \delta, q_0, F)$ as follows. Let the state set Q be $\{[\varepsilon], [x_2], \ldots, [x_N]\}$ and the starting state be $[\varepsilon]$. The transition function δ is defined by

$$\delta([x_i], a) = [x_ia] \text{ for all } [x_i] \in Q \text{ and } a \in \Sigma.$$

Finally, the accepting set F is defined by

$$F = \{[x_i] \mid x_i \in L\}.$$

From the properties (R1) and (R2) of R_L we may verify that

$$y \in [x_i] \rightarrow yR_Lx_i \rightarrow yaR_Lx_ia \rightarrow [ya] = [x_ia],$$

and hence the definitions of δ and F are unambiguous.

Denote by M the DFA defined thus far. we will show that the language accepted by M, $L(M)$, is exactly the language L. If $x = a_1 \cdots a_n \in L$, then $[x] \in F$. Beginning from the starting state $[\varepsilon]$ and reading x, the transition path of the states of DFA M will be $[\varepsilon], [a_1], [a_1a_2], \ldots, [a_1 \cdots a_n] = [x]$. Since the last state $[x] \in F$ is accepting, it means $x \in L(M)$.

On the other hand, if a string $y = [b_1 \cdots b_m] \notin L$, then its corresponding transition path of the states of M will be $[\varepsilon], [b_1], [b_1b_2], \ldots, [b_1 \cdots b_m] = [y]$. Since $y \notin L$ the final state $[y] \notin F$ and leads to $y \notin L(M)$. This finishes the proof of the "if" part of claim (1).

(c) From (a) we see that the index of R_L is less than or equal to the number of states of any DFA which accepts L. Combining this with (b) leads to the conclusion that the DFA M constructed in (b) is exactly the minimum states DFA, that is, minDFA.

(d) Now we prove that the minDFA for a given language is unique up to an isomorphism of DFA. This means that, if there is another minDFA M' accepting L, then we can establish an isomorphism between M and M'.

Since M' is a minDFA, the number of states of M' must be N. Denote its states by $q_1, \ldots, q_N$. From (a) we see that each q_i corresponds to an equivalence class of R_L or its subset. Because the number of states of M', N, is equal to the index of R_L, we conclude that there is a bijection between $q_1, \ldots, q_N$ and the equivalence classes $[x_1], \ldots, [x_N]$, the states of the minDFA M in (b), of R_L. After renaming the states q_i, if necessary, we may denote this bijection by ϕ and have

$$\phi(q_i) = [x_i] \text{ for all } i = 1, \ldots, N$$

It means that (from (a)) every input string $x \in [x_i]$ will make the DFA M' entering the state q_i after reading x.

Now denote by δ' the transition function of M'. If

$$\delta'(q_i, a) = q_j \text{ for } q_i, q_j \text{ of } M' \text{ and } a \in \Sigma,$$

then we can show that

$$[x_i a] = \delta([x_i], a) = [x_j].$$

In fact, the DFA M' enters state q_i after reading x_i, and the next symbol a makes the next state of M' being q_j, which corresponds to the class $[x_j]$. That is

$$\phi(\delta'(q_i, a)) = \delta(\phi(q_i), a).$$

This proves that the bijection ϕ is really an isomorphism between M and M' and finishes the proof of claim (2). ∎

A.3.4 Applications

Here we apply the notion of R_L and the Myhill-Nerode Theorem to Examples of section A.1.

1. Example A.1.2, where $L = \Sigma^*$. It is easy to see that the index of R_L is 1, and the DFA in Figure A.2 (a) is its minDFA.

2. Example A.1.3, where $L = \emptyset$. We have the same minDFA as above, but the accepting set $F = \emptyset$.

3. Example A.1.4, where $L = \{0^m 1^n \mid m, n \geq 0\}$. Here we may obtain the equivalence classes of R_L as follows:

$$[\varepsilon] = \{0^n \mid n \geq 0\},\ [1] = \{0^m 1^n \mid m \geq 0 \text{ and } n > 0\},\ [10] = \{x \mid x \notin L\} = L',$$

and the DFA in Figure A.2 (b) is its minDFA.

4. Example A.1.6, where L is given by the DFA of Figure A.3 (a). We have established Proposition A.1.7 there and obtain its right-linear grammar in Table A.2 and regular expression in Table A.1. Its equivalence classes are

$$[\varepsilon] = \{0^m \mid m \geq 0\},\ [1],\ [10],\ \text{and } [100],$$

and Figure A.3 (a) is its minDFA. It is easy to understand the working of this minDFA between these classes. Assume that an input string is $s = s_1 \cdots s_n$, and its prefix $s_1 \cdots s_k$ has just been read by DFA. If the state of DFA is $[\varepsilon]$ (q_0 in Figure A.3), it means simply that the prefix received consists of 0 entirely. If the state is $[1]$, then $s_k = 1$ and no substring 100 being received. If the state is $[10]$, then $s_{k-1}s_k = 10$ and no 100 being received. If the state is $[100]$, then it is certain that $s \notin L$ no matter what the symbols $s_i (i > k)$ after s_k may be. We see that in order to decide if $s \in L$ or not the information quantity required is $\log_2 3$ bit.

We discuss two new examples as follows.

Example A.3.3 $L = \{0^k 1 0^k \mid k \geq 0\}$ is not regular.

Here it is easy to show that the equivalence relation R_L is of infinite index. As a matter of fact, take $x = 0^m$ and $y = 0^n$ with $0 \leq m < n$, and $z = 10^m$. Since $xz \in L$ and $yz \notin L$, we see that $xR_L y$ is false. It means that each string in $\{0^n \mid n \geq 0\}$ belongs to a distinct equivalence class, and R_L is not of finite index. Here L itself is a equivalence class.

Example A.3.4 $L = \{0^{2^k} \mid k \geq 0\}$ is not regular.

It is easy to show that each string in L belongs to a distinct equivalence class, and R_L is of infinite index.

A.4 Properties of Regular Languages

Besides the Myhill-Nerode Theorem, more properties of regular languages are introduced in this Section.

A.4.1 Pumping Lemma of Regular Languages

Although the Myhill-Nerode theorem gives the necessary and sufficient condition of a language being regular, but in some cases the following necessary condition, the so-called *pumping Lemma* is more convenient (Hopcroft and Ullman 1979).

Lemma A.4.1 (Pumping Lemma) If L is a regular language, then there exists an integer n such that if $z \in L$ is of length $|z| \geq n$, we have a decomposition $z = uvw$ in such a way that $|uv| \leq n$, $|v| \geq 1$, and for all $i \geq 0$, $uv^iw \in L$.

Proof. Let M be DFA accepting L and n =card(Q), the number of states of M. If string $z = a_1a_2 \cdots a_k \in L$ is of length $k \geq n$, then the sequence of states of M after reading z may be denoted by $q_0q_1 \cdots q_k$, where q_0 is the starting state, and $q_i = \delta(q_{i-1}, a_i)$ for $i = 1, \ldots, k$. Since $k \geq n$, there are two integers $i < j$, $0 \leq i < j \leq n$, such that $q_i = q_j$. It implies that if we denote

$$u = a_1 \cdots a_i, \quad v = a_{i+1} \cdots a_j, \quad w = a_{j+1} \cdots a_k,$$

then the claim is true. ∎

Remark. From the proof of Lemma we really obtain

$$zR_l\, uv^iw \text{ for all } i \geq 0$$

In this form a necessary and sufficient condition can be obtained for a language being regular (Salomaa 1985).

The pumping Lemma gives a necessary condition of a language being regular. Here we apply this tool to Examples A.3.3 and A.3.4 to get new proofs that neither of their languages is regular.

1. $L = \{0^k10^k \mid k \geq 0\}$ is not regular. Assume the contrary that L is regular, then there exists the integer n asserted in the pumping Lemma. Let $z = 0^n10^n$, we have $z = uvw$ such that $|uv| \leq n$, $|v| \geq 1$, and $uv^iw \in L$ for $i \geq 0$. It implies that uv consists of symbol 0 only, and strings uv^iw cannot belong to L for $i \neq 1$.

2. $L = \{0^{2^k} \mid k \geq 0\}$ is not regular. Assume the contrary that L is regular, and let n be the integer asserted in the pumping Lemma. Let $z = 0^{2^n}$, we have $z = uvw$ such that $|uv| \leq n$, $|v| \geq 1$, and $uv^iw \in L$ for $i \geq 0$. Considering the string uv^2w, its length satisfies

$$2^n < |uv^2w| \leq 2^n + n < 2^{n+1}$$

But from $uv^2w \in L$ its length $|uv^2w|$ must be a power of 2, a contradiction.

A.4.2 Closure Properties

Denote by $\mathcal{L}$(RGL) the class of regular languages. In many applications it is important to know the set $\mathcal{L}$(RGL) is closed under some operations. For simplicity we just list some of them and give a sketch of their proofs (see Hopcroft and Ullman (1979) for details).

1. $\mathcal{L}(\mathrm{RGL})$ is closed under union, concatenation, and closure. This is a consequence of Theorem A.2.2.

2. $\mathcal{L}(\mathrm{RGL})$ is closed under complementation. That is to say, if $L \subseteq \Sigma^*$ is regular, then its complementation $L' = \Sigma^* - L$ is also regular.

   Using FA it is easy to prove this fact. Let $L = L(M)$ be the language accepted by FA $M = (Q, \Sigma, \delta, q_0, F)$, then $L' = \Sigma^* - L$ is accepted be another FA $M' = (Q, \Sigma, \delta, q_0, Q - F)$.

3. $\mathcal{L}(\mathrm{RGL})$ is closed under the operation of mirror. Here for a give string $x = a_1 \cdots a_n$ we call the string $a_n \cdots a_1$ the *mirror* of x, and denoted by x^R. For a given language L, we call $L^R = \{x^R \mid x \in L\}$ the mirror of L.

   This can be proved by using the method of regular expressions. As a matter of fact, changing all order of concatenations in the regular expression of L we obtain the expression for L^R.

4. $\mathcal{L}(\mathrm{RGL})$ is closed under the operation of substitution.

   Here we need the concept of substitution. Let $R \subseteq \Sigma^*$ be a language. If for each $a \in \Sigma$, let $R_a \subseteq \Delta^*$ be defined, where Δ may be a different alphabet from Σ. For each word $a_1 \cdots a_n \in R$ we change each a_i by a word $w_i \in R_{a_i}$ and obtain a string $w_1 \cdots w_n$. Thus we obtain a new language from R. We call this operation a *substitution* f, and denote the new language by $f(R)$. The closure property of regular languages under substitution means that if R is regular, and each R_a for $a \in \Sigma$ is regular, then $f(R)$ is also regular. It can be proved using regular expressions.

5. $\mathcal{L}(\mathrm{RGL})$ is closed under the operation of homomorphism.

   Here a *homomorphism* is a substitution where each R_a contains only one string. If we denote a homomorphism by h, then $h(a)$ is the string itself.

6. $\mathcal{L}(\mathrm{RGL})$ is closed under the operation of inverse homomorphism.

   Let h be a homomorphism, and for each $a \in \Sigma$, we have $h(a) \in \Delta^*$. For a language $L \subseteq \Delta^*$, the set

   $$h^{-1}(L) = \{x \mid h(x) \in L\}$$

   is called the *inverse homomorphism* image of L.

7. $\mathcal{L}(\mathrm{RGL})$ is closed under the operation of MIN, which is defined by

   $$\mathrm{MIN}(L) = \{x \in L \mid \text{no proper prefix of } x \text{ is in } L\}.$$

APPENDIX B
NON-REGULAR LANGUAGES

In this appendix we introduce the basic concept of Turing machine and languages associated with it. The references are those in Appendix A, and Turing (1936), Penrose (1989), Moore (1991a).

B.1 Turing Machines

B.1.1 Definitions

In order to "design" a model of machines which may accept or recognize non-regular languages, we have to go beyond the finite automata. In this aspect the most important model is the Turing machine (Turing 1936).

The simplest form of *Turing machine* is shown in Figure B.1.

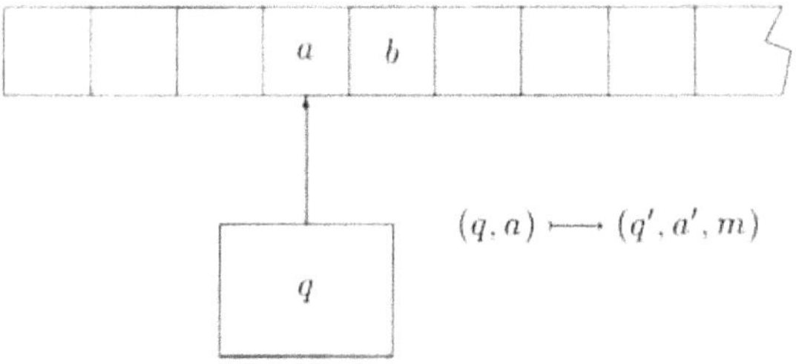

Figure B.1: A Turing machine

Comparing this machine with the finite automata shown in Figure A.1, the similarity between them is obvious. Both of them have a finite control device and an input tape, which is divided into squares and each square containing a symbol from an finite alphabet Σ. The way of working of Turing machine is also at discrete times. As implied in the case of finite automaton we also assume this input tape has its left end, but is potentially infinite to the right. It means that although each input string is of finite length, but we do not restrict its length in any way.

Now we discuss the differences between finite automata and Turing machines.

The read head in Figure B.1 has stronger ability than that of finite automata. In each move this read head scans a symbol in the current square, and then changes it with a new symbol a' into the square scanned. In other words, this head has both capability of reading and writing. In the same move the state is changed from q to q', and the head is moved left or right one square. In the Figure B.1 we see the transition

$$(q, a) \longmapsto (q', a', m),$$

where $m \in \{L, R\}$ means which way (left or right) the head will move.

As in the case of finite automata the input string of Turing machine is initially put on the tape from its left end. But in the case of finite automata, the other squares, which are not occupied initially by input symbols, are useless. For Turing machine, however, since its read head may move both way and may write as well as read, those squares, which are not used initially, are very important. As a matter of fact, they provide an infinite capacity of memory. For convenience we assume each square, which is not used initially, holding the blank, denoted by B, a special tape symbol that is not an input symbol.

There are many extensions of the simplest Turing machine above, for instance, as the nondeterministic finite automata with or without ε move (Subsection *A.1.2*) the transition $(q, a) \mapsto (q', a', m)$ may be either deterministic or nondeterministic, and we obtain nondeterministic Turing machines. We have also Turing machines with two-way infinite tape, multitrack Turing machines, multitape Turing machines, multihead Turing machines, and so on.

A basic fact is that the capability of accepting of recognizing languages is the same as the simplest Turing machine shown in Figure B.1. Thus in this book we will use Turing machine in its simplest form as above.

More precisely, we can describe a Turing machine similarly as for finite automata:

$$M = (Q, \Sigma, \Gamma, \delta, q_0, B, F),$$

where Q is the finite set of states, $\Sigma \subsetneq \Gamma$ is the finite set of input symbols, Γ is the finite set of allowable tape symbols, δ is the transition function, a mapping from $Q \times \Gamma$ to $Q \times \Gamma \times \{L, R\}$, $q_0 \in Q$ is the starting state, $B \in \Gamma$ is the blank symbol which is not in Σ, $F \subseteq Q$ is the accepting set.

The very important point about Turing machines is its nature of finite descriptions. It decides that all Turing machines, through proper coding, can be put into one-to-one correspondence with the positive integers, and hence the set of all Turing machines is countable.

B.1.2 Recursively Enumerable Languages

We now introduce the languages accepted by Turing machines.

Definition B.1.1 A language is *recursively enumerable* if it is accepted by a Turing machine.

We use REL as the abbreviation of recursively enumerable language, and $\mathcal{L}(\text{REL})$ the set of all REL. Since each finite automaton is obviously a Turing machine, we have

$$\mathcal{L}(\text{RGL}) \subsetneq \mathcal{L}(\text{REL}).$$

The fact of this inclusion being proper can be shown later.

On the other hand, if a language cannot accepted by any Turing machine, then it is referred to as a *non-recursively enumerable language*. Its existence is easily understand by the argument below.

From the definition of Turing machine we see that there are only countably many recursively enumerable languages. On the other hand, recall that languages are subsets of the free monoid Σ^*, which is always a countable set. Since for each nonempty set A, the cardinal number of power set 2^A is always higher than that of A, the set of all languages is uncountable. These considerations are summarized into the following conclusion.

Theorem B.1.2 The set of all recursively enumerable languages, $\mathcal{L}(\mathrm{REL})$, is countable, and the set of non-recursively enumerable languages (for any alphabet) is uncountable.

The adjective "enumerable" reflects the fact that an recursively enumerable language may be generated by a Turing machine as its output. As a matter of fact, since Turing machine may print symbols on its input tape, it is not surprising that a Turing machine can be used as a generator of a language. It is not difficult to prove the following theorem.

Theorem B.1.3 A language is recursively enumerable if and only if it can be generated by a Turing machine.

Here we give a sketch of its proof. But before doing it, we have to explain the *halting problem* associated with Turing machine. Let a Turing machine M be given, and consider an input string x for M. If x is of length $n = |x|$, then it is put initially on the n leftmost squares of the tape, and the read head scans the first square at beginning. If the state of M eventually enters an accepting state, then x is accepted. Without loss of generality, we may suppose M stops after accepting x. If x is not accepted by M, the situation becomes complicated. The machine M may either stop without entering accepting states or not stop and run forever. In the latter case we cannot decide if the string x will be accepted by M or not. This is an intrinsic trouble with Turing machines.

We now consider a Turing machine M as a generator (or enumerator) of strings. Adding an output tape to the simplest form of Figure B.1, we may put the strings generated by M on this tape and separate them by a special symbol, say #. Let the language generated by M be denoted by $G(M)$.

If $L = G(M_1)$ for a generator M_1, then we may construct a Turing machine M_2 as follows. This M_2 has one more input tape than M_1, on which the input string x is put initially, and it may simulates the operations of M_1 completely. After a string is generated by M_2 on its output tape, i.e., a new # appears on it, then M_2 compares it with x, if they are the same, M_2 accepts x, otherwise M_2 simulates M_1 again. It is easy to see that M_2 accepts x if and only if $x \in G(M_1)$. Note that if $x \notin G(M_1)$, then M_2 may not stop at all, even if $G(M_1)$ is finite. M_1 may not stop after it generate its all strings.

If $L = L(M_1)$ is given, then we need to construct a generator M_2 for L. Since the trouble of halting problem, we cannot simply input each string in Σ^* in some order to M_1 and generate the strings accepted. Instead we have to use an ingenious method to generate pairs (i, j) systematically, e.g., in the order of size $i + j$, and for each (i, j) input the i-th string in Σ^* to see if M_1 will accept it at the j-th step. In this way we may obtain a generator M_2, which generate exactly L, and for each string in L only once.

B.1.3 Recursive Languages

Because the trouble of halting problem for Turing machines, for a recursively enumerable language there may not exist any algorithm which can decides if a given string belongs to the language, that is to say, there is no effective procedure to solve it for each given string.

There is a proper subset of $\mathcal{L}(\mathrm{REL})$, for which there is no such trouble.

Definition B.1.4 A *recursive language* L is a recursively enumerable language, for which there exists at least one Turing machine M, such that $L = L(M)$ and M stops for every input string.

We use RL to denote recursive language, and $\mathcal{L}(\mathrm{RL})$ the set of all recursive languages. It may be shown that the inclusion relation

$$\mathcal{L}(\mathrm{RL}) \subsetneq \mathcal{L}(\mathrm{REL})$$

is proper.

Example B.1.5 There exists an recursively enumerable language L which is not a recursive language.

Since each Turing machine is of finitely describable, we may encode each Turing machine as a string in $(0 + 1)^*$. Define a language by

$$L = \{x \in (0 + 1)^* \mid x \text{ is a coding of a Turing machine, which accepts } \varepsilon\}.$$

From the halting trouble we see that if $x \in L$ and denote the corresponding Turing machine by M_x, then we may simulate the operations of M_x on input of ε and M_x will certainly stop and accept ε. But if $x \notin L$ then we cannot guarantee if M_x will stop for ε, even it x is a coding of Turing machine. This shows that L is a recursively enumerable language, but not a recursive language.

It can be proved that for a recursive language we may obtain a generator for it and which has better feature than those for recursively enumerable languages.

Theorem B.1.6 A language L is recursive if and only if there exists a generator M which enumerates exactly all strings of L and in the order of increasing size.

We only give the sketch of its proof. If a recursive language $L = L(M)$ is given, and there is no halting trouble for M, then it is easy to enumerate all strings in L by simply input each string in Σ^* to M_1 and to see if it is accepted. If the string is accepted then enumerate it, otherwise neglect it. If the input strings is put in the order of increasing order, then we obtain the required enumerator.

On the other hand, if we have a enumerator M with the required property, then we may suppose that $L = G(M)$ is infinite, otherwise we may use a finite automaton to accept L, which has no trouble of halting. For a given string x we may simulate M and enumerate strings of L one by one in the order of increasing size, then we either accept x, if we find out x is enumerated by M, or reject it if the string generated is of length beyond $|x|$ already.

From Definitions of recursively enumerable languages and recursive languages it is easy to obtain the following theorem.

Theorem B.1.7 If a language L and its complement $L' = \Sigma^* - L$ are both recursively enumerable languages, then they are also recursive languages.

B.1.4 Linear Bounded Automata

Restricting Turing machines further, we can obtain an important subclass of automata and languages.

Definition B.1.8 A *linear bounded automaton* (LBA) is a nondeterministic Turing machine satisfying the following conditions.

(1) Its input alphabet includes two special symbols ¢ and \$, the left and right end-marker, respectively.

(2) The linear bounded automaton has no moves left from ¢ or right from \$, nor may it print another symbol over ¢ and \$.

It can be shown that if restricting the Turing machine that, on each input, the part of input tape used is bounded by some linear function of the length of the input string, then the ability is the same as restricting the Turing machine to the portion of the tape containing the input—hence the name "linear bounded automaton".

Definition B.1.9 A language is *context-sensitive* if it is accepted by an linear bounded automaton.

We use CSL as the abbreviation of context-sensitive language, and $\mathcal{L}(\mathrm{CSL})$ the set of all CSL. It can be proved that

$$\mathcal{L}(\mathrm{CSL}) \subsetneq \mathcal{L}(\mathrm{RL}),$$

and the inclusion is proper.

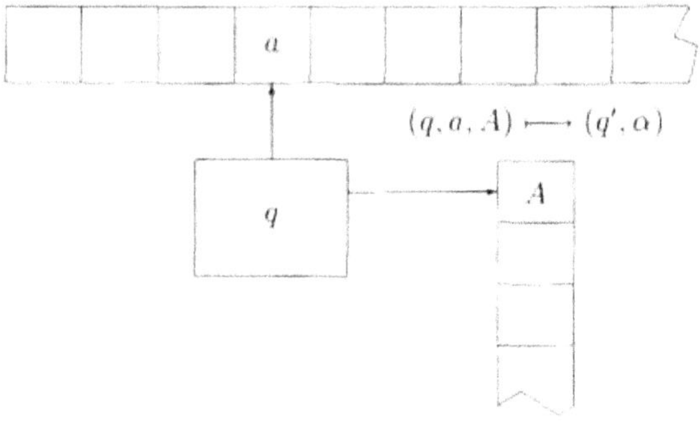

Figure B.2: A pushdown automaton

B.1.5 Pushdown Automata

A pushdown automaton (PDA) is essentially a finite automaton with an additional memory — a stack, or "first in-last out" tape, as shown in Figure B.2. Its finite control and input tape are quite the same as in Figure A.1, but the transition of states also depends on the current symbol on the top of the stack.

The stack may be considered as a list of string of symbols from an stack alphabet. Symbols may be entered or removed only at the top of the stack. When a symbol is entered at the top, the symbol at the top becomes second from the top, the symbol previously second from the top becomes third, and so on. Similarly, when a symbol is removed from the top of the stack, the symbol previously second from the top becomes the top symbol, the symbol previously third from the top becomes second, and so on.

The moves of pushdown automaton will be of two types. In the first type of move, depending on the input symbol, the top symbol of the stack, and the state of the finite control, a number of choices are possible. Each choice consists of a next state for the finite control and a (possibly empty) string of symbols to replace the top stack symbol. After selecting a choice, the input head is advanced one square right.

The second type of move, which is called an ε-move, is similar to the first, except that the input symbol is not used, and the input head is not advanced right after the move. This type of moves allows the pushdown automaton to manipulate the stack without reading input symbols.

Using pushdown automata as acceptors we obtain a new class of languages.

Definition B.1.10 A language is *context-free* if it is accepted by a pushdown automaton.

We use CFL as the abbreviation for context-free language, and $\mathcal{L}(\mathrm{CFL})$ the set of all CFL. It can be proved that

$$\mathcal{L}(\mathrm{CFL}) \subsetneq \mathcal{L}(\mathrm{CSL}),$$

and this inclusion is proper.

B.2 Languages Generated by Grammars

In the previous Section we have introduced several languages, including REL, RL, CSL, and CFL, which are accepted by different machines respectively. Now we introduce the Chomsky hierarchy to consider these languages from the viewpoint of grammars, as the right-linear grammars in Subsection *A.2.1* for regular languages.

B.2.1 The Chomsky Hierarchy

The *Chomsky hierarchy* is proposed by Chomsky (1956, 1959) to classify the languages generated from grammars into four classes to reflect the complexity of their grammars.

A general *grammar* is a 4-tuple $G = (V, \Sigma, P, S)$ (as in Subsection *A.2.1*), where V is a finite set of variables (or nonterminals), Σ is a finite set of terminals, and $V \cap \Sigma = \emptyset$, $S \in V$ is a special start variable, P is a finite set of productions, that is, grammar rules.

Each *grammar rule* is of the form $\alpha \to \beta$, where both α and β belong to $(V \cup \Sigma)^*$, and α contains at least one variable of V.

If starting from the start variable S, and applying grammar rules from P finite times, we obtain a string $x \in \Sigma^*$, that is, x consists of terminals entirely, then we say x is a *word* generated by the grammar G. The set of all words generated by S, as a subset of Σ^*, is said to be the language generated by G, and is denoted by $\mathcal{L}(G)$.

Now we explain what the application of grammar rules from P means. Let x, y be strings of $(V \cup \Sigma)^*$. If there exist decompositions $x = \alpha_1\alpha\alpha_2$ and $y = \alpha_1\beta\alpha_2$, and one grammar rule in P is $\alpha \to \beta$, then we say that the string y is obtained from string x by applying the grammar rule $\alpha \to \beta$, and denoted this process by $x \Rightarrow y$.

In generally speaking, if $x, y \in (V \cup \Sigma)^*$ are either $x = y$ or y can be obtained from x and applying grammar rules as above finite times, then we say x derives y in P, and denoted this derivation by $x \overset{*}{\Rightarrow} y$.

It is easy to understand that the grammar rules are just the rewriting rules of strings. A rule of $\alpha \to \beta$ means that the substring α of string $\alpha_1\alpha\alpha_2$ may be replaced by β. This derivation process continue until there is no variables appeared in a string, and then we obtain a word in the language $\mathcal{L}(G)$. More precisely, we may write

$$\mathcal{L}(G) = \{x \mid S \overset{*}{\Rightarrow} x\} \cap \Sigma^*$$

We called the grammar defined thus far an *unrestricted grammars*.

Definition B.2.1 A language $L = \mathcal{L}(G)$ is called of *type* i if the grammar G generating L is of type i. A grammar $G = (V, \Sigma, P, S)$ is of type i, if it satisfies the conditions below:

(1) i=0, there is no restriction.

(2) i=1, in each grammar rule $\alpha \to \beta$ the length of α is no less than the length of β, that is, $|\beta| \geq |\alpha|$. The only exception allowed is the rule

$S \to \varepsilon$, but then the start variable S cannot appear in the right hand of any rules.

(3) i=2, each grammar rule is of the form $A \to \beta$, where $A \in V$, and $\beta \in (V \cup \Sigma)^*$

(4) i=3, each grammar rule is of the form either $A \to \beta B$ or $A \to \beta$, where $A, B \in V$, and $\beta \in \Sigma^*$

We see that the grammar of type 3 is the right-linear grammar in Subsection *A.2.1*, and hence the language of type 3 is just the regular language. For the other languages of type $i(< 3)$ we have following Theorems.

Theorem B.2.2 A language is of type 0 if and only if it is recursively enumerable.

Theorem B.2.3 A language is of type 1 if and only if it is context-sensitive.

Theorem B.2.4 A language is of type 2 if and only if it is context-free.

If we denote the classes of languages of type i by $\mathcal{L}_i$, then we have

$$\mathcal{L}_3 \subsetneq \mathcal{L}_2 \subsetneq \mathcal{L}_1 \subsetneq \mathcal{L}_0,$$

or, equivalently,

$$\mathcal{L}(\mathrm{RGL}) \subsetneq \mathcal{L}(\mathrm{CFL}) \subsetneq \mathcal{L}(\mathrm{CSL}) \subsetneq \mathcal{L}(\mathrm{RL}) \subsetneq \mathcal{L}(\mathrm{REL}),$$

where the class of recursive languages is also included. An open problem is how to characterize recursive languages from their features of grammars.

From Theorem B.1.2 we know that there exist languages beyond the type 0. As a matter of fact, each class of languages above is a countable set, but the set of all languages is uncountable.

B.2.2 Context-Free Languages

A context-free language may be described either by a type 2 grammar which is referred to as a context-free grammar (CFG), or by a pushdown automaton (PDA). The class of context-free languages has found applications in many fields. But from some viewpoint the notion of context-free languages (or pushdown automata) is not an ideal model of languages (or computation). The class $\mathcal{L}(\mathrm{CFL})$ is not closed under either complementation or intersection. The ability of deterministic pushdown automata is strictly less than that of nondeterministic pushdown automata.

We now see some examples of languages of type 2.

Example B.2.5 $L = \{0^k 1 0^k \mid k \geq 0\}$ is a context-free language. This language has been discussed in Example A.1.6, where we already see that it is non-regular. In order to show that this language is of type 2, we only need to construct a grammar G of type 2, and $L = L(G)$.

Let $G = (V, \Sigma, P, S)$, where $V = \{S\}$, $\Sigma = \{0, 1\}$, $P = \{S \to 1 \mid 0S0\}$. It is easy to prove $L = L(G)$ inductively.

Example B.2.6 Dyck language is a context-free language which has both practical and theoretical values.

Let $G = (V, \Sigma, P, S)$, where $V = \{S\}$, $\Sigma = \{e_1, \cdots, e_n, e'_1, \cdots, e'_n\}$, and

$$P = \{S \to \varepsilon \mid SS \mid e_1 S e'_1 \mid \cdots \mid e_n S e'_n\}.$$

We can see that if a string $x \in L(G)$, then applying on x the rules $e_i e'_i \to \varepsilon$, $i = 1, \ldots, n$, finite times the empty word ε is obtained. Thus a Dyck language is a simulation of the different parentheses used in arithmetic expressions.

On the other hand, it can be proved that each context-free language L may be written as $h(L_D \cap R)$, where h is a homomorphism (see Subsection *A.4.2* for this notion), L_D is a Dyck language, and R a regular language.

B.2.3 Pumping Lemma of Context-Free Languages

For context-free languages there is also a pumping Lemma (Bar-Hillel, Perles and Shamir 1961), which is similar to Lemma A.4.1 for regular languages .

Lemma B.2.7 For each context-free language L there exists a positive integer $n(L)$ with the following property: each word $z \in L$ with $|z| > n(L)$, can be factorized into $z = uvwxy$, so that $|vwx| \leq p(L)$, $|vx| > 0$, and $uv^m wx^m y \in L$ for all $m \geq 0$.

Two examples are given below to show how to use this tool to prove that both of them are not context-free.

Example B.2.8 $L = \{0^{2^k} \mid k \geq 0\}$ is not a context-free language. (In Example A.3.4 we have shown that L is not regular.)

Assume the contrary that L is context-free, then we have the integer $n(L)$ asserted in Lemma. Take a string $z \in L$ with $|z| \geq n(L)$, and we have a decomposition $z = uvwxy$ such that $|vx| \geq 1$, $|vwx| \leq n(L)$, and $uv^i wx^i y \in L$ for each integer $i \geq 0$. But then every string which consists of symbol 0 entirely and of lengths $|z| + (i-1)|vx|$ belong to L. Since, however, the lengths of words of L is a geometric progression $\{2^i\}(i \geq 0)$, this leads to a contradiction.

Example B.2.9 $L = \{a^i b^i c^i \mid i \geq 0\}$ is not a context-free language.

Assume the contrary that L is a context-free language, then we have the integer $n(L)$ asserted in Lemma. Taking the word $z = a^n b^n c^n \in L$, we have a decomposition $z = uvwxy$ such that $|vx| \geq 1$, $|vwx| \leq n$, and $uv^i wx^i y \in L$ for each integer $i \geq 0$. Since $|vwx| \leq n$ the string vwx cannot contain symbols a and c simultaneously. It implies that in strings $uwy(i = 0)$ the numbers of a, b, c cannot be equal and a contradiction.

B.2.4 The Ogden Lemma

Although the pumping Lemma for context-free languages can be successfully used to prove some languages are not context-free, it is not strong enough for our purpose in this book. Here we introduce the Ogden Lemma, which is much stronger than the pumping Lemma. (see Ogden 1968 and Dömösi et al. 1994).

Lemma B.2.10 (Ogden) For each context-free language L there exists an integer $k(L)$ such that for every word $z \in L$, if any k or more distinct positions in z are designated as distinguished, then there is a decomposition of $z = uvwxy$ such that:

(i) $uv^iwx^iy \in L$ for each $i \geq 0$.

(ii) w contains at least one of the distinguished positions.

(iii) Either u and v both contain distinguished positions, or x and y both contain distinguished positions.

(iv) vwx contains at most k distinguished positions.

Remark. 1. The condition of Ogden Lemma is not sufficient for a language being context-free (Bader and Moura 1982).

2. It turns out that the weak form of Lemma in Hopcroft and Ullman (1979) is not strong enough for some applications in this book. It is also pointed out in Dömösi et al (1994) that this weak form is not equivalent to the Lemma itself.

Example B.2.11 $L = \{a^ib^jc^kd^l \mid i = 0 \text{ or } j = k = l\}$ is not a context-free language.

Here the pumping Lemma B.2.7 fails. As a matter of fact, if $i = 0$ then there is no restriction at all, otherwise we may let $uvwx = a^i$ and no contradiction happens.

Now assume that L is context-free and use the Ogden Lemma. If k is the integer in Lemma, then take a word $z = a^nb^{k+1}c^{k+1}d^{k+1}$ of L where $n > 0$, and select the positions of substring c^{k+1} as our distinguished positions:

$$a^nb^{k+1}\overbrace{c^{k+1}}^{k+1}d^{k+1}$$

Consider a decomposition of $z = uvwxy$, which satisfies the conditions listed in Lemma. From the conditions (ii) and (iv) we see that w contains symbol c and $|vwx| \leq k$, this leads to the conclusion that vwx cannot contain symbols b and d at the same time. Thus the numbers of b's and d's of strings $uv^iwx^iy \in L$ cannot be the same for all $i > 1$.

B.2.5 Two Theorems related CFL and RGL

We know that a regular language is context-free, but its converse need not be true. There are two Theorems that may be used to tell if a given context-free language is really a regular language.

Definition B.2.12 A context-free grammar $G = (V, \Sigma, P, S)$ is *self-embedding* if and only if $A \stackrel{*}{\Rightarrow} \alpha A \beta$ for some $A \in V$ and $\alpha, \beta \in (V \cup \Sigma)^*$ such that $\alpha \neq \varepsilon$ and $\beta \neq \varepsilon$. A context-free language L is *self-embedding* if and only if all context-free grammars generating L are self-embedding.

Theorem B.2.13 A language L is regular if and only if it is context-free and not self-embedding.

Since any right-linear grammar is not self-embedding, the part of "only if" is obviously true. The proof for the part of "if" can be found in Salomaa (1973).

If card $\Sigma = 1$ then we have another result.

Definition B.2.14 A set of integers is *linear* if it is of the form $\{c + pi \mid i = 0, 1, \ldots\}$. A set is *semilinear* if it is the finite union of linear sets.

Theorem B.2.15 For a language L over one symbol the following three conditions are equivalent:

1. L is regular.
2. L is context-free.
3. The set of lengths of words in L is semilinear.

Its proof may be found in Hopcroft and Ullman (1979) and Salomaa (1973).

B.2.6 Context-Sensitive Languages

We know that a context-sensitive language L may be described by either a type-1 grammar (it is called a context-sensitive grammar or CSG) G such that $L = L(G)$ or a linear bounded automaton M such that $L = L(M)$. In this Subsection we will give two examples to show how to design a context-sensitive grammar for a given context-sensitive language, and then explain the way through linear bounded automaton. Finally, we will discuss the closure properties of $\mathcal{L}(\text{CSL})$ for some operations.

If is usually not a easy task to write a context-sensitive grammar for a given context-sensitive language.

Example B.2.16 $L = \{0^{2^k} \mid k \geq 0\}$ is a context-sensitive language.

We have seen that it is neither regular nor context-free by discussion in Examples A.3.4 and B.2.8. Now let $G = (V, \{0\}, P, S)$, where

$$\begin{aligned} V &= \{S, C, D, E, F, L, R\} \text{ and} \\ P &= \{S \to 0 \mid 0^2 \mid 0^4 \mid DLCE, DL \to DCR \mid 0^2F, RC \to CCR, \\ &\quad RE \to LCCE, CL \to LC, FC \to 0^2F, FE \to 0^4\} \end{aligned}$$

Since each rule in P satisfy the condition of Definition B.2.1 that the length of right-hand side is no less than the length of left-hand side, G is a context-sensitive grammar.

Now we prove $L = L(G)$. Beginning from S we can obtain $0, 0^2, 0^4$ immediately. Generally we can have

$$S \stackrel{*}{\Rightarrow} DLC^{a_n}E,$$

where $a_0 = 1$ and $a_{n+1} = 2a_n + 3$. As a matter of fact, we have

$$\begin{aligned} DLC^{a_n}E &\Rightarrow DCRC^{a_n}E \stackrel{*}{\Rightarrow} DC^{1+2a_n}RE \\ &\Rightarrow DC^{1+2a_n}LC^2E \stackrel{*}{\Rightarrow} DLC^{2a_n+3}E \end{aligned}$$

Thus we have $a_n = 2^{n+2} - 3$. Using the rule $DL \Rightarrow 0^2F$ leads to

$$\begin{aligned} S &\stackrel{*}{\Rightarrow} DLC^{a_n}E \Rightarrow 0^2FC^{a_n}E \\ &\stackrel{*}{\Rightarrow} 0^{2a_n+2}FE \Rightarrow 0^{2a_n+6} = 0^{2n+3} \end{aligned}$$

for $n \geq 0$. This finishes the proof of $L \subseteq L(G)$. The proof of $L(G) \subseteq L$ is easy and omitted.

Example B.2.17 $L = \{a^ib^ic^i \mid i \geq 0\}$ is a context-sensitive language.

We have proved this language is not context-free in Example A.1.6. Now let $G = (V, \Sigma, P, S)$, where $V = \{S, X, Y\}$, $\Sigma = \{a, b, c\}$, and

$$\begin{aligned} P = \{&S \to abc \mid aXbc, Xb \to bX, Xc \to Ybcc, \\ &bY \to Yb, aY \to aaX \mid aa\}. \end{aligned}$$

The proof of $L = L(G)$ is easy and omitted.

Finally we point out that an open problem about linear bounded automata and context-sensitive languages is if the ability of deterministic linear bounded automata in generating languages is the same as that of nondeterministic linear bounded automata.

A new progress about CSL by Immerman (1988), Szelepcsényi (1987) is that the class $\mathcal{L}(\text{CSL})$ is closed under complementation, namely if we denote the complement language of $L \subseteq \Sigma^*$ by $L' = \Sigma^* - L$, then

$$L \in \mathcal{L}(\text{CSL}) \Longrightarrow L' \in \mathcal{L}(\text{CSL}). \tag{B.1}$$

Its proof can be found also in Davis, Sigal, and Weyuker (1994).

APPENDIX C
L SYSTEMS AND LANGUAGES

Chomsky hierarchy is not the unique way to classify languages into classes of different degrees of complexity. As we have seen in Appendices A and B that some languages seem very simple but occupy higher classes in Chomsky hierarchy. The typical example, among others, is $L = \{0^{2^k} \mid k \geq 0\}$. We have discussed this language in Examples A.3.4, B.2.8, and B.2.16. The reason is that we restrict ourselves by the grammars introduced in Definition B.2.1. Of course, the type 0 grammars, which are also known as semi-Thue, phrase structure or unrestricted grammars, is broad enough to generate every language accepted by Turing machine. From the viewpoint of computability the class $\mathcal{L}(\mathrm{REL})$ is the broadest class of computable languages. But as to type $i(> 0)$ grammars, there is no intrinsic reason that why we have to classify languages by classes of CSL, CFL and RGL.

In this appendix we introduce another grammar hierarchy, called L systems or developmental systems, which can, on one hand, generate every recursively enumerable languages, and, on the other hand, classify languages in quite different way. The standard references on these systems are Herman and Rozenberg (1975), Rozenberg and Salomaa (1981). See also Culik (1974) for relations of developmental systems with some families of languages.

C.1 Definitions of Some L Systems and Languages

In this section we give Definitions for several L systems, and some languages generated by them.

C.1.1 D0L Languages

We begin from the simplest L system—the D0L system.

Definition C.1.1 A *D0L system* is a triple

$$G = (\Sigma, h, w),$$

where Σ is an alphabet, h is a homomorphism from Σ^* to itself, and w, referred to as the starting word, is a nonempty string over S. The word sequence $S(G)$ generated by G consists of the words

$$h^0(w) = w, h^1(w) = h(w), h^2(w) = h(h(w)), \ldots$$

The language of G, which is called a *D0L language*, is defined by

$$L(G) = \{h^i(w) \mid i \geq 0\}$$

Using the notations $\Rightarrow$ and $\overset{*}{\Rightarrow}$ introduced in Subsection *B.2.1*, we have

$$a_1 \cdots a_n \Rightarrow h(a_1) \cdots h(a_n),$$

and

$$L(G) = \{x \mid w \overset{*}{\Rightarrow} x\}.$$

Some explanation of the name "D0L system": the "D" means deterministic, there is just one production rule for each symbol in Σ, the "0" stands for zero-sided context, that is, context-free or context-independent, and the "L" acknowledges A. Lindenmeyer, who first used these systems as models in developmental biology.

From the definition of D0L system and D0L language we see that there are no distinction between terminals and nonterminals (variables) as those in the Chomsky grammars, and at each step of derivation, a production rule is applied to each symbol or shorter substring of a string, rather than to just a symbol of a shorter substring. Therefore, we call such systems the *parallel rewriting systems*, while the Chomsky grammars the *sequential rewriting systems*.

Example C.1.2 $L = \{0^{2^k} \mid k \geq 0\}$ is a D0L language. As a matter of fact, let a D0L system be defined by

$$G = (\{0\}, \{0 \to 0^2\}, 0),$$

we obtain $L = L(G)$ immediately.

Thus this language L belongs to the class of D0L languages, which is the simplest class of languages generated from L systems. But from Examples A.3.4, B.2.8, and B.2.16 we know that L is a strictly context-sensitive language in Chomsky hierarchy, that is, neither a context-free language nor a regular language.

On the other hand, a finite language is in the lowest level of Chomsky hierarchy, that is, the class of regular languages, but from Definition C.1.1 we see that in general a finite language may not belong to the class of D0L languages. An example will be given after Definition C.1.6 below.

Remark. As pointed out by Salomaa (1981), D0L systems can be used to generate some useful infinite sequences. An example is the solution to the strong cube-free problem discussed there. In this book we will use D0L systems to generate many infinite sequences which have dynamical meaning in study of one-dimensional systems (see Chapter 6). Thus it is appropriate to include a Lemma as follows, in which the infinite sequence s will be referred to as the *limit* of the sequence generated by a D0L system.

Lemma C.1.3 Let A D0L system be given by $G = (\Sigma, h, w)$. If

$$h(w) = wx \text{ where } |x| > 0,$$

that is, $h(w)$ has w as its proper prefix, then there exists a unique infinite sequence s over Σ such that s has each $h^i(w)$ $(i \geq 0)$ as its prefix.

Proof. From the condition we can inductively obtain

$$h^{i+1}(w) = h^i(w)h^i(x) \text{ for all } i \geq 0,$$

thus s, the "limit" of the sequence $\{h^i(w)\}_{i\geq 0}\}$, can be defined by the condition that whose prefix of length $|h^i(w)|$ equals to $h^i(w)$ for all i. Its uniqueness is obvious. ∎

C.1.2 0L Languages

A homomorphism h of D0L system is first a mapping from Σ to Σ^* and then extended to strings and languages over Σ. This implies the D0L system is of deterministic nature. Now we consider a substitution h, which is first a mapping from Σ to 2^{Σ^*}, that is to say, for each $a \in \Sigma$, $h(a)$ is a subset of Σ^*, and then extended to strings and languages over Σ. If each $h(a)$ is finite, we call h a finite substitution. (cf. Subsection *A.4.2.*)

Definition C.1.4 A *0L system* is a triple

$$G = (\Sigma, h, w),$$

where Σ is an alphabet, h is a finite substitution from Σ^* to 2^{Σ^*}, and w, referred to as the starting word, is a nonempty string over S. The language of G, which is called a *0L language*, is defined by

$$L(G) = \{x \mid x \in h^i(w) \text{ for } i \geq 0\} = \{x \mid w \stackrel{*}{\Rightarrow} x\}.$$

It is obvious that 0L system is also a parallel rewriting system, but of nondeterministic nature.

Example C.1.5 $L = \{0^n \mid n \geq 0\}$ is a 0L language, but not a D0L language.

Let a 0L system be defined by

$$G = (\{0\}, \{0 \to \varepsilon \mid 00\}, 0),$$

it is easy to verify that $L = L(G)$.

C.1.3 T0L Languages

Definition C.1.6 A *T0L system* is a triple

$$G = (\Sigma, H, w),$$

where Σ is an alphabet, H is a finite set of substitutions, each of them is from Σ^* to 2^{Σ^*}, and w, referred to as the starting word, is a nonempty string over S. The language of G, which is called a *T0L language*, is defined by

$$L(G) = \{x \mid x = w \text{ or } x \in h_1 \cdots h_n(w), h_i \in H, i = 1, \ldots, n\} = \{x \mid w \stackrel{*}{\Rightarrow} x\}.$$

Remark. The 0L system is a special case of the T0L system when card $H = 1$.

Example C.1.7 $L = \{a^i b^i c^i \mid i \geq 0\}$ is a context-sensitive language, but not a context-free language (see Examples B.2.9 and B.2.17). We can show that L is a T0L language. Define

$$G = (\{a, b, c\}, H, abc),$$

where $H = \{h_1, h_2\}$, and
$h_1 = \{a \to aa, b \to bb, c \to cc\}$,
$h_2 = \{a \to \varepsilon, b \to \varepsilon, c \to \varepsilon\}$.
It is obvious that $L = L(G)$.

But even T0L systems are not strong enough to generate all finite languages.

Example C.1.8 $L = \{a, a^3\}$ is not a T0L language.
Assume the contrary that $L = L(G)$ for a T0L system $G = (\{a\}, H, w)$, then we must have either $w = a$ or $w = a^3$. If $w = a$ then we have $a \stackrel{*}{\Rightarrow} a^3$. But then we will have $a \stackrel{*}{\Rightarrow} a^3 \stackrel{*}{\Rightarrow} a^9$ and obtain a contradiction. On the other hand, if $w = a^3$ then form $a^3 \stackrel{*}{\Rightarrow} a$ we must have $a \stackrel{*}{\Rightarrow} \varepsilon$. But then we will have $a^3 \stackrel{*}{\Rightarrow} a^2$ and $a^2 \in L$, a contradiction again.

C.1.4 ET0L Languages

Definition C.1.9 A *ET0L system* is a 4-tuple

$$G = (V, H, w, \Sigma),$$

where $U(G) = (V, H, w)$ is a T0L system (called the *underlying system* of G), $\Sigma \subset V$ is an alphabet consisting of terminals. The language of G, which is called an *ET0L language*, is defined by

$$L(G) = L(U(G)) \cap \Sigma^*$$

It is clear that each symbol in $V - \Sigma$ plays the role of variable (as in Definition B.2.1 of Chomsky hierarchy).

The class of ET0L systems occupies a central position in the theory of L systems, it is the largest class of L systems still referred to as "zero sided" or "context-free" systems. We use notation $\mathcal{L}(\text{ET0L})$ to denote the set of all ET0L languages.

Remark. Similarly, by using a 0L system as the underlying system in Definition above, we obtain E0L systems, E0L languages, and $\mathcal{L}(\text{E0L})$, the class of all E0L languages.

C.2 Relations with Chomsky Hierarchy

C.2.1 Languages between CFL and CSL

Proposition C.2.1 $\mathcal{L}(\text{CFL}) \subsetneq \mathcal{L}(\text{E0L}) \subsetneq \mathcal{L}(\text{ET0L}) \subsetneq \mathcal{L}(\text{CSL})$.

Proof. Let $L = L(G)$ be a context-free language which is generated by a context-free grammar $G = (V, \Sigma, P, S)$. Define a 0L system by

$$G' = (V \cup \Sigma, P \cup \{a \to a \mid a \in V \cup \Sigma\}, S).$$

It is clear that each step of derivation in G is a step of parallel rewriting in G', and we have

$$L = L(G') \cap \Sigma^*$$

(Thus we have proved that each context-free language is an E0L language.) From its Definition it is easily to understand that each ET0L languages can be accepted by a linear bounded automaton, and, consequently, is a context-sensitive language. The Example C.1.2 also means that an ET0L language need not be a context-free language. The proof of inclusion relation $\mathcal{L}(\text{ET0L}) \subsetneq \mathcal{L}(\text{CSL})$ is omitted and can be found in Rozenberg and Salomaa (1981). ∎

In Figure C.1 the inclusion relations between RGL, CFL, CSL, on the one hand, and D0L, E0L, ET0L, on the other hand, are shown (here the abbreviations stand for the classes they generated).

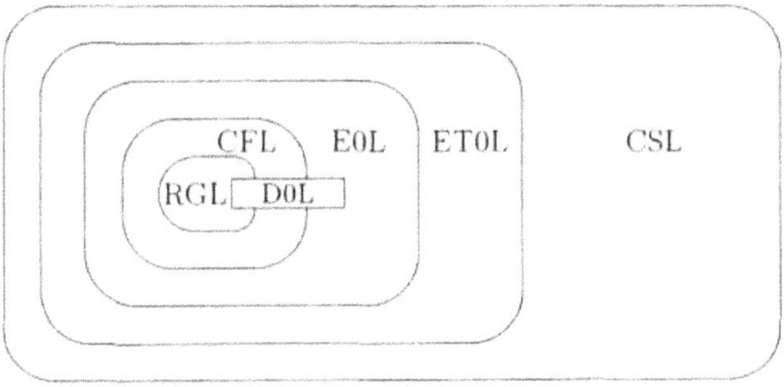

Figure C.1: Relations between Chomsky hierarchy and L systems

C.2.2 Full Abstract Family of Languages

An important feature of $\mathcal{L}(\text{ET0L})$ among L systems is that it is a full AFL, and hence it is closed under many important operations of languages.

A class of languages is said to be an *abstract family of languages* (AFL), if it is closed under union, concatenation, positive closure, ε-free homomorphism, inverse homomorphism and intersection with a regular language. (Here L^+, the positive closure of L, is $\cup_{i=1}^{\infty} L^i$.)

An AFL is called *full* if it is also closed under closure and all homomorphisms

The classes $\mathcal{L}$(RGL), $\mathcal{L}$(CFL) and $\mathcal{L}$(REL) are full AFL's, and the class $\mathcal{L}(CSL)$ is an AFL. On the other hand, $\mathcal{L}$(0L) is not closed under any one operation listed above, and $\mathcal{L}$(E0L) is not closed under homomorphism and inverse homomorphism. Hence sometimes the name of *anti-AFL* is given to $\mathcal{L}$(0L) (and similar classes).

It can be proved that there exists an increasing sequence of classes of languages $\{\mathcal{L}\}_{n\geq 0}$ such that each $\mathcal{L}_n$ is a full AFL, and $\mathcal{L}$(ET0L) is the smallest full AFL among them:

$$\mathcal{L}(\text{ET0L}) = \mathcal{L}_0 \subsetneq \mathcal{L}_1 \subsetneq \cdots \subsetneq \mathcal{L}_n \subsetneq \cdots \subsetneq \mathcal{L}(\text{CSL}).$$

Therefore, in the sense of grammatical complexity we can consider an ET0L language is much simpler than a CSL.

REFERENCES

1. Adler, R. L., Konheim, A. G. and McAndrew, M. H., Topological entropy, *Trans. Amer. Math. Soc.* **114** (1965) 309–319.

2. Adler, R. L., Geodesic flows, interval maps, and symbolic dynamics, in *Ergodic Theory, Symbolic Dynamics and Hyperbolic Spaces*, eds. Bedford, T., Keane, M., Series, C. (Oxford University Press, Oxford 1991) 93–123.

3. Alekseev, Y. M. and Yakobson, M. V., Symbolic dynamics and hyperbolic dynamic systems, *Physics Reports* **75** (1981) 287–325.

4. Alsedà, L., Llibre, J. and Misiurewicz, M., *Combinatorial Dynamics and Entropy in Dimension One* (World Scientific, Singapore 1993).

5. Auerbach, D., Dynamical Complexity of Strange Sets, in *Measures of Complexity and Chaos*, eds. N. B. Abraham, A. M. Albano, A. Passamante and P. E. Rapp, (Plenum, New York 1990), 203–207.

6. Auerbach, D. and Procaccia, I., Grammatical complexity of strange sets, *Phys. Rev.* **A41** (1990) 6602–6614.

7. Bader, Ch. and Moura, A., A generalization of Ogden's lemma, *Journal of the ACM* **29** (1982) 404–407.

8. Bar-Hillel, Y., Perles, M., Shamir, E., On formal properties of simple phrase structure grammars, *Z. Phonetik. Sprachwiss. Komm.* **14** (1961) 143–172.

9. Block, L. S. and Coppel, W. A., *Dynamics in One Dimension* (Lecture Notes in Math., vol. 1513, Springer-Verlag, Berlin 1992).

10. Bowen, R., Topological entropy and axiom *A*, *Global Analysis, Proc. Symp. Pure Math.* **14** (1970a) 23–41.

11. Bowen, R., Markov partition for axiom *A* diffeomorphisms, *Amer. J. Math.* **92** (1970b) 725–747.

12. Bowen, R. and Frank, J., The periodic points of maps of the disk and the interval, *Topology* **15** (1976) 337–342.

13. Boyle, M., Kitchens, B. and Marcus, B., A note on minimal covers for sofic systems, *Proc. Amer. Math. Soc.* **95** (1985) 403–411.

14. Brudno, A. A., Entropy and the complexity of the trajectories of a dynamical systems, *Trans. Moscow Math. Soc.* **2** (1983) 127–151.

15. Çambel, A. B., *Applied Chaos Theory, A Paradigm for Complexity* (Academic Press, Boston 1993), 94.

16. Chaitin, G. J., On the length of programs for computing finite binary sequences, *J. ACM* **13** (1966) 547–569.

17. Chen, X., Lu, Q-H. and Xie, H-M., Grammatical Complexity of one-dimensional dynamical systems, *J. Suzhou Univ.* **9** (1993a) 68–72 (in Chinese).

18. Chen, X., Lu, Q-H. and Xie, H-M., Grammatical Complexity of Feigenbaum Attractor, *Preprint* (An English abstract is in *Advances in Mathematics (China)* **22** (1993b) 185–186).

19. Chomsky, N., Three models for the description of language, *IRE Trans. on Information Theory* **2** (1956) 113–124.

20. Chomsky, N., On certain formal properties of grammars, *Information and Control* **2** (1959) 137–167.

21. Collet, P. and Eckmann, J-P., *Iterated Maps on the Interval as Dynamical Systems* (Birkhäuser, Boston, 1980).

22. Coven, E. M. and Paul, M. E., Finite procedures for sofic systems, *Mh. Math.* **83** (1977) 265–278.

23. Cramer, F., *Chaos and Order, the Complex Structure of Living Systems* (VCH, Weinheim, 1993), 140.

24. Crutchfield, J. P. and Young, K., Computation at the onset of chaos, in *Complexity, Entropy and the Physics of Information*, ed. W. H. Zurek (SFI Studies in the Sciences of Complexity, Vol. VIII, Addison-Wesley, Reading MA 1990), 223–267.

25. Crutchfield, J. P., The calculi of emergence: computation, dynamics and induction, *Physica* **D75** (1994) 11–54.

26. Čulik II, K., On some families of languages related to developmental systems, *Intern. J. Computer Math.* **4** (1974) 31–42.

27. Csizar, I. and Komlos, J., On the equivalence of two models of finite-state noiseless channels from the point of view of the output, in *Proceedings of the colloquium of information theory*, ed. A. Renyi (J. Bolyai Math. Soc., Budapest 1968).

28. D'Alessandro, G. and Politi, A., Hierarchical approach to complexity with applications to dynamical systems, *Phys. Rev. Let.* **64** (1990) 1609–1612.

29. Davis, M. D., Sigal, R. and Weyuker, E. J., *Computability, Complexity and Languages*, 2nd edition (Academic Press, Boston 1994).

30. de Luca, A. and Varricchio, S., Some combinatorial properties of factorial languages, in *Sequence*, ed. R. M. Capocelli (Springer-Verlag, Berlin 1990), 258–266.

31. de Melo, W. and van Strien, S., *One-Dimensional Dynamics* (Springer-Verlag, Berlin 1993).

32. Derrida, B., Gervois, A. and Pomeau, Y., Iteration of endomorphisms on the real axis and representation of numbers, *Ann. Inst. Henri Poincaré* **29A** (1978) 305–356.

33. Devaney, R. L., *An Introduction to Chaotic Dynamical Systems*, 2nd edition (Addison-Wesley, Reading MA 1989).

34. Dömösi, P., Horváth, S., Ito, M., Kászonyi, L. and Katsura, M., Some combinatorial properties of words, and the Chomsky-hierarchy, In *Words, Languages and Combinatorics II*, eds. M. Ito and H. Jürgensen (World Scientific, Singapore 1994), 105–123.

35. Douady, A., Topological entropy of unimodal maps, In *Real and Complex Dynamical Systems*, eds. B. Branner and P. Hjorth (Kluwer Academic Publishers, 1995), 65–87.

36. Feigenbaum, M. J., Quantitative universality for a class of nonlinear transformations, *J. Stat. Phys.* **19** (1978) 25-52.

37. Fischer, R., Sofic systems and graphs, *Mh. Math.* **80** (1975a) 179–186.

38. Fischer, R., Graphs and symbolic dynamics, *Transactions of the Colloquium on Information Theory* (Keszthely, Hungary 1975b), 229–244.

39. Friedmann, E. J., Structure and uncomputability in one-dimensional maps, *Complex Systems* **5** (1991) 335–349.

40. Gambaudo, J-M., Lanford III, O. and Tresser, C., Dynamique symbolique des rotations, *C. R. Acad. Sc. Paris*, Sr 1 **299** (1984) 823–826.

41. Gantmacher, F. R., *Applications of the Theory of Matrices* (Interscience, New York 1959).

42. Golze, U., Differences between 1- and 2-dimensional cell spaces, in *Automata, Languages, Development*, eds. A. Lindenmayer and G. Rozenberg (North-Holland, Amsterdam 1976), 369–384.

43. Gottschalk, W. H. and Hedlund, G. A., *Topological Dynamics*, Amer. Math. Soc. Colloq. Publ., **36** (Amer. Math. Soc., Providence 1955).

44. Grassberger, P., Toward a quantitative theory of self-generated complexity, *Int. J. Theor. Phys.* **25** (1986) 907–936.

45. Grassberger, P., On symbolic dynamics of one-humped maps of the interval, *Z. Naturforsch.* **43a** (1988a) 671–680.

46. Grassberger, P., Complexity and forecasting in dynamical systems, in *Measures of Complexity, Lecture Notes in Physics* **314**, eds. L. Peliti and A. Vulpiani (Springer-Verlag, Berlin 1988b), 1–21.

47. Guckenheimer, J. and Holmes, P., *Nonlinear Oscillations, Dynamical Systems, and Bifurcations of Vector Fields* (Springer-Verlag, Berlin 1983).

48. Günther, R., Schapiro, B. and Wagner, P., Complex systems, complexity measures, grammars and model-inferring, *Chaos, Solitons & Fractals* **4** (1994) 635–651.

49. Gutowitz, H., ed., *Cellular Automata: Theory and Experiment, Physica* **D45** (1990).

50. Hadamard, J., Les surfaces à courbure poopsées et leur lignes géodésiques, *Journal de Mathématiques* (5th series) **4** (1898) 27–73.

51. Hanson, J. E. and Crutchfield, J. P., The attractor-basin portrait of a cellular automaton, *J. Stat. Phys.* **66** (1992) 1415–1462.

52. Hao, B-L., *Elementary Symbolic Dynamics and Chaos in Dissipative Systems* (World Scientific, Singapore 1989).

53. Hao, B-L., Symbolic dynamics and characterization of complexity, *Physica* **D51** (1991) 161–176.

54. Hardy, G. H. and Wright, E. M., *An Introduction to the Theory of Numbers* (5th edition, Oxford, 1981).

55. Hedlund, A., Remarks of the work of Axel Thue, *Nordisk Mat. Tidskr.* **15** (1967) 148–150.

56. Herman, G. T. and Rozenberg, G., *Developmental Systems and Languages* (North-Holland, Amsterdam, 1975).

57. Herzel, H., Complexity of symbolic sequences, *Syst. Anal. Model. Simul.* **5** (1988) 435–444.

58. Hofbauer, F., The topological entropy of the transformation $x \mapsto ax(1-x)$, *Mh. Math.* **90** (1980) 117–141.

59. Hopcroft, J. E. and Ullman, J. D., *Introduction to Automata Theory, Languages, and Computation* (Addison-Wesley, Reading MA 1979).

60. Hurd, L. P., The application of formal language theory to the dynamical behavior of cellular automata. (doctoral thesis, Department of Mathematics, Princeton University, 1988).

61. Hurd, L. P., Recursive cellular automata invariant sets, *Complex Systems* **4** (1990a) 119–129.

62. Hurd, L. P., Nonrecursive cellular automata invariant sets, *Complex Systems* **4** (1990b) 131–138.

63. Hurd, L. P., Kari, J and Culik II, K., The topological entropy of cellular automata is uncomputable, *Ergodic Theory and Dynamical Systems* **12** (1992) 255–265.

64. Immerman, N., Nondeterministic Space is closed under Complementation, *SIAM J. Comput.* **17** (1988) 935–938.

65. Jackson, E. A., *Perspectives of Nonlinear Dynamics*, Vols. 1 and 2 (Cambridge University Press, Cambridge 1990).

66. Jonker, L. and Rand, D., The periodic orbits and entropy of certain maps of the unit interval, *J. London Math. Soc.* **22** (1980) 175–181.

67. Kari, J., The nilpotency problem of one-dimensional cellular automata, *SIAM J. Comp.* **21** (1992) 571–586.

68. Kaspar, K. and Schuster, H. G., Easily calculable measure for the complexity of spatiotemporal patterns, *Physical Rev.* **A36** (1987) 842–848.

69. Keller, G., Lyapunov exponents and complexity for interval maps. (Lecture Notes in Math., vol. 1486, Springer-Verlag, Berlin 1991) 216–226.

70. Ko, K-I., *Complexity Theory of Real Functions* (Birkhäuser, Boston 1991).

71. Kolmogorov, A. N., Three approaches to the definition of the concept 'quantity of information', *Problem of Information Transmission* **1** (1965) 1–11.

72. Lempel, A. and Ziv, J., On the complexity of finite sequences, *IEEE Trans.* **IT22** (1976) 75–81.

73. Li, M. and Vitányi, P., *An Introduction to Kolmogorov Complexity and its Applications* (Springer-Verlag, Berlin 1993).

74. Li, T-Y. and Yorke, J. A., Period three implies chaos, *Am. Math. Monthly* **82** (1975) 985–992.

75. Li, W., On the relationship between complexity and entropy for Markov chain and regular languages, *Complex Systems* **5** (1991) 381–399.

76. Lind, D. A., Application of ergodic theory and sofic systems to cellular automata, *Physica* **D10** (1984) 36–44.

77. Lindgren, K. and Nordahl, M., Complexity measures and cellular automata, *Complex Systems* **2** (1988) 409–440.

78. Lorenz, B. N., Deterministic nonperiodic flow, *J. Atmos. Sci.* **20** (1963) 130–141.

79. Lothaire, M., *Combinatorics on Words* (Addison-Wesley, Reading MA, 1983).

80. Lyubich, M. and Milnor, J., The Fibonacci unimodal map, *Journal A.M.S.* **6** (1993) 425–457.

81. Mandelbrot, B. B., On recurrent noise limiting coding, *Proc. Symposium on Information Networks, Polytechnic Institute of Brocklyn* (1954) 205–221.

82. May, R. M., Simple mathematical models with very complicated dynamics, *Nature* **261** (1976) 459–467.

83. Merry, U., *Coping with Uncertainty, Insights for the New Sciences of Chaos, Self-Organization, and Complexity* (Praeger, Westport CT 1995).

84. Metropolis, M., Stein, M. L., Stein, P. R., On finite limit sets for transformations of the unit interval, *J. Combinatorial Theory* **15** (1973) 25–44.

85. Milnor, J. and Thurston, W., On iterated maps of the interval, (Lecture Notes in Math., vol. 1342, Springer-Verlag, Berlin 1988) 465–563.

86. Mira, C., *Chaotic Dynamics* (World Scientific, Singapore 1987).

87. Misiurewicz, M. and Szlenk, W., Entropy of piecewise monotone mapping, *Astérisque* **50** (1977) 299–310.

88. Misiurewicz, M. and Szlenk, W., Entropy of piecewise monotone mapping, *Studia Mathematica* **67** (1980) 45–63.

89. Moore, C., Generalized shifts: unpredictability and undecidability in dynamical systems, *Nonlinearity* **4** (1991a) 199–230.

90. Moore, C., Generalized one-sided shifts and maps of the intervals, *Nonlinearity* **4** (1991b) 727–745.

91. Morse, M., A one-to-one representation of geodesics on a surface of negative curvature, *Amer. J. Math.* **43** (1921a) 33–51.

92. Morse, M., Recurrent geodesics on a surface of negative curvature, *Trans. Amer. Math. Soc.* **22** (1921b) 84–100.

93. Morse, M. and Hedlund, G. A., Symbolic Dynamics, *Amer. J. of Math.* **60** (1938) 815–866.

94. Morse, M. and Hedlund, G. A., Symbolic Dynamics II. Sturmian trajectories, *American J. of Math.* **60** (1940) 1–42.

95. Nasu, M., An invariant for bounded-to-one factor maps between transitive sofic subshifts, *Ergodic Theory and Dynamical Systems* **5** (1985) 89–105.

96. Ogden, W., A helpful result for proving inherent ambiguity, *Math. Systems Theory* **2** (1968) 191-194.

97. Ostlund, S. and Kim, S-H., Renormalization of Quasiperiodic Mappings, *Physica Scripta* **T9** (1985) 193–198.

98. Parry, W., Intrinsic Markov chains, *Trans. Amer. Math. Soc.* **112** (1964) 55–66.

99. Parry, W., Symbolic dynamics and transformations of the unit interval, *Trans. Amer. Math. Soc.* **122** (1966) 368–378.

100. Peliti, L. and Vulpiani, A., *Measures of Complexity* (Springer-Verlag, New York 1988).

101. Penrose, R., *The Emperor's New Mind, Concerning Computers, Minds and The Laws of Physics* (Oxford University Press, Oxford 1989).

102. Pólya, G. and Szegö, G., *Problems and Theorems in Analysis*, Vol 1 (Springer-Verlag, Berlin 1972).

103. Pour-El, M. B. and Richards, J. I., *Computability in Analysis and Physics* (Springer-Verlag, Berlin 1989).

104. Preston, C., *Iterates of Maps on an Interval* (Lecture Notes in Math., vol. 999, Springer-Verlag, Berlin 1983).

105. Preston, C., *Iterates of Piecewise Monotone Mappings of an Interval* (Lecture Notes in Math., vol. 1347, Springer-Verlag, Berlin 1988).

106. Procaccia, I., Thomae, S. and Tresser, C., First-return maps as a unified renormalization scheme for dynamical systems, *Physical Review* **A35** (1987) 1884–1900.

107. Queffélec, M., *Substitution Dynamical Systems—Spectral Analysis* (Lecture Notes in Math., vol. 1294, Springer-Verlag, Berlin 1987).

108. Rasband, S. N., *Chaotic Dynamics of Nonlinear Systems* (Wiley, New York 1990).

109. Roetzheim, W., *Enter the Complexity Lab* (Sams Publishing, Indianapolis 1994).

110. Rozenberg, G. and Salomaa, A., *The Mathematical Theory of L Systems* (Academic Press, New York 1980).

111. Salomaa, A., *Formal Languages* (Academic Press, New York 1973).

112. Salomaa, A., *Jewels of Formal Language Theory* (Computer Science Press, Rockville, 1981).

113. Salomaa, A., *Computation and Automata* (Cambridge Univ. Press, Cambridge, 1985).

114. Sarkovskii, A., Coexistence of cycles of a continuous map of the line into itself, *Ukrain. Math. Zh.* **16** (1964) 61–71.

115. Shannon, C. T. and Weaver, W., *The Mathematical Theory of Communication* (The University of Illinois Press, Urbana, 1949).

116. Shyr, H-J., *Free Monoids and Languages* (Hon Min Book Co., Taichung, Taiwan, 1991).

117. Siegel, R. M., Tresser, C. and Zettler, G., A decoding problem in dynamics and in number theory, *Chaos* **2** (1992) 473–493.

118. Smale, S., Differentiable dynamical systems, *Bulletin Amer. Math. Soc.* **73** (1967) 747–817.

119. Solomonoff, R. J., A formal theory of inductive inference, Part I, *Inform. Control* **7** (1964) 1–22.

120. Stefan, P., A theorem of Šarkovskii on the existence of periodic orbits of continuous endomorphisms of the real line, *Commun. Math. Phys.* **54** (1977) 237–248.

121. Stein, P. R. and Ulam, S. M., Non-linear transformation studies on electronic computers, *Rozprawy Mat* **39** (1964) 1–66.

122. Szelepcsényi, R., The method of forcing for nondeterministic automata, *Bull. Europ. Assoc. Theor. Comp. Sci.* **33** (1987) 96–100.

123. Thue, A., Über unendliche Zeichenreihen, *Norske Vid. Selsk. Skr. I. Mat. Nat. Kl.*, Cristiania no. 7 (1906) 1-22.

124. Thue, A., Über die gegenseitige Lage gleicher Teile gewisser Zeichenreihen, *Norske Vid. Selsk. Skr. I. Mat. Nat. Kl.*, Cristiania no. 1 (1912) 1-67.

125. Titchmarsh, E. C., *The Theory of Functions* (2nd edition, Oxford University Press, Oxford 1939).

126. Troll, G., Truncated horseshoes and formal languages in chaotic scattering *Chaos* **3** (1993) 459–473.

127. Turing, A., On computable numbers with an application to the Entscheidungsproblem, *Proc. London Math. Soc.* **2** (1936) 230–265.

128. Ulam, S. M. and von Neumann, J., On combinations of stochastic and deterministic processes, *Bull. Am. Math. Soc.* **53** (1947) 1120.

129. Vul, E.B., Sinai, Y. G. and Khanin, K. M., Feigenbaum university and thermodynamic formation, *Usp. Mat. Nauk* **39** (1984) 3–37.

130. Wackerbauer, R., Witt, A., Atmanspacher, H., Kurths, J. and Scheingraber, H., A comparative classification of complexity measures, *Chaos, Solitons & Fractals* **4** (1994) 133–173.

131. Walters, P., *An Introduction to Ergodic Theory* (Springer-Verlag, New York 1982).

132. Wang, Y. and Xie, H-M., Grammatical complexity of unimodal maps with eventually periodic kneading sequences, *Nonlinearity* **7** (1994) 1419–1436.

133. Weisbuch, G., *Complex Systems Dynamics, An Introduction to Automata Networks* (Addison-Wesley, New York 1991).

134. Weiss, B., Subshifts of finite type and sofic systems, *Mh. Math.* **77** (1973) 462–474.

135. Wolfram, S., Computation theory of cellular automata, *Commun. Math. Phys* **96** (1984) 15–57.

136. Wolfram, S., *Theory and Applications of Cellular Automata* (World Scientific, Singapore 1986).

137. Xie, H-M., The finite automata of eventually periodic unimodal maps on the interval, *J. Suzhou Univ.* **9** (1993a) 112–118.

138. Xie, H-M., On formal languages of one-dimensional dynamical systems, *Nonlinearity* **6** (1993b) 997–1007.

139. Xie, H-M., *Complexity and Dynamical Systems* (Shanghai Scientific and Technological Education Publishing House, Shanghai 1994) (in Chinese).

140. Xie, H-M., Distinct excluded blocks and grammatical complexity of dynamical systems, *Complex Systems* **9** (1995a) 73–90.

141. Xie, H-M., Fibonacci sequences and homomorphisms of free submonoid for unimodal maps (1995b) (Preprint).

142. Xie, H-M., Complexity of symbolic behaviors of circle homeomorphisms (1995c) (Preprint).

143. Xu, J-H., Liu, Z-R. and Liu, R., The Measures of sequence complexity for EEG studies, *Chaos, Solitons & Fractals* **4** (1994) 2111–2119.

144. Zheng, W-M., Construction of median itineraries without using the anti-harmonic, *Int. J. Mod. Phys.* **B3** (1989a) 1703–1711.

145. Zheng, W-M., Generalized composition law for symbolic itineraries, *J. Phys.* **A22** (1989b) 3307–3313.

146. Zheng, W-M., The W sequence for circle maps and misbehaved itineraries, *J. Phys. A: Math. Gen.* **22** (1989c) 3647–3652.

147. Zheng, W-M., Symbolic dynamics for the circle map, *J. of Modern Physics* **B5** (1991) 481–495.

Symbols Index

Subject Index